William Stanley Jevons

The Principles of Science - A Treatise on Logic and Scientific Method

Vol. I

William Stanley Jevons

The Principles of Science - A Treatise on Logic and Scientific Method
Vol. I

ISBN/EAN: 9783744747417

Printed in Europe, USA, Canada, Australia, Japan

Cover: Foto ©berggeist007 / pixelio.de

More available books at **www.hansebooks.com**

THE LOGICAL MACHINE.

THE PREDICATE-KEY C IS DEPRESSED.

THE PRINCIPLES OF SCIENCE:

A TREATISE ON

SCIENTIFIC METHOD.

BY

W. STANLEY JEVONS, M.A., F.R.S.,

FELLOW OF UNIVERSITY COLLEGE, LONDON;

PROFESSOR OF LOGIC AND POLITICAL ECONOMY IN THE OWENS COLLEGE, MANCHESTER.

VOL. I.

London

MACMILLAN AND CO.

1874

OXFORD:
E. PICKARD HALL, AND J. H. STACY,
PRINTERS TO THE UNIVERSITY.

PREFACE.

It may be truly asserted that the rapid progress of the physical sciences during the last three centuries has not been accompanied by a corresponding advance in the theory of reasoning. Physicists speak familiarly of Scientific Method, but they could not readily describe what they mean by that expression. Profoundly engaged in the study of particular classes of natural phenomena, they are usually too much engrossed in the immense and ever-accumulating details of their special sciences, to generalize upon the methods of reasoning which they unconsciously employ. Yet few will deny that these methods of reasoning ought to be studied, especially by those who endeavour to introduce scientific order into less successful and methodical branches of knowledge.

The application of Scientific Method cannot be restricted to the sphere of lifeless objects. We must sooner or later have strict sciences of those mental and social phenomena, which, if comparison be possible, are of more interest to us than purely material phenomena. But it is the proper course of reasoning to proceed from the known to the unknown—from the evident to the obscure—from the material and palpable to the subtle and refined. The physical sciences may therefore be properly made the practice-ground of the reasoning powers, because they furnish us with a great body of precise and successful investigations. In these sciences we meet with happy instances of unquestionable deductive reasoning, of extensive generalization, of happy prediction, of satisfactory verification, of nice calculation of probabilities. We can note how the slightest analogical clue has been followed up to a glorious discovery, how a rash generalization has at length been exposed, or a conclusive *experimentum crucis* has decided the long-continued strife between two rival theories.

In following out my design of detecting the general methods of inductive investigation, I have found that the more elaborate and interesting processes of quantitative induction have their necessary foundation in the simpler science of Formal Logic. The earlier, and probably by far the least attractive part of this work, consists, therefore, in a statement of the so-called Fundamental Laws of Thought, and of the all-important Principle of Substitution, of which, as I think, all reasoning is a development. The whole procedure of inductive inquiry, in its most complex cases, is foreshadowed in the combinational view of Logic, which arises directly from these fundamental principles. Incidentally I have described the mechanical arrangements by which the use of the important form called the Logical Abecedarium, and the whole working of the combinational system of Formal Logic, may be rendered evident to the eye, and easy to the mind and hand.

The study both of Formal Logic and of the Theory of Probabilities, has led me to adopt the opinion that there is no such thing as a distinct method of induction as contrasted with deduction, but that induction is simply an inverse employment of deduction. Within the last century a reaction has been setting in against the purely empirical procedure of Francis Bacon, and physicists have learnt to advocate the use of hypotheses. I take the extreme view of holding that Francis Bacon, although he correctly insisted upon constant reference to experience, had no correct notions as to the logical method by which, from particular facts, we educe laws of nature. I endeavour to show that hypothetical anticipation of nature is an essential part of inductive inquiry, and that it is the Newtonian method of deductive reasoning combined with elaborate experimental verification, which has led to all the great triumphs of scientific research.

In attempting to give an explanation of this view of Scientific Method, I have first to show that the sciences of number and quantity repose upon and spring from the simpler and more general science of Logic. The Theory of Probability, which enables us to estimate and calculate quantities of knowledge, is then described, and especial attention is drawn to the Inverse Method of Probabilities, which involves, as I conceive, the true principle of inductive procedure. No inductive conclusions are more than probable, and I adopt the opinion that the theory of probability is an essential part of logical method, so that the logical value of every inductive result must be determined consciously or unconsciously, according to the principles of the inverse method of probability.

The phenomena of nature are commonly manifested in quantities of time, space, force, energy, &c., and the observation, measurement, and analysis of the various quantitative conditions or results involved, even in a simple experiment, demand much employment of systematic procedure. I devote a book, therefore, to a simple and general description of the devices by which exact measurement is effected, errors eliminated, a probable mean result attained, and the probable error of that mean ascertained. I then proceed to the principal, and probably the most interesting, subject of the book, illustrating successively the conditions and precautions requisite for accurate observation, for successful experiment, and for the sure detection of the quantitative laws of nature. As it is impossible to comprehend aright the value of quantitative laws without constantly bearing in mind the degree of quantitative approximation to the truth probably attained, I have devoted a special chapter to the Theory of Approximation, and however imperfectly I may have treated this subject, I must look upon it as a very essential part of a work on Scientific Method.

It then remains to illustrate the sound use of hypothesis, to distinguish between the portions of knowledge which we owe to empirical observation, to accidental discovery, or to scientific prediction. Interesting questions arise concerning the accordance of quantitative theories and experiments, and I point out how the successive verification of an hypothesis by distinct methods of experiment yields conclusions approximating to but never attaining certainty. Additional illustrations of the general procedure of inductive investigations are given in a chapter on the Character of the Experimentalist, in which I endeavour to show, moreover, that the inverse use of deduction was really the logical method of such great masters of experimental inquiry as Newton, Huyghens, and Faraday.

In treating Generalization and Analogy, I consider the precautions requisite in inferring from one case to another, or from one part of the universe to another part, the validity of all such inferences resting ultimately upon the inverse method of probabilities. The treatment of Exceptional Phenomena appeared to afford an interesting subject for a further chapter illustrating the various modes in which an outstanding fact may eventually be explained. The formal part of the book closes with the subject of Classification, which is, however, very inadequately treated.

I have, in fact, almost restricted myself to showing that all classification is fundamentally carried out upon the principles of Formal Logic and the Logical Abecedarium described at the outset.

In certain concluding remarks I have expressed the conviction which the study of Logic has by degrees forced upon my mind, that serious misconceptions are entertained by some scientific men as to the logical value of our knowledge of nature. We have heard much of what has been aptly called the Reign of Law, and the necessity and uniformity of natural forces has been not uncommonly interpreted as involving the non-existence of an intelligent and benevolent Power, capable of interfering with the course of natural events. Fears have been expressed that the progress of Scientific Method must therefore result in dissipating the fondest beliefs of the human heart. Even the 'Utility of Religion' is seriously proposed as a subject of discussion. It seemed to be not out of place in a work on Scientific Method to allude to the ultimate results and limits of that method. I fear that I have very imperfectly succeeded in expressing my strong conviction that before a rigorous logical scrutiny the Reign of Law will prove to be an unverified hypothesis, the Uniformity of Nature an ambiguous expression, the certainty of our scientific inferences to a great extent a delusion. The value of science is of course very high, while the conclusions are kept well within the limits of the data on which they are founded, but it is pointed out that our experience is of the most limited character compared with what there is to learn, while our mental powers seem to fall infinitely short of the task of comprehending and explaining fully the nature of any one object. I draw the conclusion that we must interpret the results of Scientific Method in an affirmative sense only. Ours must be a truly positive philosophy, not that false negative philosophy which, building on a few material facts, presumes to assert that it has compassed the bounds of existence, while it nevertheless ignores the most unquestionable phenomena of the human mind and feelings.

I have to thank my colleague, Professor Barker, for carefully revising several of the sheets most abounding in mathematical considerations. It is approximately certain that in freely employing illustrations drawn from many different sciences, I have frequently fallen into errors of detail. In this respect I must throw myself upon the indulgence of the reader, who will bear in mind, as I hope, that the scientific facts are generally mentioned purely for the purpose of illustration, so that inaccuracies of detail will not in the majority of cases affect the truth of the general principles illustrated.

December 15th, 1873.

CONTENTS.

BOOK I.

FORMAL LOGIC, DEDUCTIVE AND INDUCTIVE.

CHAPTER I.

INTRODUCTION.

CHAPTER II.

TERMS.

CHAPTER III.

PROPOSITIONS.

CHAPTER IV.

DEDUCTIVE REASONING.

CHAPTER V.

DISJUNCTIVE PROPOSITIONS.

CHAPTER VI.

THE INDIRECT METHOD OF INFERENCE.

CHAPTER VII.

INDUCTION.

BOOK II.

NUMBER, VARIETY, AND PROBABILITY.

CHAPTER VIII.

PRINCIPLES OF NUMBER.

CHAPTER IX.

THE VARIETY OF NATURE, OR THE DOCTRINE OF COMBINATIONS AND PERMUTATIONS.

BOOK III.

METHODS OF MEASUREMENT.

CHAPTER XIII.

THE EXACT MEASUREMENT OF PHENOMENA.

CHAPTER XIV.

UNITS AND STANDARDS OF MEASUREMENT.

CHAPTER XV.

ANALYSIS OF QUANTITATIVE PHENOMENA.

THE PRINCIPLES OF SCIENCE.

CHAPTER I.

INTRODUCTION.

SCIENCE arises from the discovery of Identity amid Diversity. The process may be described in many different words, but our language must always imply the presence of one common and necessary element. In every act of inference or scientific method we are engaged about a certain identity, sameness, similarity, likeness, resemblance, analogy, equivalence or equality apparent between two objects. It is doubtful whether an entirely isolated phenomenon could present itself to our notice, since there must always be a contrast between object and object to awaken our consciousness. But in any case an isolated phenomenon could be studied to no useful purpose. The whole value of science consists in the power which it confers upon us of applying to one object the knowledge acquired from like objects; and it is only so far, therefore, as we can discover and register resemblances or differences that we can turn our observations to account.

Nature is a spectacle continually exhibited to our senses, in which phenomena are mingled in combinations of endless variety and novelty. Wonder fixes the mind's attention; memory stores up a record of each distinct impression; the powers of association bring forth the record when the like is felt again. By the higher faculties of judgment and reasoning the mind compares the new with the old, recognises essential identity, even when disguised by diverse circumstances, and expects to find again what was before experienced. It must be the ground of all reasoning and inference that *what is true of one thing will be true of its equivalent*, and that under carefully ascertained conditions *Nature repeats herself.*

Were this indeed a Chaotic Universe, the powers of mind employed in science would be useless to us. Did Chance wholly take the place of order, and did all phenomena come out, not of one same *Infinite Lottery*, to use Condorcet's expression, but out of lotteries ever changing in their conditions, there could be no reason to expect the like result in like circumstances. It is possible to conceive a world in which no two things should be associated more often, in the long run, than any other two things. The frequent conjunction of any two events would then be purely fortuitous, and if we expected conjunctions to recur continually we should be disappointed. In such a world we might recognise the same phenomenon as it appeared from time to time, just as we might recognise a marked ball as it was occasionally drawn from a ballot-box; but the approach of any one phenomenon would be in no way indicated by what had gone before, nor would it be at all a sign of what was to come after. In such a world knowledge would be no more than the memory of past coincidences, and the reasoning powers, if they existed at all, would give no clue to the nature of the present, and no presage of the future.

Happily the Universe in which we dwell is not the result of chance, and where chance seems to work it is our own deficient faculties which prevent us from recognising the operation of Law and of Design. In the material framework of this world, substances and forces present themselves in definite and stable combinations. All things are not in perpetual flux, as ancient philosophers held. Element remains element; iron changes not into gold, nor oxygen into hydrogen. With suitable precautions we can calculate upon finding the same thing again endowed with the same properties. The constituents of the globe, indeed, appear in almost endless combinations; but each combination bears its fixed character, and when resolved is found to be the compound of definite substances. Misapprehensions must continually occur, owing to the limited extent of our experience. We can never have examined and registered possible existences so thoroughly as to be sure that no new ones will occur and frustrate our calculations. The same outward appearances may cover any amount of hidden differences which we have not yet suspected. To the variety of substances and powers diffused through nature at its creation, we must not suppose that our brief experience can assign a limit; and the necessary imperfection of our knowledge should be ever borne in mind.

Yet there is much to give us confidence in science. The wider our experience, the more minute our examination of the globe, the greater the accumulation of well-reasoned knowledge,—the fewer must become the failures of inference compared with the successes. Exceptions to the prevalence of Law are gradually reduced to Law themselves. Certain deep similarities have been detected among the objects around us, and have never yet been found wanting. As the means of examining distant parts of the universe have been acquired, those similarities have been traced there as here. Other worlds and stellar systems may be almost incomprehensively different from ours in magnitude, condition and disposition of parts, and yet we detect there the same elements of which our own limbs are composed. The same natural laws can be detected in operation in every part of the universe within the scope of our instruments ; and doubtless these laws are obeyed irrespective of distance, time and circumstance.

It is the prerogative of Intellect to discover what is uniform and unchanging in the phenomena around us. So far as object is different from object, knowledge is useless and inference impossible. But so far as object resembles object, we can pass from one to the other. In proportion as resemblance is deeper and more general, the commanding powers of knowledge become more wonderful. Identity in one or other of its phases is thus always the bridge by which we pass in inference from case to case ; and it is my purpose in this treatise to trace out the various forms in which the one same process of reasoning presents itself in the ever-growing achievements of Scientific Method.

The Powers of Mind concerned in the Creation of Science.

It is no part of the purpose of this work to investigate the nature of mind, except so far as its powers are requisite to the formation of Science. In this place I need only point out that the mental powers engaged in knowledge are probably three in number. They are substantially as Mr. Bain has stated them[a] : —

    1. The Power of Discrimination.
    2. The Power of Detecting Identity.
    3. The Power of Retention.

We exert the first power in every act of perception. Hardly can we have a sensation or feeling unless we discriminate it from something else which preceded.

---

[a] 'The Senses and the Intellect,' Second Ed., pp. 5, 325, &c.

Consciousness would almost seem to consist in the break between one state of mind and the next, just as an induced current of electricity arises from the beginning or the ending of the primary current. We are always engaged in discrimination ; and the rudiment of thought which exists in the lower animals probably consists in their power of feeling difference and being agitated by its occurrence.

But had we power of discrimination only, Science could not be created. To know that one feeling differs from another gives purely negative information. It cannot teach us what will happen. Each sensation would stand out distinct from any other, and there would be no tie, no bridge of affinity between them. We want a unifying power by which the present and the future may be linked to the past ; and this seems to be accomplished by a different power of mind. Francis Bacon has pointed out that different men possess in very different degrees the powers of discrimination and identification. It may be said indeed that discrimination necessarily implies the opposite process of identification ; and so it doubtless does in superficial points. But there is a rare property of mind which consists in penetrating the disguise of variety and seizing the common elements of sameness ; and it is this property which furnishes the true measure of intellect. The very name of intellect (*interligo*) expresses the action, not of separating, but of uniting and binding together the particular and various into the general and like. Logic is but another name for the same process [b], the peculiar work of reason ; and Plato said of this unifying power, that if he met the man who could detect *the one in the many*, he would follow him as a god.

[b] Max Müller, 'Lectures on Language,' Second Series, vol. ii. p. 63.

Laws of Identity and Difference.

At the basis of all thought and science must lie the laws which express the very nature and conditions of the discriminating and identifying powers of mind. These are the so-called Fundamental Laws of Thought, usually stated as follows :—

1. The Law of Identity. *Whatever is, is.*
2. The Law of Contradiction. *A thing cannot both be and not be.*
3. The Law of Duality. *A thing must either be or not be.*

The first of these statements may perhaps be regarded as a description of identity itself, if so fundamental a notion can admit of description. A thing at any moment is perfectly identical with itself, and if any person were unaware of the meaning of the word 'identity' we could not better describe it than by such an example.

The second law points out that contradictory attributes can never be joined together. The same object may vary in its different parts; here it may be black, and there white; at one time it may be hard and at another time soft: but at the same time and place an attribute cannot be both present and absent. Aristotle truly described this law as the first of all axioms[c]—one of which we need not seek for any demonstration. All truths cannot be proved, otherwise there would be an endless chain of demonstration; and it is in self-evident truths like this that we find the fittest foundation.

The third of these laws completes the other two. It asserts that at every step there are two possible alternatives—presence or absence, affirmation or negation. Hence I propose to name this law the Law of Duality,

[c] 'Metaphysics,' Bk. III. chap. iii. 9–12.

for it gives to all the formulæ of reasoning a dual character. It asserts also that between presence or absence, existence or non-existence, affirmation or negation, there is no third alternative. As Aristotle said, there can be no mean between opposite assertions : we must either affirm or deny. Hence the somewhat inconvenient name by which it has been generally known— The Law of Excluded Middle.

It may be held that these laws are not three independent and distinct laws, they rather express three different aspects of the same truth, and each law doubtless presupposes and implies the other two. But it has not hitherto been found possible to state these characters of identity and difference in less than the three-fold formula. The reader may perhaps desire some information as to the mode in which these laws have been stated, or the way in which they have been regarded, by philosophers in different ages of the world. Abundant information on this and many other points of logical history will be found in Ueberweg's 'System of Logic,' of which an excellent translation has been published by Mr. T. M. Lindsay[d]. I must confess however that the history of logical doctrines has seemed to me one of the most confusing and least beneficial studies in which a person can engage ; and over-abundant attention perhaps has been paid to it by Hamilton, Mansel, and many German logicians.

The Nature and Authority of the Laws of Identity and Difference.

I must at least allude to the profoundly difficult question concerning the nature and authority of these

---

[d] Ueberweg's 'System of Logic,' transl. by Lindsay, London, 1871, pp. 228-281.

Laws of Identity or Difference. Are they Laws of Thought or Laws of Things? Do they belong to mind or to material nature? On the one hand it may be said that science is a purely mental existence, and must therefore conform to the laws of that which formed it. Science is in the mind and not in the things, and the properties of mind are therefore all important. It is true that these laws are verified in the observation of the exterior world; and it would seem that they might have been gathered and proved by generalisation, had they not already been in our possession. But on the other hand, it may well be urged that we cannot prove these laws by any process of reasoning or observation, because the laws themselves are presupposed, as Leibnitz acutely remarked, in the very notion of a proof. They are the prior conditions of all thought and all knowledge, and even to question their truth is to allow them true. Hartley ingeniously refined upon this argument, remarking that if the fundamental laws of logic be not certain, there must exist a logic of a second order whereby we may determine the degree of uncertainty: if the second logic be not certain, there must be a third, and so on *ad infinitum.* Thus we must suppose either that absolutely certain laws of thought exist, or that there is no such thing as certainty whatever [c].

Logicians, indeed, appear to me to have paid insufficient attention to the fact that mistakes in reasoning are always likely to occur. The Laws of Thought are often called necessary laws, that is, laws which cannot but be obeyed. Yet as a matter of fact who is there that does not often fail to obey them? They are the laws which the mind ought to obey rather than what it always does obey. Our thoughts cannot be the criterion of truth, for we often have to acknowledge

c Hartley on Man, vol. i. p. 359.

mistakes in arguments of very moderate complexity, and we sometimes only discover our mistakes by a collision between our mental expectations and the events of objective nature.

Mr. Herbert Spencer holds that the laws of logic are objective laws [f], and he regards the mind as being in a state of constant education, each act of false reasoning or miscalculation leading to results which are likely to prevent similar mistakes from being again committed. I am quite inclined to accept such ingenious views; but at the same time it is necessary to distinguish between the accumulation of knowledge and experience, and the constitution of the mind which allows of the acquisition of knowledge. Before the mind can perceive or reason at all it must have the conditions of thought impressed upon it. Before a mistake can be committed, the mind must clearly distinguish the mistaken conclusion from all other assertions. Are not the Laws of Identity and Difference the prior conditions of all consciousness and all existence? Must they not hold true, alike of things material and immaterial? and if so, can we say that they are only subjectively true or objectively true? I am inclined, in short, to regard them as true both 'in the nature of thought and things,' as I expressed it in my first logical essay [g], and I hold that they belong to the common basis of all existence. But this is one of the most profound and difficult questions of psychology and metaphysics which can be raised, and it is hardly one for the logician to decide. As the mathematician does not inquire into the nature of unity and plurality, but developes the formal laws of plurality, so the logician, as I conceive, must assume the truth of the Laws of

[f] 'Principles of Psychology,' Second Ed., vol. ii. p. 86.

[g] 'Pure Logic, or the Logic of Quality apart from Quantity,' London (Stanford), 1864, pp. 10, 16, 22, 29, 36, &c.

Identity and Difference, and occupy himself in developing the variety of forms of reasoning in which their truth may be manifested.

Again, I need hardly dwell upon the question whether logic treats of language, notions, or things. As reasonably might we debate whether a mathematician treats of symbols, quantities, or things. A mathematician certainly does treat of symbols, but only as the instruments whereby to facilitate his reasoning concerning quantities; and as the axioms and rules of mathematical science must be verified in concrete objects in order that the calculations founded upon them may have any validity or utility, it follows that the ultimate objects of mathematical science are the things themselves. In like manner I conceive that the logician treats of language so far as it is essential for the embodiment and exhibition of thought. Even if reasoning can take place in the inner consciousness of man without the use of any signs, at any rate it cannot become the subject of discussion until by some system of material signs it is manifested to other persons. The logician then uses words and symbols as instruments of reasoning, and leaves the nature and peculiarities of existing language to the grammarian. But signs again must correspond to the thoughts and things expressed, in order that they shall serve their intended purpose. We may therefore say that logic treats ultimately of thoughts and things, and immediately of the signs which stand for them. Signs, thoughts and exterior objects may be regarded as parallel and analogous series of phenomena, and to treat one series is equivalent to treating either of the other series [h].

---

[h] See also 'Elementary Lessons in Logic,' Second Ed., p. 10.

The Process of Inference.

The fundamental action of our reasoning faculties consists in inferring or carrying to a new instance of a phenomenon whatever we have previously known of its like, analogue, equivalent or equal. Sameness or identity presents itself in all degrees, and is known under various names; but the great rule of inference embraces all degrees, and affirms that *so far as there exists sameness, identity or likeness, what is true of one thing will be true of the other.* The great difficulty of reasoning doubtless consists in ascertaining that there does exist a sufficient degree of likeness or sameness to warrant an intended inference; and it will be our main task to investigate the conditions under which the inference is valid. In this place I wish to point out that there is something common to all acts of inference however different their apparent forms. The one same rule lends itself to the most diverse applications.

The simplest possible case of inference, perhaps, occurs in the use of a *pattern, example*, or, as it is commonly called, a *sample*. To prove the exact similarity of two portions of commodity, we need not bring one portion beside the other. It is sufficient that we cut a sample which exactly represents the texture, appearance, and general nature of one portion, and according as this sample agrees or not with the other, so will the two portions of commodity agree or differ. Whatever is true as regards the colour, texture, density, material of the sample will be true of the goods themselves. In such cases likeness of quality is the condition of inference.

Exactly the same mode of reasoning holds true of magnitude and figure. To compare the size of two objects, we need not lay them alongside each other. A staff, string, or other kind of measure may be employed to represent the length of one object, and according as it agrees or not with the other, so must the two objects agree or differ. In this case the proxy or sample represents length; but the fact that lengths can be added and multiplied renders it unnecessary that the proxy should always be as large as the object. Any standard of convenient length, such as a common foot-rule, may be made the medium of comparison. The height of a church in one town may be carried to that in another, and objects existing immoveably at opposite sides of the earth may be vicariously measured against each other. We obviously employ the rule that whatever is true of a thing as regards its length, is true of its equal.

To every other simple phenomenon in nature the same principle of substitution is applicable. We may compare weights or densities or degrees of hardness, and all other qualities, in like manner. To ascertain whether two sounds are in unison we need not compare them directly, but a third sound may be the go-between. If a tuning-fork is in unison with the middle C of York Minster organ, and we afterwards find it to be in unison with the same note of the organ in Westminster Abbey, then it follows that the two organs are tuned in unison. The rule of inference now is that what is true, as regards pitch, of the tuning-fork, is true of any sound in unison with it.

The skilful employment of this substitutive process enables us to make measurements beyond the powers of our senses. No one can count the vibrations, for instance, of an organ pipe. But we can construct an instrument called the *syren,* so that while producing a sound of any pitch it shall register the number of vibrations constituting the sound. Adjusting the sound of the syren in unison with an organ pipe, we measure indirectly the number of vibrations belonging to a sound of that pitch. To measure a sound of the same pitch is as good as to measure the sound itself.

Sir David Brewster, in a somewhat similar manner, succeeded in measuring the refractive index of irregular fragments of transparent minerals. It was a troublesome, and sometimes impracticable work to grind the minerals into prisms, so that their powers of refracting light could be directly observed ; but he fell upon the ingenious device of forming a liquid possessing exactly the same refractive power as the transparent fragment under examination. The moment when this equality was attained could be known by the fragments ceasing to reflect or refract light when immersed in the liquid, so that they became almost invisible in it. The refractive power of the liquid being then measured gives that of the solid ; and a more beautiful instance of representative measurement, depending immediately upon the principle of inference, could not be found [i].

Throughout the various logical processes which we are about to consider—Deduction, Induction, Generalisation, Analogy, Classification, Quantitative Reasoning—we shall find the one same principle operating in a more or less disguised form.

Deduction and Induction.

The processes of inference always depend on the one same method of substitution ; but they may nevertheless be distinguished according as the results are inductive or deductive. As generally stated, deduction consists in

[i] Brewster, 'Treatise on New Philosophical Instruments,' p. 273. See also Whewell, ' Philosophy of the Inductive Sciences,' vol. ii. p. 355 ; Tomlinson, 'Philosophical Magazine,' Fourth Series, vol. xl. p. 328 ; Tyndall, in Youman's 'Modern Culture,' p. 16.

passing from more general to less general truths; induction is the contrary process from less to more general truths. We may however describe the difference in another manner. In deduction we are engaged in developing the consequences of a law or identity. We learn the meaning, contents, results or inferences, which attach to any given proposition. Induction is the exactly inverse process. Given certain results or consequences, we are required to discover the general law from which they flow.

In a certain sense all knowledge is inductive. We can only learn the laws and relations of things in nature by observing those things. But the knowledge gained from the senses is knowledge only of particular facts, and we require some process of reasoning by which we may construct out of the facts the laws obeyed by them. Experience gives us the materials of knowledge : induction digests those materials, and yields us general knowledge. Only when we possess such knowledge, in the form of general propositions and natural laws, can we usefully apply the reverse process of deduction to ascertain the exact information required at any moment. In its ultimate origin or foundation, then, all knowledge is inductive—in the sense that it is derived by a certain inductive reasoning from the facts of experience.

But it is nevertheless true,—and this is a point to which insufficient attention has been paid,—that all reasoning is founded on the principles of deduction. I call in question the existence of any method of reasoning which can be carried on without a knowledge of deductive processes. I shall endeavour to show that *induction is really the inverse process of deduction.* There is no mode of ascertaining the laws which are obeyed in certain phenomena, except we previously have the power of determining what results would follow from a given law. Just as the process of division necessitates a prior knowledge of multi-

plication, or the integral calculus rests upon the observation and remembrance of the results of the differential
calculus, so induction requires a prior knowledge of
deduction. An inverse process is the undoing of the
direct process. A person who enters a maze must either
trust to chance to lead him out again, or he must carefully
notice the road by which he entered. The facts furnished
to us by experience are a maze of particular results ; we
might by chance observe in them the fulfilment of a law,
but this is scarcely possible, unless we thoroughly learn
the effects which would attach to any particular law.

Accordingly, the importance of deductive reasoning is
doubly supreme. Even when we gain the results of induction they would be of little or no use without we
could deductively apply them. But before we can gain
them at all we must understand deduction, since it is the
inversion of deduction which constitutes induction. Our
first task then, in this work, must be to trace out fully the
nature of identity in all its forms of occurrence. Having
given any series of propositions we must be prepared to
develop the whole meaning embodied in them, and the
whole of the consequences which flow from them.

Symbolic Expression of Logical Inference.

In developing the results of the Principle of Inference
we require to use an appropriate language of signs. It
would indeed be quite possible to explain the processes of
reasoning merely by the use of words found in the ordinary
grammar and dictionary. Special examples of reasoning,
too, may seem to be more readily apprehended than general
and symbolic forms. But it has been abundantly proved
in the mathematical sciences that the attainment of truth
depends greatly upon the invention of a clear, brief, and
appropriate system of symbols. Not only is such a

language convenient, but it is essential to the expression
of those general truths which are the very soul of science.
To apprehend the truth of special cases of inference does
not constitute logic ; we must apprehend them as cases of
more general truths.   The object of all science is the
separation of what is common and general from what is
accidental and different.   In a system of logic, if anywhere,
we should esteem this generality, and strive to exhibit
clearly what is similar in very diverse cases.   Hence the
great value of *general symbols* by which we can represent
the form and character of a reasoning process, disentangled
from any consideration of the special subject to which it is
applied.

The signs required in logic are of a very simple kind.
As every sameness or difference must exist between two
things or notions, we need signs or terms to indicate the
things or notions compared, and other signs to denote the
relation between them.   We shall need, then, (1) symbols
for terms, (2) a symbol for sameness, (3) a symbol for difference, and (4) one or two symbols to take the place of
conjunctions.

Ordinary nouns substantive, such as *Iron, Metal, Electricity, Undulation*, might serve as terms, but for the
reasons explained above it is better to adopt blank letters,
devoid of special signification, such as A, B, C, D, E, &c.
Each letter must be understood to represent a noun, and,
so far as the conditions of the argument allow, *any noun.*
Just as in Algebra, x, y, z, p, q, r, &c. are used for *any
quantities*, undetermined or unknown, except when the
special conditions of the problem are taken into account,
so will our letters stand for undetermined or unknown
things.

These letter-terms will be used indifferently for nouns
substantive and adjective.   Between these two kinds of
nouns there may be important differences in a metaphysical

or grammatical point of view. But grammatical usage readily sanctions the free conversion of adjectives into substantives, and *vice versâ;* we may avail ourselves of this latitude without in any way prejudging the metaphysical difficulties which may be involved. Here, as throughout this work, I shall devote my attention to truths which I can exhibit in a clear and formal manner, believing that, in the present condition of logical science, this will lead to much greater advantage than discussion upon the metaphysical questions which may underlie any part of the subject.

Every noun or term denotes an object, and usually implies the possession by that object of certain qualities or circumstances common to all the objects denoted. There are certain terms, however, which imply the absence of qualities or circumstances attaching to other objects. It will be convenient to employ a special mode of indicating these *negative terms,* as they are called. If the general name *A* denotes an object or class of objects possessing certain defined qualities, then the term *Not-A* will denote any object which does not possess the whole of those qualities; in short, *Not-A* is the sign for anything which differs from *A* in regard to any one or more of the assigned qualities. If *A* denote ' transparent object,' *Not-A* will denote ' not transparent object.' Brevity and facility of writing and reading are of no slight importance in a system of notation, and it will therefore be desirable to substitute for the negative term *Not-A* a briefer mode of expression. The late Prof. de Morgan represented negative terms by small Roman letters, or sometimes by small italic letters [k], and as the latter seem to be highly convenient, I shall use *a, b, c, d, e, ... p, q, r,* &c., as the negative terms corresponding to A, B, C, D, E, ... P, Q, R, &c. Thus if A means ' fluid,' *a* will mean ' not-fluid,' and so on.

[k] ' Formal Logic,' p. 38.

To denote the relation of sameness or identity I unhesitatingly adopt the sign =, so long used by mathematicians to denote equality. This symbol was originally appropriated by Robert Recorde in his 'Whetstone of Wit,' to avoid the tedious repetition of the words 'is equal to'; and he chose a pair of parallel lines, because no two things can be more equal[1]. The meaning of the sign has however been gradually extended beyond that of common equality; mathematicians have themselves used it to indicate equivalency of operations. The force of analogy has been so great that writers in all other branches of science have more or less employed the same sign. The philologist indicates by it equivalency of meaning of words: chemists adopt it to signify the identity in kind and equality in weight of the elements which form two different compounds. Not a few logicians, for instance Ploucquet, Condillac[m], George Bentham[n], Boole, have employed it as the copula of propositions. Prof. de Morgan declined to use it for this purpose, but still further extended its meaning so as to include the equivalency of a proposition with the premises from which it can be inferred[o], and Herbert Spencer has applied it in a like manner[p].

Many persons may think that the choice of a symbol is a matter of slight importance or of mere convenience, but I hold that the common use of this sign = in so many different meanings is really founded upon a generalisation

[1] Hallam's 'Literature of Europe,' First Ed. vol. ii. p. 444.

[m] Condillac, 'Langue des Calculs,' p. 157.

[n] 'Outline of a New System of Logic,' London, 1827, pp. 133, &c.

[o] 'Formal Logic,' pp. 82, 106. In his later work, 'The Syllabus of a New System of Logic,' he discontinued the use of the sign.

[p] 'Principles of Psychology,' Second Ed., vol. ii. pp. 54, 55.

of the widest character and of the greatest importance—
one indeed which it is a principal object of this work to
endeavour to explain. The employment of the same sign
in different cases would be wholly unphilosophical unless
there were some real analogy between its diverse meanings.
If such analogy exist, it is not only allowable, but highly
desirable and even imperative, to use the symbol of equivalency with a generality of meaning corresponding to the
generality of the principles involved. Accordingly Prof.
de Morgan's refusal to use the symbol in logical propositions indicated his opinion that there was a want of analogy
between logical propositions and mathematical equations.
I use the sign because I hold the contrary opinion.

I conceive that the sign $=$ always denotes some form
or degree of sameness or equivalency, and the particular
form is usually indicated by the nature of the terms joined
by it. Thus '6720 pounds $=$ 3 tons' is evidently an
equation of quantities. The formula $- \times - = +$ expresses the equivalency of operations. 'Exogens $=$ Dicotyledons' is a logical identity expressing a profound
truth concerning the character of vegetables.

We have great need in logic of a distinct sign for the
copula, because the little verb *is*, hitherto used both in
logic and ordinary discourse, is thoroughly ambiguous.
It sometimes denotes identity, as in 'St. Paul's is the
chef-d'œuvre of Sir Christopher Wren,' but it more
commonly indicates inclusion of class within class, or
partial identity, as in 'Bishops are members of the House
of Lords.' This latter relation involves identity, but requires careful discrimination from simple identity, as will
be shown further on.

When with this sign of equality we join two nouns or
logical terms, as in

Hydrogen $=$ The least dense element,
we signify that the object or group of objects denoted by

one term is identical with that denoted by the other in everything except the names. The general formula

$$A = B$$

must be taken to mean that A and B are symbols for the same object or group of objects. This identity may sometimes arise from the mere imposition of names, but it may also arise from the deepest laws of the constitution of nature ; as when we say

Gravitating matter = Matter possessing inertia,
Exogenous plants = Dicotyledonous plants,
Plagihedral quartz crystals = Quartz crystals rotating
the plane of polarisation of light.

We shall need carefully to distinguish between relations of terms which can be modified at our own will and those which are fixed as expressing the laws of nature ; but at present we are considering only the mode of expression.

We may sometimes, but much less frequently, require a symbol to indicate difference or the absence of complete sameness. For this purpose we may generalise in like manner the symbol $\sim$, which was introduced by Wallis to signify difference between two numbers or quantities. The general formula

$$B \sim C$$

denotes that B and C are the names of some two objects or groups of objects which are not identical with each other. Thus we may say

Acrogens $\sim$ Flowering plants.
Snowdon $\sim$ The highest mountain in Great Britain.

I shall also occasionally use the sign ∞ to signify in the most general manner the existence of any relation between the two terms connected by it. Thus ∞ might mean not only the relations of equality or inequality, sameness or difference, but any special relation of time, place, size, causation, &c. in which one thing may stand to another. By $A \infty B$ I mean, then, any two objects of thought related to each other in any matter whatsoever.

General Formula of Logical Inference.

The one supreme rule of inference consists, as I have said, in the direction to affirm of anything whatever is known of its like, equal or equivalent. The *Substitution of Similars* is a phrase which seems aptly to express the power of mutual replacement existing between any two objects which are to a sufficient degree like or equivalent. It is a matter for further investigation to point out when and for what purposes a degree of similarity less than complete identity is sufficient to warrant substitution. For the present we think only of the exact sameness expressed in the form

$$A = B.$$

Now if we take the letter C to denote any third conceivable object, and use the sign ∞ in its stated meaning of *indefinite relation*, then the general formula of all inference may be thus exhibited :—

From $\qquad A = B \infty C$

we may infer $\qquad A \infty C$

or, in words—*In whatever relation a thing stands to a second thing, in the same relation it stands to the like or equivalent of that second thing.* The identity between A and B allows us indifferently to place A where B was or B where A was, and there is no limit to the variety of special meanings which we can bestow upon the signs used in this formula consistently with its truth. Thus if we first specify only the meaning of the sign ∞, we may say that if *C is the weight of B,* then *C is also the weight of A.* Similarly

> If C is the father of B, C is the father of A ;
> If C is a fragment of B, C is a fragment of A ;
> If C is a quality of B, C is a quality of A ;
> If C is a species of B, C is a species of A ;
> If C is the equal of B, C is the equal of A ;

and so on *ad infinitum.*

We may also endow with special meanings the letter-
terms A, B and C, and the process of inference will never
be false. Thus let the sign ∽ mean 'is height of,' and
let

A = Snowdon,
B = Highest mountain in England or Wales,
C = 3590 feet ;

then it obviously follows that since '3590 feet is the
height of Snowdon,' and 'Snowdon = the highest mountain
in England or Wales,' then '3590 feet is the height of the
highest mountain in England or Wales.'

One result of this general process of inference is that
we may in any aggregate or complex whole replace any
part by its equivalent without altering the whole. To
alter is to make a difference, but if in replacing a part I
make no difference, there is no alteration of the whole.
Many inferences which have been very imperfectly included in logical formulæ at once follow. I remember the
late Prof. de Morgan remarking that all Aristotle's logic
could not prove that 'Because a horse is an animal, the
head of a horse is the head of an animal.' I conceive that
this amounts merely to replacing in the complete notion
head of a horse, the term 'horse' by its equivalent *some
animal* or *an animal*. Similarly, since

The Lord Chancellor = The Speaker of the House of
Lords,

it follows that

The death of the Lord Chancellor = The death of the
Speaker of the House of Lords ;

and any event, circumstance or thing which stands in a
certain relation to the one will stand in like relation to
the other. Milton reasons in this way when he says, in
his Areopagitica, 'Who kills a man, kills a reasonable creature, God's image.' If we may suppose him to mean

God's image = man = some reasonable creature,

it follows that 'The killer of a man is the killer of some reasonable creature,' and also 'The killer of God's image.'

This replacement of equivalents may be repeated over and over again to any extent. Thus if *person* is identical in meaning with *individual*, it follows that

Meeting of persons = meeting of individuals ;

and if *assemblage = meeting*, we may make a new replacement and show that

Meeting of persons = assemblage of individuals.

We may in fact found upon this principle of substitution a most general axiom in the following terms [q] :—

Same parts samely related make same wholes.

If, for instance, exactly similar bricks be used to build two houses, and they be similarly placed in each house, the two houses must be similar. There are millions of cells in a human body, but if each cell of one person were represented by an exactly similar cell similarly placed in another body, the two persons would be undistinguishable, and would be only *numerically* different. It is upon this principle, as we shall see, that all accurate processes of measurement depend. If for a weight in a scale of a balance we substitute another weight, and the equilibrium remains entirely unchanged, then the weights must be exactly equal. The general test of equality is substitution. Objects are equally bright when on replacing one by the other the eye perceives no difference. Two objects are equal in dimensions when tested by the same gauge they fit in the same manner. Generally speaking, two objects are alike so far as when substituted one for another no alteration is produced, and *vice versâ* when alike no alteration is produced by the substitution.

[q] 'Pure Logic, or the Logic of Quality,' p. 14.

The Propagating Power of Identity.

The relation of identity or sameness in all its degrees is reciprocal. So far as things are alike, either may be substituted for the other ; and this may perhaps be considered the very meaning of the relation. But it is well worth notice that there is in identity a peculiar power of extending itself among all the things which are identical. To render a number of things similar to each other we need only render them similar to one standard object. Each coin struck from a pair of dies not only exactly resembles the matrix or original pattern from which the dies were struck, but exactly resembles every other coin manufactured from the same original pattern. Among a million such coins there are not less than 499,999,500,000 of *pairs of coins* exactly resembling each other. Similars to the same are similars to all. It is one great advantage of printing that all copies of a document taken from the same type are necessarily identical each with each, and whatever is true of one copy will be true of every copy. Similarly, if fifty rows of pipes in an organ be tuned in perfect unison with one row, usually the Principal, they must be in unison with each other. Identity can also reproduce or propagate itself *ad infinitum ;* for if a number of tuning-forks be adjusted in perfect unison with one standard fork, all instruments tuned to any one fork will agree with any instrument tuned to any other fork. Standard measures of length, capacity, or weight, or any other measureable quality, are propagated in the same manner. So far as copies of the original standard, or copies of copies, or copies again of those copies, are accurately executed, they must all agree each with every other.

It is the power of mutual substitution which gives

such great value to the modern methods of mechanical construction, according to which all the parts of a machine are exact facsimiles of a fixed pattern. All the rifles used in the British army are constructed on the interchangeable system, so that any one part of any one rifle can be substituted indifferently for the same part of another. A bullet fitting one rifle will fit all others of the same bore. Sir J. Whitworth has extended the same system to the screws and screw-bolts used in connecting together the parts of machines, by establishing a series of standard screws.

Anticipations of the Principle of Substitution.

In such a subject as logic it is hardly possible to put forth any opinions or principles which have not been in some degree previously entertained. The germ at least of every doctrine will be found in earlier writers, and novelty must arise chiefly in the mode of harmonising and developing ideas. When I first proposed to employ the process and name of *substitution* in logic [r], I believe that I was led to do so from analogy with the familiar mathematical process of substituting for a symbol its value as given in an equation. In writing my first logical essay I had a most imperfect conception of the importance and generality of the process, and I described, as if they were of equal importance, a number of other laws which now seem to be but particular cases of the one general rule of substitution.

My second essay, the Substitution of Similars, was written shortly after I had become aware of the great simplification which may be effected by a proper application of the principle of substitution. I was not then acquainted with the fact that the German logician

[r] 'Pure Logic,' pp. 18–19.

Beneke had employed the principle of substitution, and had used the word itself in forming a theory of the syllogism. My imperfect acquaintance with the German language had prevented me from acquiring a complete knowledge of Beneke's views, but there is no doubt that Mr. Lindsay is right in saying that he, and probably other previous logicians, were in some degree familiar with the principle[s]. Even Aristotle's dictum may be regarded as an imperfect statement of the principle of substitution; and, as I have pointed out, we have only to modify that dictum in accordance with the quantification of the predicate in order to arrive at the complete process of substitution[t]. The Port-Royal logicians appear to have entertained nearly equivalent views, for they considered that all moods of the syllogism might be reduced under one general principle[u]. Of two premises they regard one as the *containing proposition* (propositio continens), and the other as the *applicative proposition*. The latter proposition must always be affirmative, and represents that by which a substitution is made; the former may or may not be negative, and is that in which a substitution is effected. They also show that this method will embrace certain cases of complex reasoning which had no place in the Aristotelian syllogism. Their views probably constitute the greatest improvement in logical doctrine made up to that time since the days of Aristotle. But a true reform in logic must consist, not in explaining the syllogism in one way or another, but in doing away with all the narrow restrictions of the Aristotelian system, and in showing that there exists

[s] Ueberweg's 'System of Logic,' transl. by Lindsay, pp. 442-446, 57[1], 57[2].

[t] 'Substitution of Similars,' p. 9.

[u] 'Port-Royal Logic,' transl. by Spencer Baynes, pp. 212-219. Part III. chap. x. and xi.

an indefinite extent of logical arguments immediately deducible from the principle of substitution of which the ancient syllogism forms but a small and not even the most important part.

The Logic of Relatives.

There is a difficult and important branch of logic which may be called the Logic of Relatives. If I argue, for instance, that because Daniel Bernoulli was the son of John, and John the brother of James, therefore Daniel was the nephew of James, it is not possible to prove this conclusion by any simple logical process. We require at any rate to assume that the son of a brother is a nephew. A simple logical relation is that which exists between properties and circumstances of the same object or class. But objects and classes of objects may also be related according to all the properties of time and space. I believe it may be shown, indeed, that where an inference concerning such relations is drawn, a process of substitution is really employed and an identity must exist; but I will not undertake to prove the assertion in this work. The relations of time and space are logical relations of a complicated character demanding much abstract and difficult investigation. The subject has been treated with such great ability by Professors Peirce [x], De Morgan [y], Ellis [z], and Harley, that I will not in the

[x] 'Description of a Notation for the Logic of Relatives, resulting from an Amplification of the Conceptions of Boole's Calculus of Logic.' By C. S. Peirce. 'Memoirs of the American Academy,' vol. ix. Cambridge, U.S., 1870.

[y] 'On the Syllogism, No. IV, and on the Logic of Relations.' By Augustus De Morgan. 'Transactions of the Cambridge Philosophical Society,' vol. x. part ii. 1860.

[z] 'Observations on Boole's Laws of Thought.' By the late R. Leslie Ellis; communicated by the Rev. Robert Harley, F.R.S. 'Report

present work attempt any review of their writings, but merely refer to the publications in which they are to be found.

of the British Association,' 1870.   'Report of Sections,' p. 12.   Also, 'On Boole's Laws of Thought.'   By the Rev. Robert Harley, F.R.S., ibid. p. 14.

CHAPTER II.

EVERY proposition expresses the resemblance or difference of the things denoted by its terms. As reasoning
or inference treats of the relation between two or more
propositions, so a proposition consists in a relation
between two or more terms. In the portion of this
work which treats of deduction it will be convenient
to follow the usual order of exposition, and consider in
succession the various kinds of terms, propositions, and
arguments, and we commence in this chapter with terms.

The simplest and most palpable meaning which can
belong to a term consists of some single material object,
such as Westminster Abbey, the Sun, Sirius, Stonehenge,
&c. It is probable that in the earliest stages of intellect
only concrete and palpable things are the objects of
thought. The youngest child knows the difference
between a hot and a cold body. The dog can recognise
his master among a hundred other persons, and animals
of much lower intelligence know and discriminate their
haunts. In all such acts there is judgment concerning
the likeness or unlikeness of physical objects, but there
is little or no power of analysing each object and regarding it as a group of qualities or circumstances.

The dignity of intellect begins with the power of
separating points of agreement from those of difference.
Comparison of two objects may lead us to perceive that

they are at once like and unlike. Two fragments of
rock may differ entirely in outward form, yet they may
have the same colour, hardness, and texture. Flowers
which agree in colour may differ in odour. The mind
learns to regard each object as an aggregate of qualities,
and acquires the power of dwelling at will upon one or
other of those qualities to the exclusion of the rest.
Logical abstraction, in short, comes into play, and the
mind becomes capable of reasoning, not merely about
objects which are physically complete and concrete, but
about things which may be thought of separately in
the mind though they exist not separately in nature.
We can think of the hardness of a rock, or the colour
of a flower, and thus produce abstract notions, denoted
by abstract terms which will form a subject for further
consideration.

At the same time arise general notions and classes of
objects. We cannot fail to observe that the quality
hardness exists in many objects, for instance in many
fragments of rock ; and mentally joining these we create
the class *hard object*, which will include, not only the
actual objects examined, but all others which may
happen to agree with them as they agree with each
other. As our senses cannot possibly report to us all
the contents of space, we cannot usually set any limits
to the number of objects which may fall into any such
class. At this point we begin to perceive the power and
generality of thought which enables us at once to treat
of indefinitely or even infinitely numerous objects. We
can safely assert that whatever is true of any one object
coming under a general notion or class is true of any of
the other objects so far as they possess the common
qualities implied in their belonging to the class. We
must not place an individual thing in a class unless we
are prepared to believe of it all that is believed of the

class in general ; but it remains as a matter of important consideration how far and in what manner we can safely undertake thus to assign the place of objects in that general system of classification which constitutes the whole body of science.

Twofold Meaning of General Names.

Etymologically the *meaning* of a name is what we are caused to think of when the name is used. Now every general name causes us to think of some one or more of the objects belonging to a class; it may also cause us to think of the common qualities possessed by those objects. A name is said to *denote* the distinct object of thought to which it may be applied ; it *implies* at the same time the possession of certain qualities or circumstances. The number of objects denoted forms the *extent* of meaning of the term ; the number of qualities implied forms the *intent* of meaning. Crystal is the name of any substance of which the molecules are arranged in a regular geometrical manner. The substances or objects in question form the extent of meaning ; the circumstance of having the molecules so arranged forms the intent of meaning.

When we compare a variety of general terms it may often be found that the meaning of one is included in the meaning of another. Thus all *crystals* are included among *material substances,* and all *opaque crystals* are included among *crystals:* here the inclusion is in extension. We may also have inclusion of meaning in regard to intension. For as all crystals are material substances, the qualities implied by the term material substance must be among those implied by crystal. Again, it is obvious that while in extension of meaning opaque crystals are but a part of crystals, in intension of meaning crystal is but part of opaque crystal. We increase the intent of meaning of a term by joining adjectives, or phrases equivalent to adjectives, to it, and the removal of such adjectives of course decreases the intensive meaning. Now concerning such changes of meaning the following all-important law holds universally true. *When the intent of meaning of a term is increased the extent is decreased;* and vice versâ, *when the extent is increased the intent is decreased.* In short, as one is increased the other is decreased.

This law refers only to logical changes. The number of steam engines in the world may be undergoing a rapid increase without the intensive meaning of the name being altered. The law will only be verified again when there is a real change in the intensive meaning, and an adjective may often be joined to a noun without making a change. *Elementary metal* is identical with *metal*; *mortal man* with *man*; it being a *property* of all metals to be elements, and all men to be mortals.

There is no limit to the amount of meaning which a term may have. A term may denote one object, or many, or an infinite number; it may imply a single quality, if such there be, or a group of any number of qualities, and yet the law connecting the extension and intension will infallibly apply. Taking the general name *planet*, we increase its intension and decrease its extension by prefixing the adjective *exterior*; and if we further add *nearest to the earth*, there remains but one planet *Mars*, to which the name can then be applied. Singular terms, which denote a single individual only, come under the same law of meaning as general names. They may be regarded as general names of which the meaning in extension is reduced to a minimum. Logicians have erroneously asserted, as it seems to me, that singular terms are devoid of meaning in intension, the

fact being that they exceed all other terms in that kind
of meaning, as I have elsewhere tried to show [a].

Abstract Terms.

Comparison of different objects, and analysis of the
complex resemblances and differences which they present,
lead us to the conception of *abstract qualities*. We learn
to think of one object as not only different from another,
but as differing in some particular point, such as colour,
or weight, or size. We may then convert points of
agreement or difference into separate objects of thought
called *qualities,* and denoted by *abstract terms*. Thus
the term *redness* means something in which a number
of objects agree as to colour, and in virtue of which they
are called red. Redness forms, in fact, the intensive
meaning of the term red.

Abstract terms are strongly distinguished from general
terms by possessing only one kind of meaning; for as
they denote qualities there is nothing which they can in
addition imply. The adjective 'red' is the name of red
objects, but it implies the possession by them of the
quality *redness*; but this latter term has one single
meaning—the quality alone. Thus it arises that abstract
terms are incapable of number or plurality. Red objects
are numerically distinct each from each, and there are a
multitude of such objects; but redness is a single existence which runs through all those objects, and is the
same in one as it is in another. It is true that we may
speak of *rednesses,* meaning different kinds or tints of
redness, just as we may speak of *colours,* meaning different kinds of colours. But in distinguishing kinds,

[a] J. S. Mill, 'System of Logic,' Book I. chap. ii. section 5. Jevons'
'Elementary Lessons in Logic,' pp. 41–43; 'Pure Logic,' p. 6. See
also Shedden's 'Elements of Logic,' London, 1864, pp. 14, &c.

D

degrees, or other differences, we render the terms so far
concrete. In that they are merely red there is but a
single nature in red objects, and so far as things are
merely coloured, colour is a single indivisible quality.
Redness, so far as it is redness merely, is one and the
same everywhere, and possesses absolute oneness or unity.
In virtue of this unity we acquire the power of treating
all instances of such quality as we may treat any one.
We possess, in short, general knowledge.

Substantial Terms.

Logicians appear to have taken very little notice of a
large class of terms which partake in certain respects of
the character of abstract terms and yet are undoubtedly
the names of concrete existing things. These terms are
the names of substances, such as gold, carbonate of lime,
nitrogen, &c. We cannot speak of two golds, twenty
carbonates of lime, or a hundred nitrogens. There is no
such distinction between the parts of a uniform substance as will allow of a discrimination of numerous
individuals. The qualities of colour, lustre, malleability,
density, &c., by which we recognise gold, extend through
its substance irrespective of particular size or shape. So
far as a substance is gold, it is one and the same everywhere; so that terms of this kind, which I propose to call
substantial terms, possess the peculiar unity of abstract
terms. Yet they are not abstract; for gold is of course
a tangible visible body, entirely concrete, and existing
physically independent of other bodies.

It is only when we break up, by actual mechanical
division, the uniform whole which forms the meaning of
a substantial term, that we introduce the notion of
number. *Piece of gold* is a term capable of plurality;
for there may be an endless variety of pieces discriminated

from each other, either by their various shapes and sizes, or, in the absence of such marks, by occupying simultaneously different parts of space. In substance they are one ; as regards the properties of space they are many. We need not further pursue this distinction between unity and plurality until we come to consider the principles of number in a subsequent chapter.

Collective Terms.

We must clearly distinguish between the *collective* and the *general meaning* of terms. The same name may be used to denote the whole body of existing objects of a certain kind, or any one of those objects taken separately. ' Man ' may mean the aggregate of existing men, which we sometimes describe as *mankind* ; it is also the general name applying to any man. The vegetable kingdom is the name of the whole aggregate of *plants*, but ' plant ' itself is a general name applying to any one or other plant. Every material object may be conceived as divisible into parts, and is therefore collective as regards those parts. The animal body is made up of cells and fibres, a crystal of molecules ; wherever physical division, or as it has been called *partition*, is possible, there we deal in reality with a collective whole. Thus the greater number of general terms are at the same time collective as regards each individual whole which they denote.

It need hardly be pointed out that we must not infer of a collective whole what we know of the parts, nor of the parts what we know only of the whole. The relation of whole and part is not one of identity, and does not allow of substitution. There may nevertheless be qualities or circumstances which are true alike of the whole and its parts. Thus a number of organ pipes tuned in unison produce an aggregate of sound which is of exactly the same pitch as each separate sound. In the case of substantial terms, certain qualities may be present equally in each minutest part as in the whole. The chemical nature of the largest mass of pure carbonate of lime in existence is the same as the nature of the smallest particle. In the case of abstract terms, again, we cannot draw a distinction between whole and part; what is true of redness in any case is always true of redness, so far as it is merely red.

Synthesis of Terms.

We continually combine simple terms together so as to form new terms of more complex meaning. Thus, to increase the intension of meaning of a term we write it with an adjective or a phrase of adjectival nature. By joining 'brittle' to 'metal,' we obtain a combined term, 'brittle metal,' which denotes a certain portion of the metals, namely such as are selected on account of possessing the quality of *brittleness*. As we have already seen, 'brittle metal' possesses less extension and greater intension than metal. Nouns, prepositional phrases, participial phrases and subordinate propositions may also be added to terms so as to increase their intension and decrease their extension.

In our symbolic language we need some mode of indicating this junction of terms, and the most convenient device will be the simple juxtaposition of the distinct letter-terms. Thus if A mean brittle, and B mean metal, then AB will mean brittle metal. Nor need there be any limit to the number of letters thus joined together, or the complexity of the notions which they may represent.

Thus if we take the letters

P = metal,
Q = white,
R = monovalent,

$$S = \text{of specific gravity } 10\cdot5,$$
$$T = \text{melting above } 1000°C,$$
$$V = \text{good conductor of heat and electricity,}$$

then we can form a combined term $PQRSTV$, which will denote 'a white monovalent metal, of specific gravity $10\cdot5$, melting above $1000°C$, and a good conductor of heat and electricity.'

There are many grammatical rules or usages concerning the junction of words and phrases to which we need pay no attention in logic. We can never say in ordinary language 'of wood table,' meaning 'a table of wood,' but we may consider 'of wood' as logically an exact equivalent of 'wooden'; so that if

$$X = \text{of wood,}$$
$$Y = \text{table,}$$

there is no reason why, in our symbols, XY should not be the correct term for 'table of wood.' In this case indeed we might substitute the corresponding adjective 'wooden,' but we should often fail to find any adjective answering exactly to a phrase. There is no single word which could express the notion 'of specific gravity $10\cdot5$': but logically we may consider these words as forming an adjective; and denoting this by S and metal by P, we may say that SP means 'metal of specific gravity $10\cdot5$.' It is one of many advantages in these blank letter-symbols that they enable us completely to abstract all grammatical peculiarities and fix our attention solely on the purely logical relations involved. Investigation will probably show that the rules of grammar are mainly founded upon traditional usage and have little logical signification. This indeed is sufficiently proved by the wide grammatical differences which exist between languages where the logical foundation must be the same.

Symbolic Expression of the Law of Contradiction.

The synthesis of terms is subject to the all-important Law of Thought, described in a previous section (p. 6) and called the Law of Contradiction. It is self-evident that no quality or circumstance can be both present and absent at the same time and place. This fundamental condition of all thought and all existence is expressed symbolically by a rule that a term and its negative shall never be allowed to come into combination. Such combined terms as Aa, Bb, Cc, &c. are self-contradictory and devoid of all meaning. If they represented anything, it would be what cannot exist, and cannot even be imagined in the mind. They can therefore only enter into our consideration to suffer immediate exclusion. The criterion of false reasoning, as we shall find, is that it involves self-contradiction, the affirming and denying of the same statement. Thus we might represent the object of all reasoning as the separation of the consistent and possible from the inconsistent and impossible ; and we cannot make any inference without implying that certain combinations of terms are contradictory and excluded from thought. To conclude that 'all A's are B's' is equivalent to the assertion that 'A's which are not B's cannot exist.'

It will be convenient to have the means of indicating this exclusion of the self-contradictory ; and we may use the familiar sign for *nothing*, the cipher o. Thus the second law of thought may be symbolised in the forms

$$Aa = o \qquad ABb = o \qquad ABCa = o.$$

We may variously describe the meaning of o in logic as the *non-existent*, the *impossible*, the *self-inconsistent*, the *inconceivable*. Close analogy exists between this meaning and its mathematical signification.

Certain Special Conditions of Logical Symbols.

In order that we may argue and infer truly we must treat our logical symbols according to the fundamental laws of Identity and Difference. But in thus using our symbols we shall frequently meet with combinations of which the meaning will not at first be apparent. In some cases, for instance, we may learn that an object is 'yellow and round,' in other cases that it is 'round and yellow': there arises the question whether these two descriptions are identical in meaning or not. Or again, if we proved that an object was 'round round' the meaning of such an expression would be open to doubt. Accordingly we must take notice, before proceeding further, of certain special laws which govern the combination of logical terms.

In the first place the combination of a logical term with itself is without effect, just as the repetition of a statement does not alter the meaning of the statement: 'a round round object' is simply 'a round object.' What is yellow yellow is merely yellow; metallic metals cannot differ from metals, nor elementary elements from elements. In our symbolic language we may similarly hold that AA is identical with A, or

$$A = AA = AAA = \&c.$$

The late Professor Boole is the only logician in modern times who has drawn attention to this remarkable property of logical terms[b]; but in place of the name which he gave to the law, I have proposed to call it The Law of Simplicity[c]. Its high importance will only become apparent when we attempt to determine the relations of logical and mathematical science. Two symbols of quantity, and only

[b] 'Mathematical Analysis of Logic,' Cambridge, 1847, p. 17. 'An Investigation of the Laws of Thought,' London, 1854, p. 29.

[c] 'Pure Logic,' p. 15.

two, seem to obey this law ; we may say that $1.1 = 1$, and $0.0 = 0$ (taking o to mean absolute zero or $1 - 1$) ; there is apparently no other number which combined with itself gives an unchanged result. I shall point out, however, in the chapter upon Number, that in reality all numerical symbols obey this logical principle.

It is curious that this Law of Simplicity, though almost unnoticed in modern times, was known to Boëthius, who makes a singular remark in his treatise ' De Trinitate et Unitate Dei' (p. 959). He says, ' If I should say sun, sun, sun, I should not have made three suns, but I should have named one sun so many times [d].' Ancient discussions concerning the doctrine of the Trinity drew more attention to subtle questions concerning the nature of unity and plurality than has ever since been given to them.

It is a second law of logical symbols that order of combination is a matter of indifference. ' Rich and rare gems' are the same as 'rare and rich gems,' or even as 'gems, rich and rare.' Grammatical, rhetorical or poetic usage may give considerable significance to order of expression. The limited power of our minds prevents our grasping many ideas at once, and thus the order of statement may produce some effect, but not in a strictly logical manner. All life proceeds in the succession of time, and we are obliged to write, speak, or even think of things and their qualities one after the other ; but between the things and their qualities there need be no such relation of order in time or space. The sweetness of sugar is neither before nor after its weight and solubility. The hardness of a metal, its colour, weight, opacity, malleability, electric and chemical properties, are all coexistent and coextensive, pervading the metal and every part of it

___

[d] ' Velut si dicam Sol, Sol, Sol, non tres soles effecerim, sed uno toties prædicaverim.'

in perfect community, none before nor after the others. In our words and symbols we cannot observe this natural condition; we must name one quality first and another second, just as some one must be the first to sign a petition, or to walk foremost in a procession. In nature there is no such precedence.

A little reflection will show that knowledge in the highest perfection would consist in the *simultaneous* possession of a multitude of facts. To comprehend a science perfectly we should have every fact present with every other fact. We must write a book and we must read it successively word by word, but how infinitely higher would be our powers of thought if we could grasp the whole in one collective act of consciousness. Compared with the brutes we do possess some slight approximation to such power, and it is just conceivable that in the indefinite future mind may acquire a vast increase of capacity, and be less restricted to the piecemeal examination of a subject. But I wish here to make plain that there is no logical foundation for the successive character of thought and reasoning unavoidable under our present mental conditions. The fact that we must think of one thing first, and another second, is a logical weakness and imperfection. We must describe metal as 'hard and opaque,' or 'opaque and hard,' but in the metal itself there is no such difference of order; the properties are simultaneous and coextensive in existence.

Setting aside all grammatical peculiarities which render a substantive less moveable than an adjective, and disregarding any meaning indicated by emphasis or marked order of words, we may state, as a general law of logic, that AB is identical with BA.

$$AB = BA$$
$$ABC = ACB = BCA = \&c.$$

The late Professor **Boole** first drew attention, so far as I know, to this property of logical terms, and he called it the property of Commutativeness [e]. He not only stated the law with the utmost clearness, but pointed out that it is a Law of Thought rather than a Law of Things. I shall have in various parts of the following pages to show how the necessary imperfection of our symbols expressed in this law clings to our modes of expression, and introduces complication into the whole body of mathematical formulæ, which are really founded on a logical basis.

It is of course apparent that the power of commutation belongs only to terms related in the simple logical mode of synthesis. No one can confuse 'a house of bricks,' with 'bricks of a house,' 'twelve square feet' with 'twelve feet square,' 'the water of crystallization' with 'the crystallization of water.' All relations which involve differences of time and space are inconvertible; the higher must not be made to change place with the lower, or the first with the last. For the parties concerned there is all the difference in the world between A killing B and B killing A. The law of commutativeness simply asserts that difference of order does not attach to the connection between the properties and circumstances of a thing—to what I shall call *simple logical relations*.

[e] 'Laws of Thought,' p. 29.

CHAPTER III.

We now proceed to consider the variety of forms of propositions in which the truths of science must be expressed. I shall endeavour to show that, however diverse these forms may be, they all admit the application of the one same principle of influence, that what is true of one thing or circumstance is true of the like or same. This principle holds true whatever be the kind or manner of the likeness, provided proper regard be had to its degree. Propositions may assert an identity of time, space, manner, quantity, degree, or any other circumstance in which things may agree or differ.

We find an instance of a proposition concerning time in the following :—'The year in which Newton was born, was the year in which Galileo died.' This proposition expresses an approximate identity of time between two events; hence whatever is true of the year in which Galileo died is true of that in which Newton was born, and *vice versâ*. 'Tower Hill is the place where Raleigh was executed' expresses an identity of place; and whatever is true of the one spot is true of the spot otherwise defined, but in reality the same. In ordinary language we have many propositions obscurely expressing identities of number, quantity, or degree. 'So many men, so many minds,' is a proposition concerning number or an equation; whatever is true of the number of men is true of the number of minds, and *vice versâ*. 'The density of Mars is (nearly) the same as that of the Earth,' 'The force

of gravity is directly as the product of the masses, and
inversely as the square of the distance,' are propositions
concerning magnitude or degree. Logicians have not paid
adequate attention to the great variety of propositions
which can be stated by the use of the little conjunction
as, together with *so*. 'As the home so the people,' is a
proposition expressing identity of manner ; and a great
number of similar propositions all indicating some kind of
resemblance might be quoted. Whatever be the special
kind or form of identity, all such expressions of identity
are subject to the great principle of inference ; but as we
shall in later parts of this work treat more particularly
of inference in cases of number and magnitude, we will
here confine our attention to the logical propositions
which involve only notions of quality.

Simple Identities.

The most important class of propositions consists of
those which fall under the formula

$$A = B,$$

and may be called *simple identities*. I may instance, in
the first place, those most elementary propositions which
express the exact similarity of a quality encountered in
two or more objects. I may compare by memory or
otherwise the colour of the Pacific ocean with that of
the Atlantic, and declare them identical. I may assert
that 'the smell of a rotten egg is that of hydrogen
sulphide,' 'the taste of silver hyposulphite is that of
cane sugar,' 'the sound of an earthquake is that of distant
artillery.' Such are propositions stating, accurately or
otherwise, the identity or non-identity of simple physical
sensations. Judgments of this kind are necessarily presupposed in more complex judgments. If I declare that
'this coin is made of gold,' I must base the judgment upon

the exact likeness of the substance in several qualities to other pieces of substance which are undoubtedly of gold. I must make a judgment of the colour, the specific gravity, the hardness, sound, and chemical properties; and each of these judgments might be expressed in an elementary proposition, 'the colour of this coin is the colour of gold,' and so on. Even when we establish the identity of a thing with itself under a different name or aspect, it is by distinct judgments concerning single circumstances. To prove that the Homeric χαλκός is copper we must show the identity of each quality recorded of χαλκός with a quality of copper. To establish Deal as the landing-place of Cæsar, every circumstance must be shown to agree. If the modern Wroxeter is the ancient Uriconium, there must be the like agreement of all features of the country not subject to alteration by time.

All such identities may be expressed in the form $A = B$. We may say

Colour of Pacific Ocean = Colour of Atlantic Ocean.

Smell of rotten egg = Smell of hydrogen sulphide.

In these and similar propositions we assert identity of single qualities or sensations. But in the same form we may express identity of any group of qualities, as in

χαλκός = Copper.

Deal = Landing-place of Cæsar.

A multitude of propositions involving singular terms fall into the same form, as in

The Pole star = The slowest-moving star.

Jupiter = The greatest of the planets.

The ringed planet = The planet having seven satellites.

The Queen of England = The Queen of India.

The number two = The even prime number.

Honesty = The best policy.

In mathematical and scientific theories we often meet with simple identities capable of expression in the same form. Thus in mechanical science ' The process for finding the resultant of forces = the process for finding the resultant of simultaneous velocities [a].' Theorems in geometry often give results in this form, as—

Equilateral triangles = Equiangular triangles.

Circle = Finite plane curve of constant curvature.

Circle = Curve of least perimeter.

The more profound and important laws of nature are often expressible in the form of identities ; in addition to some instances which have already been given I may suggest—

Crystals of cubical system = Crystals incapable of double refraction.

All definitions are necessarily of this form of simple identity, whether the objects defined be many, few, or singular. Thus we may say—

Common salt = Sodium chloride.

Chlorophyl = Green colouring matter of leaves.

Square = Equal-sided rectangle.

It is an extraordinary fact that propositions of this elementary form, all-important and very numerous as they are, had no recognised place in Aristotle's system of Logic. Accordingly their importance was overlooked until very recent times, and logic was the most deformed of sciences. But it is quite impossible that Aristotle or any other person should avoid constantly using them ; not a term could be defined without their use. In one place at least Aristotle actually notices a proposition of the kind. He observes :—' We sometimes say that that white thing is Socrates, or that the object approaching is Callias [b].' Here we certainly have simple identity of terms ; but he

---

[a] Thomson and Tait, ' Treatise on Natural Philosophy,' vol. i. p. 182.

[b] ' Prior Analytics,' I. cap. xxvii. 3.

considered such propositions purely accidental, and came
to the extraordinary conclusion, that 'Singulars cannot be
predicated of other terms.'

Propositions may also express the identity of extensive
groups of objects taken collectively or in one connected
whole ; as when we say—

> 'The Queen, Lords, and Commons = The Legislature
> of the United Kingdom.'

When Blackstone asserts, 'The only true and natural
foundation of society are the wants and fears of individuals,' we must interpret him as meaning that the whole
of the wants and fears of individuals in the aggregate form
the foundation of society. But many propositions which
might seem to be collective are but groups of singular propositions or identities. When we say 'Potassium and sodium
are the metallic bases of potash and soda,' we obviously
mean—

> Potassium = Metallic base of potash ;
> Sodium = Metallic base of soda.

It is the work of grammatical analysis to separate the
various propositions often combined in a single sentence.
Logic cannot be properly required to interpret the forms
and devices of language, but to treat the meaning or
information when clearly exhibited.

Partial Identities.

However numerous and important may be propositions
expressing *simple identity* of one term or class with
another, there is an almost equally important kind of
proposition which I propose to call *a partial identity*.
When we say that 'All mammalia are vertebrata,' we do
not mean that mammalian animals are identical with
vertebrate animals, but only that the mammalian form a
part of the class vertebrata. Such a proposition was
regarded in the old logic as asserting the inclusion of one

class in another, or of an object in a class. It was called a universal affirmative proposition, because the attribute *vertebrate* was affirmed of the whole subject *mammalia;* but the attribute was said to be *undistributed,* because not all vertebrata were of necessity involved in the proposition. Aristotle, overlooking the importance of simple identities, and indeed almost denying their existence, unfortunately founded his system upon the notion of inclusion in a class, in place of identity. He regarded inference as resting upon the rule that what is true of the containing class is true of the contained, instead of the vastly more general rule that what is true of a class or thing is true of the like. Thus he not only reduced logic to a fragment of its proper self, but destroyed the deep analogies which bind together logical and mathematical reasoning. Hence a crowd of defects, difficulties and errors which will long disfigure the first and simplest of the sciences.

It is surely evident that the relation of inclusion rests upon a relation of identity. Mammalian animals cannot be included among vertebrates unless they be identical with part of the vertebrates. Cabinet Ministers are included almost always in the class Members of Parliament, because they are identical with some who sit in Parliament. We may indicate this identity with a part of the larger class in various ways; as for instance—

Mammalia = part of the vertebrata

Diatoms = species of plants.

Cabinet Ministers = some Members of Parliament.

Iron = a metal.

In ordinary language the verbs *is* or *are* express mere inclusion more often than not. *Men are mortals,* means that *men* form a part of the class *mortal,* but great confusion exists between this sense of the verb and that in which it expresses identity, as in ' The sun is the centre of the planetary system.' The introduction of the indefinite article *a* often seems to express partiality, as when we say 'Iron is a metal' we clearly mean *one only* of several metals.

Certain eminent recent logicians have proposed to avoid the indefiniteness in question by what is called the Quantification of the Predicate, and they have generally used the little word *some* to show that only a part of the predicate is identical with the subject[c]. *Some* is an *indeterminate adjective;* it implies unknown qualities by which we might select the part in question if they were known, but it gives no hint as to their nature. I might make extensive use of such an indeterminate sign to express partial identities in this work. Thus, taking the special symbol $V = $ some, the general form of a partial identity would be $A = VB$, and in Boole's Logic expressions of the kind were freely used. But I find that indeterminate symbols only introduce complexity, and destroy the beauty and simple universality of the system which may be created without their use. A vague word like *some* is only used in ordinary language by *ellipsis,* and to avoid the trouble of attaining accuracy. We can always substitute for it more definite expressions if we like ; but when once the indefinite *some* is introduced we cannot replace it by the special description. We do not know whether *some* colour is red, yellow, blue, or what it is ; but on the other hand *red* colour is certainly *some* colour ; as is also yellow, blue, &c.

Throughout this system of logic I shall usually dispense with all such indefinite expressions ; and this can readily be done by substituting one of the other terms. To express the proposition 'All A's are some B's' I shall not use the form $A = VB$, but

$$A = AB.$$

[c] 'Elementary Lessons in Logic,' p. 183. 'Substitution of Similars,' p. 7.

This formula expresses that the class A is identical with the class AB; and as the latter must be a part at least of the class B, it implies the inclusion of the class A in that of B. Thus we might represent our former example thus—

Mammalia = Mammalian vertebrata.

This proposition asserts identity between a part of the vertebrata and the mammalia. If it is asked What part? the proposition affords no answer except that it is the part which is mammalian; but the assertion 'mammalia = some vertebrata' tells us no more.

It is quite likely that some readers may think this mode of representing the universal affirmative proposition of the old logic artificial and complicated. I will not undertake to convince them of the opposite at this point of the system. My justification for it will be found, not in the immediate treatment of this proposition, but in the general harmony which it enables us to discover between all parts of reasoning. I have no doubt that this is the point of critical difficulty in the relation of logical to other forms of reasoning. Grant this mode of denoting that 'all A's are B's,' and I fear no further difficulties; refuse it, and we find want of analogy and endless complication in every direction. For instance —Aristotle, in accepting inclusion of class in class as the fundamental relation of logic, was at once obliged to ignore the existence of the very extensive and all-important class of propositions denoting the similarity of one thing with another. It is on general grounds that I hope to show overwhelming reasons for seeking to reduce every kind of proposition to the form of an identity.

I may add that not a few previous logicians have accepted this view of the universal affirmative proposition. Boole often employed this mode of expression, and

Spalding [d] distinctly says that the proposition 'all metals are minerals' might be described as an assertion of *partial identity* between the two classes. Hence the name which I have adopted for the proposition.

Limited Identities.

A highly important class of propositions have the general form

$$AB = AC,$$

expressing the identity of the class AB with the class AC. In other words, 'Within the sphere of the class of things A, all the B's are all the C's,' or 'The B's and C's, which are A's, are identical.' But it will be observed that nothing is asserted concerning things which are outside of the class A; and thus the identity is of limited extent. It is the proposition B = C limited to the sphere of the class A. Thus if we say 'Plants are devoid of locomotive power,' we must limit the statement to large plants, since minute microscopic plants often have very remarkable powers of motion. When we say 'Metals possess metallic lustre,' we mean in their uncombined state.

A barrister may make numbers of most general statements concerning the relations of persons and things in the course of an argument, but it is of course to be understood that he speaks only of persons and things under the English Law. Even mathematicians make statements which are not true with absolute generality. They say that imaginary roots enter into equations by pairs; but this is only true under the tacit condition that the equations in question shall not have imaginary coefficients.[e]

[d] 'Encyclopædia Britannica,' Eighth Ed. art. Logic, sect. 37, note. 8vo reprint, p. 79.

[e] De Morgan 'On the Root of any Function.' Cambridge Philosophical Transactions, 1867, vol. xi. p. 25.

The universe, in short, within which they habitually discourse, is that of equations with real coefficients. These
implied limitations form part of that great mass of tacit
knowledge which accompanies all special arguments.

It is worthy of inquiry whether almost all identities
are not really limited to an implied sphere of meaning.
When we make such a plain statement as ' Gold is malleable' we obviously speak of gold only in its solid state ;
when we say that ' Mercury is a liquid metal' we must
be understood to exclude the frozen condition to which it
may be reduced in the Arctic regions. Even when we
take such a fundamental law of nature as ' All substances
gravitate,' we must mean by substance, material substance, not including that basis of heat, light and electrical
undulations which occupies space and possesses many
mechanical properties, but not gravity. The proposition
then is really of the form

Material substance = Material gravitating substance.

To De Morgan is due the remark, that we do usually
think and argue in a limited universe or sphere of notions
even when it is not expressly stated[f].

Negative Propositions.

In every act of intellect, as we have seen, we are engaged with a certain degree of identity or difference
between certain things or sensations compared together.
Hitherto I have treated only of identities ; and yet
it might seem that the relation of difference must be
infinitely more common than that of likeness. One
thing may resemble a great many other things, but
then it differs from all remaining things in the world.
Difference or diversity may almost be said to constitute
life, being to thought what motion is to a river. The

[f] 'Syllabus of a Proposed System of Logic,' §§ 122, 123.

very perception of an object involves its discrimination from all other objects. But we may nevertheless be said to detect resemblance as often as we detect difference. We cannot, in fact, assert the existence of a difference, without at the same time implying the existence of an agreement.

If I compare mercury, for instance, with other metals, and decide that it is *not solid*, here is a difference between mercury and solid things, expressed in a negative proposition ; but there must be implied, at the same time, an agreement between mercury and the other substances which are not solid. As it is impossible in the alphabet to separate the vowels from the consonants without at the same time separating the consonants from the vowels, so I cannot select as the object of thought *solid things*, without thereby throwing together into another class all things which are *not solid*. The very fact of not possessing a quality, constitutes a new quality or circumstance which may equally be the ground of judgment and classification. In this point of view, agreement and difference are ever the two sides of the same act of intellect, and it becomes equally possible to express the same judgment in the one or other aspect.

Between affirmation and negation there is accordingly a perfect balance or equilibrium. Every affirmative proposition implies a negative one, and *vice versâ*. It is even a matter of indifference, in a logical point of view, whether a positive or negative term be used to denote a given quality and the class of things possessing it. If the ordinary state of man's body be called *good health*, then in other circumstances he is said *not to be in good health ;* but we might equally describe him in the latter state as *sickly*, and in his normal condition he would be *not sickly*. Animal and vegetable substances are now called *organic*, so that the other substances, forming an immensely greater

part of the globe, are described negatively as *inorganic*.
But we might, with at least equal logical correctness,
have described the preponderating class of substances as
mineral, and then vegetable and animal substances would
have been *non-mineral*.

It is plain that any positive term, and its corresponding
negative divide between them the whole universe of
thought: whatever does not fall into one must fall into
the other, by the third fundamental Law of Thought,
the Law of Duality. It follows at once that there are
two modes of representing a difference. Suppose that
the things or classes represented by A and B are found
to differ, we may indicate the result of the judgment by
the notation (see p. 20)

$$A \sim B.$$

But we may now represent the same judgment by the
assertion that A agrees with those things which differ from
B, or that A agrees with the not-B's. Using our notation
for negative terms (see p. 17), we obtain

$$A = Ab$$

as the expression of the ordinary negative proposition.
Thus if we take A to mean quicksilver, and B solid, then
we have the following proposition :—

Quicksilver = Quicksilver not-solid.

There may also be several other classes of negative
propositions, of which no notice was taken in the old logic.
We may have cases where all A's are not-B's, and at the
same time all not-B's are A's ; there may, in short, be a
simple identity between A and not-B, which may be
expressed in the form

$$A = b.$$

An example of this form would be

Conductors of electricity = non-electrics.

We shall also frequently have to deal as results of

deduction, with simple, partial, or limited identities between negative terms, in the forms

$$a = b, \quad a = ab, \quad aC = bC.$$

It would be equally possible to represent affirmative propositions in the negative form. Thus 'Iron is solid,' might be expressed as 'Iron is not not-solid,' or 'Iron is not fluid'; or, taking A and b for the terms 'iron,' and 'not-solid,' the form would be

$$A \sim b.$$

But there are very strong reasons why we should employ all propositions in their affirmative form. All inference proceeds by the substitution of equivalents, and a proposition expressed in the form of an identity is ready to yield all its consequences in the most direct manner. As will be more fully shown, we can infer *in* a negative proposition, but not *by* it. Difference is incapable of becoming the ground of inference ; it is only the implied agreement with other differing objects, which admits of deduction ; and it will always be found advantageous to employ propositions in the form which exhibits clearly all the implied agreements.

Conversion of Propositions.

The old books of logic contain many rules concerning the conversion of propositions, that is, the transposition of the subject and predicate in such a way as to obtain a new proposition which will be equally true with the original. The reduction of every proposition to the form of an identity renders all such rules and processes needless. Identity is essentially reciprocal. If the colour of the Atlantic Ocean is the same as that of the Pacific Ocean, that of the Pacific must be the same as that of the Atlantic. Sodium chloride being identical with common salt, common salt must be identical with sodium

chloride.   If the number of windows in Salisbury
Cathedral equals the number of days in the year, the
number of days in the year must equal the number of
the windows.   Lord Chesterfield was not wrong when
he said, 'I will give anybody their choice of these two
truths, which amount to the same thing; He who loves
himself best is the honestest man; or, The honestest man
loves himself best[g].'   Scotus Erigena exactly expresses this
reciprocal character of identity in saying, 'There are not
two studies, one of philosophy and the other of religion;
true philosophy is true religion, and true religion is true
philosophy.'

A mathematician would not think it worth mention
that if $x = y$ then also $y = x$.   He would not consider
these to be two equations at all, but one same equation
accidentally written in two different manners.   In written
symbols one of two names must come first, and the other
second, and a like succession must perhaps be observed in
our thoughts: but in the relation of identity there is no
need for succession in order; each is simultaneously equal
and identical to the other.   These remarks will hold true
equally of logical and mathematical identity; so that I
shall consider the two forms

$$A = B \text{ and } B = A$$

to express exactly the same identity differently written.
All need for rules of conversion disappears, and there
will be no single proposition in the system which may
not be written with either term foremost.   Thus $A = AB$
is the same as $AB = A$, $AB = AC$ as $AC = AB$, and so on.

The same remarks are partially true of differences or
inequalities, which are also reciprocal to the extent that
one thing cannot differ from a second without the second
differing from the first.   Mars differs in colour from

[g] Chesterfield's Letters, 8vo, 1744; vol. i. p. 302.

Venus, and Venus must differ from Mars. The Earth differs from Jupiter in density; therefore Jupiter must differ from the Earth. Speaking generally, if $A \sim B$ we shall also have $B \sim A$, and these two forms may be considered expressions of the same difference. But the reader will notice that the relation of differing things is not wholly reciprocal. The density of Jupiter does not differ from that of the Earth in the same way that that of the Earth differs from that of Jupiter. The change of sensation which we experience in passing from Venus to Mars is not the same as what we experience in passing back to Venus, but just the opposite in nature. The colour of the sky is lighter than that of the ocean; therefore that of the ocean cannot be lighter than that of the sky, but darker. In these and all similar cases we gain a notion of *direction* or character of change, and results of immense importance may be shown to rest on this notion. For the present we shall be concerned with the mere fact of identity existing or not existing.

Twofold Interpretation of Propositions.

Terms, as we have seen (p. 31), may have a meaning either in extension or intension; and according as one or the other meaning is attributed to the terms of a proposition, so may a different interpretation be assigned to the proposition itself: When the terms are abstract we must read them in intension, and a proposition connecting such terms must denote the identity or non-identity of the qualities respectively denoted by the terms. Thus if we say

$$\text{Equality} = \text{Identity of magnitude,}$$

the assertion means that the circumstance of being equal

exactly corresponds with the circumstance of being identical in magnitude. Similarly in

Opacity = Incapability of transmitting light,

the quality of being incapable of transmitting light is declared to be the same as the intended meaning of the word opacity.

When general names form the terms of a proposition we may apply a double interpretation. Thus

Exogens = Dicotyledons

means either that the qualities which belong to all exogens are the same as those which belong to all dicotyledons, or else that every individual falling under one name falls equally under the other. Hence it may be said that there are two distinct fields of logical thought. We may argue either by the qualitative meaning of names or by the quantitative, that is, the extensive meaning. Every argument involving concrete plural terms might be converted into one involving only abstract singular terms, and *vice versâ*. But there are many reasons for believing that the intensive or qualitative form of reasoning is the primary and fundamental one. It is sufficient to point out that we may use abstract terms which contain no reference to an extensive meaning; and when there is a mode which we must sometimes and may always adopt, it is higher in importance than a mode which we never need adopt necessarily.

CHAPTER IV.

THE general principle of inference having been explained in the previous chapters, and a suitable system of symbols provided, we have now before us the comparatively easy task of tracing out the most common and important forms of deductive reasoning. The general problem of deduction is as follows :—*From one or more propositions called premises to draw such other propositions as will necessarily be true when the premises are true.* By deduction we investigate and unfold the information contained in the premises ; and this we can do by one single rule—*For any term occurring in any proposition or expression substitute the expression which is asserted in any premise to be identical with it.* To obtain certain deductions, especially those involving negative conclusions, we shall require to bring into use the second and third Laws of Thought, and the process of reasoning will then be called *Indirect Deduction.* In the present chapter, however, I shall confine my attention to those results which can be obtained by the process of *Direct Deduction,* that is, by applying to the premises themselves the rule of substitution. It will be found that we can combine in one harmonious system, not only the various moods of the ancient syllogism, but a great number of equally important forms of reasoning, which had no distinct place in the old logic. We can at the same time dispense entirely with the elaborate apparatus of logical rules and mnemonic lines, which were requisite so long as the vital principle of reasoning was not clearly expressed.

Immediate Inference.

Probably the simplest of all forms of inference is that which has been called *Immediate Inference*, because it can be performed upon a single proposition. It consists in joining an adjective, or other qualifying clause of the same nature, to both sides of an identity, and asserting the equivalence of the terms thus produced. For instance, since

Conductors of electricity = Non-electrics,

it follows that

Liquid conductors of electricity = Liquid non-electrics.

If we suppose that

Plants = Bodies decomposing carbonic acid,

it follows that

Microscopic plants = Microscopic bodies decomposing
    carbonic acid.

In general symbols, from the identity

$$A = B$$

we can infer the identity

$$AC = BC.$$

This is but a case of plain substitution; for by the first Law of Thought it must be admitted that

$$AC = AC,$$

and if in the second side of this identity we substitute for A its equivalent B, we obtain

$$AC = BC.$$

In like manner from the partial identity

$$A = AB$$

we may obtain

$$AC = ABC$$

by an exactly similar form of substitution; and in every

other case the rule will be found capable of verification by
the principle of inference. The process when performed as
here described will be found free from the liability to
error which I have shown [a] to exist in Immediate Inference
by added Determinants, as described by Dr. Thomson[b].

Inference with Two Simple Identities.

One of the most common forms of inference, and one to
which I shall especially direct attention, is practised with
two simple identities. From the two statements that
'London is the capital of England' and 'London is the
most populous city in the world,' we instantaneously draw
the conclusion that 'The capital of England is the most
populous city in the world.' Similarly, from the identities

Hydrogen = Substance of least density

Hydrogen = Substance of least atomic weight,

we infer

Substance of least density = Substance of least atomic
weight.

The general form of the argument is exhibited in the
symbols

$$B = A \qquad (1)$$
$$B = C \qquad (2)$$
$$\text{hence} \qquad A = C. \qquad (3)$$

We may describe the result by saying that terms
identical with the same term are identical with each
other; and it is impossible to overlook the analogy to the
first axiom of Euclid that 'things equal to the same thing
are equal to each other.' It has been very commonly supposed that this was a fundamental principle of thought
incapable of reduction to anything simpler. But I entertain no doubt that this form of reasoning is only one case

[a] 'Elementary Lessons in Logic,' p. 86.
[b] 'Outline of the Laws of Thought,' § 87.

of the general rule of inference. We have two propositions, $A = B$ and $B = C$, and we may for a moment consider the second one as affirming a truth concerning B
while the former one informs us that B is identical with
A ; hence by substitution we may affirm the same truth
of A. It happens in this particular form that the truth
affirmed is identity to C, and we might, if we had preferred,
have considered the substitution as made by means of the
second identity in the first. Having two identities we
have a choice of the mode in which we will make the
substitution, though the result is exactly the same in
either case.

Now compare the three following formulæ

(1) $A = B = C$ hence $A = C$
(2) $A = B \sim C$ hence $A \sim C$
(3) $A \sim B \sim C$, no inference.

In the second formula we have an identity and a difference, and we are able to infer a difference ; in the third
we have two differences and are unable to make any
inference at all. Because A and C both differ from B, we
cannot tell whether they will or will not differ from each
other. The flowers and leaves of a plant may both differ
in colour from the earth in which the plant grows, and
yet they may differ from each other ; in other cases the
leaves and stem may both differ from the soil and yet agree
with each other. Where we have difference only we can
make no inference ; where we have identity we can infer.
This fact gives great countenance to my assertion that
inference proceeds always through an identity, but may
be indifferently effected in a difference or an identity.

Deferring a more complete discussion of this point, I
will only mention now that arguments from double
identity occur very frequently, and are usually taken
for granted owing to their extreme simplicity. In the
equivalency of words it must be constantly employed. If

the ancient Greek χαλκός is our *copper*, then it must be
the French *cuivre*, the German *kupfer*, the Latin *cuprum*,
because these are words, in one sense at least, equivalent
to copper. Whenever we can give two definitions or
expressions for the same term, the formula applies; thus
Senior defined wealth as 'whatever is transferable, limited
in supply, and productive of pleasure or preventive of
pain;' it is also equivalent to 'whatever has value in
exchange;' hence obviously 'Whatever has value in exchange' = 'Whatever is transferable, limited in supply, and
productive of pleasure or preventive of pain.' Two expressions for the same term are often given in the same
sentence, and their equivalency implied. Thus Thomson
and Tait say[c], 'The naturalist may be content to know
matter as that which can be perceived by the senses, or as
that which can be acted upon by or can exert force.' I
take this to mean—

Matter = what can be perceived by the senses;

Matter = what can be acted upon by or can exert force.

For the term 'matter' in either of these identities we
may substitute its equivalent given in the other definition.
Elsewhere they often employ sentences of the form exemplified in the following[d]; 'The integral curvature, or whole
change of direction of an arc of a plane curve, is the angle
through which the tangent has turned as we pass from
one extremity to the other.' This sentence is certainly of
the form—

The integral curvature = the whole change of direction,
&c. = the angle through which the tangent has
turned, &c.

Disguised cases of the same kind of inference occur
throughout all sciences, and a remarkable instance is
found in algebraic geometry. Mathematicians readily

[c] 'Treatise on Natural Philosophy,' vol. i. p. 161.

[d] *Ibid.* vol. i. p. 6.

show that every equation of the form $y = mx + c$ is equivalent to or represented by a straight line ; it is also easily proved that the same equation is equivalent to one of the form $Ax + By + C = 0$, and *vice versâ.* Hence it follows that every equation of the first degree is equivalent to or represents a straight line [e].

Inference with a Simple and a Partial Identity.

A form of reasoning somewhat different from that last considered consists in inference between a simple and a partial identity. If we have two propositions of the form

$$A = B,$$
$$B = BC,$$

we may then substitute for B in either proposition its equivalent in the other, getting in both cases $A = BC$; in this we may if we like make a second substitution for B, getting

$$A = AC.$$

Thus, since ' Mont Blanc is the highest mountain in Europe, and Mont Blanc is deeply covered with snow,' we infer by an obvious substitution that ' The highest mountain in Europe is deeply covered with snow.' These propositions when rigorously stated fall into the form above exhibited.

This form of inference is constantly employed when for a term we substitute its definition, or *vice versâ.* The very purpose of a definition is in fact to allow a single term to be employed in place of a long descriptive phrase. Thus when we say ' Circles are curves of the second degree,' we may substitute the definition of a circle, getting ' A plane curve, all points of whose perimeter are at equal distances from a certain fixed point, is a curve of

[e] Todhunter's ' Plane Co-ordinate Geometry,' chap. ii. pp. 11-14.

the second degree.' The real forms of the propositions
here given are exactly those shown in the symbolic statement, but in this and many other cases it will be sufficient
to state them in ordinary elliptical language for sake of
brevity. In scientific treatises a term and its definition
are often both given in the same sentence, as in 'The
weight of a body in any given locality, or the force with
which the earth attracts it, is proportional to its mass.'
The conjunction *or* in this statement gives the force of
equivalence to the parenthetic definition, so that the
propositions really are

> Weight of a body = force with which the earth at-
> tracts it.
> Weight of a body = weight, &c. proportional to its
> mass.

A slightly different case of inference consists in substituting in a proposition of the form $A = AB$ a definition of the term B. Thus from $A = AB$ and $B = C$
we get $A = AC$. For instance, we may say that 'Metals
are elements' and 'Elements are incapable of decomposition.'

> Metal = metal element.
> Element = what is incapable of decomposition.

Hence

> Metal = metal incapable of decomposition.

It is almost needless to point out that the form of these
arguments would not suffer any real modification if some
of the terms happened to be negative; indeed in the last
example 'incapable of decomposition' may be treated as
a negative term. Taking

> A = metal
> B = element
> C = what is capable of decomposition
> c = what is incapable of decomposition (p. 17);

the propositions are of the form

$$A = AB$$
$$B = c\,;$$

whence, by substitution,

$$A = Ac.$$

Inference of a Partial from Two Partial Identities.

However common be the cases of inference already noticed, there is a form occurring almost more frequently, and which deserves much attention because it occupied a prominent place in the ancient syllogistic system. That system strangely overlooked all the kinds of argument we have as yet considered, and selected as the type of all reasoning one which employs two partial identities as premises. Thus from the propositions

Sodium is a metal	(1)
Metals conduct electricity,	(2)

we may conclude that

Sodium conducts electricity.	(3)

Taking A, B, C, respectively to represent the three terms, the premises are of the form

$A = AB$	(1)
$B = BC.$	(2)

Now for B in (1) we can substitute its description as given in (2), obtaining

$A = ABC,$	(3)

or, in words, from

Sodium = sodium metal	(1)
Metal = metal conducting electricity,	(2)

we infer

Sodium = sodium metal conducting electricity, (3)

which in the elliptical language of common life becomes

'Sodium conducts electricity.'

The above is a syllogism in the mood called Barbara [f] in the truly barbarous language of ancient logicians ; and the first figure of the syllogism alone contained three other moods which were esteemed distinct forms of argument. But it is worthy of notice that without any real change in our form of inference we readily include these three other moods under it. The negative mood Celarent will be represented by the example

	Neptune is a planet	(1)
	No planet has retrograde motion,	(2)
hence	Neptune has not retrograde motion.	(3)

If we put A for Neptune, B for planet, and C for ' having retrograde motion,' then by the corresponding negative term c, we denote 'not having retrograde motion.' The premises now fall into the form

$$A = AB \qquad (1)$$
$$B = Bc, \qquad (2)$$

and by substitution for B, exactly as before, we obtain

$$A = ABc. \qquad (3)$$

What is called in the old logic a particular conclusion may be deduced without any real variation in the symbols. Particular quantity is indicated, as before mentioned (p. 49), by joining to the term an indefinite adjective of quantity, such as *some, a part, certain,* &c., meaning that an unknown part of the term enters into the proposition as subject. Considerable doubt and ambiguity arise out of the question whether the part may not in some cases be the whole, and in the syllogism at least it must be understood in this sense [g]. Now if we take a letter to represent this indefinite part, we need make no change in

---

[f] An explanation of this and other technical terms of the old logic will be found in my 'Elementary Lessons in Logic,' Second Ed. 1871. Macmillan & Co.

[g] 'Elementary Lessons in Logic,' pp. 67. 79.

our formulæ to express either of the syllogisms Darii or
Ferio. Consider the example—

>   Some metals are of less density than water   (1)
>   All bodies of less density than water will float
>   upon its surface   (2)
>   Some metals will float upon its surface.   (3)

Let   A = some metals
      B = body of less density than water
      C = floating on the surface of water ;
then the propositions are evidently as before,

$$A = AB \qquad\qquad (1)$$
$$B = BC ; \qquad\qquad (2)$$

hence $A = ABC.$ (3)

Thus the syllogism Darii does not really differ from Barbara. If the reader prefer it, we can readily employ a
distinct symbol for the indefinite sign of quantity.

Let   P = some
          Q = metal,

B and C having the same meanings as before.   Then the
premises become

$$PQ = PQB \qquad\qquad (1)$$
$$B = BC ; \qquad\qquad (2)$$

hence, by substitution, as before,

$$PQ = PQBC. \qquad\qquad (3)$$

Except that the formulæ look a little more complicated
there is no difference whatever.

The mood Ferio is of exactly the same character as
Darii or Barbara, except that it involves the use of a
negative term. Take the example—

>   Bodies which are equally elastic in all directions do
>   not doubly refract light,
>   Some crystals are bodies equally elastic in all direc-
>   tions; therefore some crystals do not doubly
>   refract light.

Assigning the letters as follows—

> A = some crystals.
> B = bodies equally elastic in all directions
> C = doubly refracting light
> c = not doubly refracting light.

Our argument is of the same form as before, and may be concisely stated in one line

$$A = AB = ABc.$$

If we take PQ for the indefinite *some crystals*, we have

$$PQ = PQB = PQBc.$$

The only difference is that the negative term *c* occurs instead of C in the mood Darii (p. 68).

On the Ellipsis of Terms in Partial Identities.

The reader will probably have noticed that the conclusion which we obtain from premises is often more full than that drawn by the old Aristotelian processes. Thus from 'Sodium is a metal,' and 'Metals conduct electricity,' we inferred (p. 66) that 'Sodium = sodium metal, conducting electricity,' whereas the old logic simply concludes that 'Sodium conducts electricity.' Symbolically, from $A = AB$, and $B = BC$, we get $A = ABC$, whereas the old logic gets at the most $A = AC$. It is therefore well to show that without employing any other principles of inference than those already described, we may infer $A = AC$ from $A = ABC$, though we cannot infer the latter more full and accurate result from the former. We may show this most simply as follows :—

By the first law of thought it is evident that

$$AA = AA ;$$

and if we have given the proposition $A = ABC$, we may substitute for both the A's in the second side of the above, obtaining

$$AA = ABC . ABC.$$

But from the property of logical symbols expressed in the

Law of Simplicity (p. 39) some of the repeated letters may be made to coalesce, and we have

$$A = ABC . C.$$

Substituting again for ABC its equivalent A, we obtain

$$A = AC,$$

the desired result.

By a similar process of reasoning it may be shown that we can always drop out any term appearing in one member of a proposition, provided that we substitute for it the whole of the other member. This process was described in my first logical Essay[h], as *Intrinsic Elimination*, but it might perhaps be better entitled the *Ellipsis of Terms*. It enables us to get rid of needless terms by strict substitutive reasoning.

Inference of a Simple from Two Partial Identities.

Two terms may be connected together by two partial identities in yet another manner, and a case of inference then arises which is of the highest importance. In the two premises

$$A = AB \tag{1}$$
$$B = AB, \tag{2}$$

the second member of each is the same; so that we can by obvious substitution obtain

$$A = B.$$

Thus in plain geometry we readily prove that 'Every equilateral triangle is also an equiangular triangle,' and we can with equal ease prove that 'Every equiangular triangle is an equilateral triangle.' Thence by substitution, as explained above, we pass to the simple identity—

Equilateral triangle = equiangular triangle.

We thus prove that one class of triangles is entirely identical with another class; that is to say, they differ only in our way of naming and regarding them.

[h] 'Pure Logic,' p. 19.

The great importance of this process of inference arises from the fact that the conclusion is more simple and general than either of the premises, and contains as much information as both of them put together. It is on this account constantly employed in inductive investigation, as will afterwards be more fully explained, and it is the natural mode by which we arrive at a conviction of the truth of simple identities as existing between classes of numerous objects.

Inference of a Limited from Two Partial Identities.

We have just considered arguments which are of the type treated by Aristotle in the first figure of the syllogism. But there are two other types of argument which employ a pair of partial identities. If our premises are, as shown in these symbols,

$$B = AB \qquad\qquad (1)$$
$$B = CB, \qquad\qquad (2)$$

we may substitute for B either by (1) in (2) or by (2) in (1), and by both modes we obtain the conclusion

$$AB = CB, \qquad\qquad (3)$$

a proposition of the kind which we have called a limited identity (p. 51). Thus, for example,

Potassium = potassium metal $\qquad\qquad$ (1)

Potassium = potassium floating on water ; $\qquad$ (2)

hence

Potassium metal = potassium floating on water. $\quad$ (3)

Now this is really a syllogism of the mood Darapti in the third figure, except that we obtain a conclusion of a much more exact character than the old syllogism gives. From the premises 'Potassium is a metal' and 'Potassium floats on water,' Aristotle would have inferred that 'Some metals float on water.' But if inquiry were made what the some metals are, the answer would certainly be 'Metal which is potassium.' Hence Aristotle's conclusion simply leaves out some of the information afforded in the premises; it even leaves us open to interpret the *some metals* in a wider sense than we are warranted in doing. From these distinct defects of the syllogism the process of substitution is free, and it only incurs the possible objection of being tediously minute and accurate.

Miscellaneous Forms of Deductive Inference.

The more simple and common forms of deductive reasoning having been exhibited and demonstrated on the principle of substitution, there remain many, in fact an indefinite number, which may be explained with nearly equal ease. Such as involve the use of disjunctive propositions will be deferred to a later chapter, and several of the syllogistic moods which include negative terms will be more conveniently treated after we have introduced the symbolic use of the second and third laws of thought.

We sometimes meet with a chain of propositions which allow of repeated substitution and form an argument called in the old logic a Sorites. Take, for instance, the premises

Iron is a metal	(1)
Metals are good conductors of electricity	(2)
Good conductors of electricity are useful for telegraphic purposes.	(3)

It obviously follows that

Iron is useful for telegraphic purposes.	(4)

Now if we take our letters thus—

A = Iron,   B = metal,   C = good conductor of electricity,   D = useful for telegraphic purposes, the premises will assume the form—

$$A = AB \qquad (1)$$
$$B = BC \qquad (2)$$
$$C = CD \qquad (3)$$

For B in (1) we can substitute its equivalent in (2), and

for C in (2) we can substitute its equivalent in (3). We shall obtain as an intermediate result,

$$A = ABC,$$

and from this the complete conclusion

$$A = ABCD. \qquad (4)$$

The full interpretation is that *Iron is iron, metal, good conductor of electricity, useful for telegraphic purposes,* which is abridged in common language by the ellipsis of the circumstances which are not of immediate importance.

Instead of all the propositions being of one type, as in the last example, we may have a series of premises of various character ; for instance

Common salt is sodium chloride	(1)
Sodium chloride crystallizes in a cubical form	(2)
What crystallizes in a cubical form does not possess the power of double refraction ;	(3)

it will follow that

Common salt does not possess the power of double refraction.	(4)

Taking our letter-terms thus—

A = Common salt,
B = Sodium chloride,
C = Crystallizing in a cubical form,
D = Possessing the power of double refraction,

we may state the premises in the form

$$A = B, \qquad (1)$$
$$B = BC, \qquad (2)$$
$$C = Cd. \qquad (3)$$

Substituting by (2) in (1) and by (3) in (2) we obtain

$$A = BCd, \qquad (4)$$

which is a more precise version of the common conclusion.

We often meet with a series of propositions describing the qualities or circumstances of one same thing, and we may if we like combine them all into one proposition by the process of substitution. This case is, in fact, that which Archbishop Thomson has called 'immediate inference by the sum of several predicates,' and his example will serve my purpose well[i]. He describes copper as 'A metal, of a red colour, and disagreeable smell and taste, all the preparations of which are poisonous, which is highly malleable, ductile, and tenacious, with a specific gravity of about 8.83.' If we assign the letter A to copper, and the succeeding letters of the alphabet in succession to the series of predicates, we have nine distinct statements, of the form

$$A = AB \ (1) \quad A = AC \ (2) \quad A = AD \ (3) \ldots \ldots A = AK \ (9).$$

We can readily combine these propositions into one by substituting for A in the second side of (1) its expression in (2). We thus get

$$A = ABC,$$

and by repeating the process over and over again we obtain the single proposition

$$A = ABCDEFGHIJK.$$

But Dr. Thomson is mistaken in supposing that we can obtain in this manner a definition of copper. Strictly speaking, the above proposition is only a description of copper, and all the ordinary descriptions of substances in scientific works may be summed up in this form. Thus we may assert of the organic substances called Paraffins that they are all saturated hydrocarbons, incapable of uniting with other substances, produced by heating the alcoholic iodides with zinc, and so on. It may be shown that no amount of ordinary description can be equivalent to definition.

Fallacies.

I have hitherto been engaged in showing that all the forms of reasoning of the old syllogistic logic, and an indefinite number of other forms in addition, may be

[i] 'An Outline of the Laws of Thought,' Fifth Ed. p. 161.

readily and clearly explained on the single principle of substitution. It is now desirable to show that the same principle would prevent us falling into fallacies. So long as we exactly observe the one rule of substitution of equivalents it will be impossible to commit a *paralogism,* or to break any one of the elaborate rules of the ancient system. One rule is thus proved to be as powerful as the six, eight, or more rules by which the correctness of syllogistic reasoning was guarded.

It was a fundamental rule, for instance, that two negative premises could give no conclusion. If we take the propositions—

Granite is not a sedimentary rock, (1)

Basalt is not a sedimentary rock, (2)

we ought not to be able to draw any inference concerning the relation of granite and basalt. Taking our letter-terms thus

$$A = \text{granite}$$
$$B = \text{sedimentary rock}$$
$$C = \text{basalt,}$$

the premises may be expressed in the form

$$A \smallsmile B \qquad (1)$$
$$C \smallsmile B. \qquad (2)$$

We have in this form two statements of difference; but the principle of inference can only work with a statement of agreement or identity (p. 62). Thus our rule gives us no power whatever of drawing any inference.

It is to be remembered, indeed, that we claim the power of always turning a negative proposition into an affirmative one; and it might seem that the old rule of negative premises would be thus circumvented. Let us try. The premises (1) and (2) when affirmatively stated (see p. 54), will take the form

$$A = Ab \qquad (1)$$
$$C = Cb. \qquad (2)$$

The reader will find it impossible by the rule of substitution to discover a relation between A and C. Three terms occur in these premises, namely A, b, and C; but they are so combined that no term occurring in one has its exact equivalent stated in the other. No substitution can therefore be made, and the principle holds true. Fallacy is impossible.

It would be a mistake to suppose that the mere occurrence of negative terms in both premises render them incapable of yielding a conclusion. The old rules of logic informed us that from two negative premises no conclusion could be drawn, but it is a fact that the rule in this bare form does not hold universally true; and I am not aware that any precise explanation has been given of the conditions under which it is or is not imperative. Consider the following example—

Whatever is not metallic is not capable of power-
    ful magnetic influence,   (1)

Carbon is not metallic,   (2)

Therefore, carbon is not capable of powerful mag-
    netic influence.   (3)

Here we have two distinctly negative premises (1) and (2), and yet they yield a perfectly valid negative conclusion (3). The syllogistic rule is actually falsified in its bare and general statement. In this and many other cases we can convert the propositions into affirmative ones which yield a conclusion. To show this let

    A = carbon,   B = metallic,

    C = capable of powerful magnetic influence.

The premises readily take the form

$$b = bc \qquad (1)$$
$$A = Ab, \qquad (2)$$

and substitution for b in (2) by means of (1), gives the conclusion

$$A = Abc \qquad (3)$$

Our principle of inference then includes the rule of negative premises whenever it is true, and discriminates correctly between the cases where it does and does not apply.

The paralogism, anciently called *Undistributed Middle,* is also easily exhibited and infallibly avoided by our system. Let the premises be

Hydrogen is an element, (1)

All metals are elements. (2)

According to the syllogistic rules the middle term *element* is here undistributed, and no conclusion can be obtained; we cannot tell then whether hydrogen is or is not a metal. Represent the terms as follows—

$$A = \text{hydrogen}$$
$$B = \text{element}$$
$$C = \text{metal.}$$

The premises then become

$$A = AB \qquad (1)$$
$$C = CB. \qquad (2)$$

The reader will here, as in a former page (p. 75), find it impossible to make any substitution. The only term which occurs in both premises is B, but it is combined with different letters. For CB we cannot substitute the equivalent of AB. We have no right to decompose combinations; and if we adhere rigidly to the rule given, that if two terms are stated to be equivalent we may substitute one for the other, we cannot commit the fallacy. It is apparent that the form of premises given above is the same as that which we obtained by translating two negative premises into the affirmative form.

The old fallacy, technically called the Illicit Process of the Major Term, is more easy to commit and more difficult to detect than any other breach of the syllogistic rules. In our system it could hardly occur. From the premises

All planets are subject to gravity, (1)

Fixed stars are not planets, (2)

we might inadvertently but fallaciously infer that, 'Fixed stars are not subject to gravity.' To reduce the premises to symbolic form, let

$$A = \text{planet}$$
$$B = \text{fixed star}$$
$$C = \text{subject to gravity};$$

then we have the propositions

$$A = AC \qquad\qquad (1)$$
$$B = Ba. \qquad\qquad (2)$$

The reader will try in vain to produce from these premises by legitimate substitution any relation between B and C; he could not then commit the fallacy of asserting that B is not C.

There remain two other kinds of paralogism, commonly known as the fallacy of Four Terms and the Illicit Process of the Minor term. They are so evidently impossible while we obey the rule of the substitution of equivalents, that it is not necessary to give any illustrations. When there are four distinct terms in two propositions there could be no opening for a substitution. As to the Illicit Process of the Minor it consists in a flagrant substitution for a term of another wider term which is not known to be equivalent to it, and which is therefore forbidden by our rule to be substituted for it.

CHAPTER V.

In the previous chapter I have exhibited various forms of deductive reasoning by the process of substitution, so far as they can be treated without the use of disjunctive propositions ; but we cannot long defer the consideration of this more complex class of identities. General terms arise, as we have seen (p. 29), from classifying or mentally uniting together all objects which agree in certain qualities, the value of this union consisting in the fact that the power of knowledge is multiplied thereby. In forming such classes or general notions, we overlook or abstract the points of difference which exist between the objects joined together, and fix our attention only on the points of agreement. But every process of thought may be said to have its inverse process, which consists in undoing the effects of the direct process. Just as division undoes multiplication, and evolution undoes involution, so we must have a process which undoes abstraction, or the operation of forming general notions. This inverse process will consist in distinguishing the separate objects or minor classes which are the constituent parts of any wider class. When we mentally unite together certain objects visible in the sky and call them planets, we shall afterwards need to distinguish the contents of this general notion, which we do in the disjunctive proposition —

A planet is either Mercury or Venus or the Earth or ..... or Neptune.

Having formed the very wide class 'vertebrate animal,'

we may specify its subordinate classes thus :—'A vertebrate animal is either a mammalian, bird, reptile, or fish.' Nor is there any limit to the number of possible alternatives. 'An exogenous plant is either a ranunculus, a poppy, a crucifer, a rose, or it belongs to some one of the other seventy natural orders of exogens at present recognised by botanists.' A cathedral church in England must be either that of London, Canterbury, Winchester, Salisbury, Manchester, or of one of about twenty-four cities possessing such churches. And if we were to attempt to specify the meaning of the term 'star,' we should require to enumerate as alternatives, not only the many thousands of stars recorded in catalogues, but the many millions yet unnamed.

Whenever we thus distinguish the parts of a general notion we employ a disjunctive proposition, in at least one side of which are several alternatives joined by the so called disjunctive conjunction *or*, a contracted form of *other*. There must be some relation between the parts thus connected in one proposition; we may call it the *disjunctive* or *alternative* relation, and we must carefully inquire into its nature and results. This relation is that of doubt and ignorance, giving rise to choice or uncertainty. Whenever we classify and abstract we must open the way to such uncertainty. By fixing our attention on certain attributes to the exclusion of others we necessarily leave it doubtful what those other attributes are. The term 'molar tooth' bears upon the face of it that it is a part of the wider term 'tooth.' But if we meet with the simple term 'tooth' there is nothing to indicate whether it is an incisor, a canine, or a molar tooth. This doubt, however, may be resolved by other information, and we have to consider what are the appropriate logical processes for treating disjunctive propositions in connection with other propositions disjunctive or otherwise.

Expression of the Alternative Relation.

In order to represent disjunctive propositions with convenience we require a sign of the alternative or disjunctive relation, equivalent to one meaning at least of the little conjunction *or* so frequently used in common language. I propose to use for this purpose the symbol ┿. In my first logical Essay I followed the example of Dr. Boole and adopted the common sign + ; but this sign should not be employed unless there exists exact analogy between mathematical addition and logical alternation. We shall find that the analogy is of a very partial character, and that there is such profound difference between a logical and a mathematical term as should prevent our uniting them by the same symbol. Accordingly I have chosen a sign ┿, which seems aptly to suggest whatever degree of analogy may exist without implying more. The exact meaning of the symbol we will now proceed to investigate and determine.

Nature of the Alternative Relation.

Before treating disjunctive propositions it is indispensable to decide whether the alternatives shall be considered exclusive or unexclusive. By *exclusive alternatives* we mean those which cannot contain the same things. Thus

Matter is solid, or liquid, or gaseous ;

but the same portion of matter cannot be at once solid and liquid, properly speaking ; still less can we suppose it to be solid and gaseous, or solid, liquid and gaseous all at the same time. Many examples on the other hand can readily be suggested in which two or more alternatives may hold true of the same object. Thus

Luminous bodies are self-luminous or luminous by reflection.

It is undoubtedly possible by the laws of optics, that the
same surface may at one and the same moment give off
light of its own and reflect the light from other bodies.
We speak familiarly of *deaf or dumb* persons, knowing
that the majority of those who are deaf from birth are
also dumb.

There can be no doubt that in a great many cases,
perhaps the greater number of cases, alternatives are
exclusive as a matter of fact. Any one number is incompatible with any other; one point of time or place is
exclusive of all others. Roger Bacon died either in 1284
or 1292; it is certain that he could not die in both years.
Henry Fielding was born either in Dublin or Somerset-
shire; he could not be born in both places. There is so
much more precision and clearness in the use of exclusive
alternatives that we ought doubtless to select them
when possible. Old works on logic accordingly contained
a rule directing that the *Membra dividentia*, the parts of
a division or the constituent species of a genus should be
exclusive of each other.

It is no doubt owing to the great prevalence and
convenience of exclusive divisions that the majority of
logicians have held it necessary to make every alternative
in a disjunctive proposition exclusive of every other one.
Aquinas considered that when this was not the case the
proposition was actually *false*, and Kant adopted the same
opinion[a]. A multitude of statements to the same effect
might readily be quoted, and if the question were to be
determined by the weight of historical evidence, it would
.certainly go against my view. Among recent logicians
Sir W. Hamilton, as well as Dr. Boole, took the exclusive
side. But there are authorities to the opposite effect.
Whately, Mansel, and J. S. Mill, have all pointed out that

---

[a] Mansel's 'Aldrich,' p. 103, and 'Prolegomena Logica,' p. 221.

we may often treat alternatives as *Compossible*, or true at the same time. Whately gives as an example[b], 'Virtue tends to procure us either the esteem of mankind, or the favour of God,' and he adds, 'Here both members are true, and consequently from one being affirmed we are not authorized to deny the other. Of course we are left to conjecture in each case, from the context, whether it is meant to be implied that the members are or are not exclusive.' Mansel says[c], '*We may happen to know* that two alternatives cannot be true together, so that the affirmation of the second necessitates the denial of the first; but this, as Boethius observes, is a *material*, not a *formal* consequence.' Mr. J. S. Mill has also pointed out the absurdities which would arise from always interpreting alternatives as exclusive. 'If we assert,' he says[d], 'that a man who has acted in some particular way must be either a knave or a fool, we by no means assert, or intend to assert, that he cannot be both.' Again, 'to make an entirely unselfish use of despotic power, a man must be either a saint or a philosopher. . . . . . . Does the disjunctive premise necessarily imply, or must it be construed as supposing, that the same person cannot be both a saint and a philosopher? Such a construction would be ridiculous.'

I discuss this subject fully because it is really the point which separates my logical system from that of the late Dr. Boole. In his 'Laws of Thought' (p. 32) he expressly says, 'In strictness, the words "and," "or," interposed between the terms descriptive of two or more classes of objects, imply that those classes are quite distinct, so that no member of one is found in another.' This I altogether

---

[b] 'Elements of Logic,' Book II. chap. iv. sect. 4.

[c] Aldrich, 'Artis Logicæ Rudimenta,' p. 104.

[d] 'Examination of Sir W. Hamilton's Philosophy,' pp. 452–454.

dispute. In the ordinary use of these conjunctions we do not necessarily join distinct terms only ; and when terms so joined do prove to be logically distinct, it is by virtue of a *tacit premise,* something in the meaning of the names and our knowledge of them, which teaches us they are distinct. And when our knowledge of the meanings of the words joined is defective it will often be impossible to decide whether terms joined by conjunctions are exclusive or not.

Take, for instance, the proposition 'A peer is either a duke, or a marquis, or an earl, or a viscount, or a baron.' If expressed in Professor Boole's symbols, it would be implied that a peer cannot be at once a duke and marquis, or marquis and earl. Yet many peers do possess two or more titles, and the Prince of Wales is Duke of Cornwall, Earl of Dublin, and Baron Renfrew. If it were enacted by parliament that no peer should have more than one title, this would be the tacit premise which Professor Boole assumes to exist. Nor is the restriction true of more common terms.

In the sentence 'Repentance is not a single act, but a habit or virtue,' it cannot be implied that a virtue is not a habit ; by Aristotle's definition it is.

Milton has the expression in one of his sonnets, 'Unstain'd by gold or fee,' where it is obvious that if the fee is not always gold, the gold is meant to be a fee or bribe.

Tennyson has the expression 'wreath or anadem.' Most readers would be quite uncertain whether a wreath may be an anadem, or an anadem a wreath, or whether they are quite distinct or quite the same.

From Darwin's 'Origin,' I take the expression, 'When we see any *part or organ* developed in a remarkable *degree or manner.*' In this, *or* is used twice, and neither time disjunctively. For if *part* and *organ* are not synonymous, at any rate an organ is a part. And it is obvious that a part may be developed at the same time both in an extraordinary degree and manner, although such cases may be comparatively very rare.

From a careful examination of ordinary writings, it will thus be found that the meanings of terms joined by 'and' 'or' vary from absolute identity up to absolute contrariety. There is no logical condition of distinctness at all, and when we do choose exclusive expressions, it is because our subject demands it. The matter, not the form of an expression, points out whether terms are exclusive[e].

The question, as we shall afterwards see, is one of the greatest theoretical importance, because it furnishes the true distinction between the sciences of Logic and Mathematics. It is the very foundation of number that every unit shall be distinct from every other unit; but Dr. Boole imported the conditions of number into the science of Logic, and produced a system which, though wonderful in its results, was not a system of logic at all.

Laws of the Disjunctive Relation.

In considering the combination or synthesis of terms (p. 39), we found that certain laws, those of Simplicity and Commutativeness, must be observed. In uniting terms by the disjunctive symbol we shall find that the same or closely similar laws hold true. The alternatives of either member of a disjunctive proposition are certainly commutative. Just as we cannot properly distinguish between *rich and rare gems* and *rare and rich gems,* so we must consider as identical the expression *rich or rare gems,* and *rare or rich gems.* In our symbolic language we may say generally

$$A + B = B + A.$$

[e] 'Pure Logic,' pp. 76, 77.

The order of statement, in short, has no effect upon the meaning of an aggregate of alternatives, so that the Law of Commutativeness holds true of the disjunctive symbol.

As we have admitted the possibility of joining as alternatives terms which are not really different, the question arises, How shall we treat two or more alternatives when they are clearly shown to be the same? If we have it asserted that P is Q or R, and it is afterwards proved that Q is but another name for R, the result is that P is either R or R. How shall we interpret such a statement? What would be the meaning, for instance, of 'wreath or anadem' if, on referring to a dictionary, we found *anadem* described as a wreath? I take it to be self-evident that the meaning would then become simply 'wreath.' Accordingly we may affirm the general law

$$A + A = A.$$

Any number of identical alternatives may always be reduced to, and are logically equivalent to, any one of those alternatives. This is a law which distinguishes mathematical terms from logical terms, because it obviously does not apply to the former. I propose to call it the *Law of Unity*, because it must really be involved in any definition of a mathematical unit. This law is closely analogous to the Law of Simplicity, $AA = A$; and the nature of the connection is worthy of attention.

I am not aware that logicians have in any adequate way noticed the close relation between combined and disjunctive terms, namely that every disjunctive term is the negative of a corresponding combined term, and *vice versâ*. Consider the term

Malleable dense metal.

How shall we describe the class of things which are not malleable-dense-metals? Whatever is included under that term must have all the qualities of malleability, denseness, and metallic nature. Wherever any one or more of the qualities is wanting, the combined term will not apply. Hence the negative of the whole term is

Not-malleable or not-dense or not-metallic.

In the above the conjunction *or* must clearly be interpreted as unexclusive; for there may readily be objects which are both not-malleable, and not-dense, and perhaps not-metallic at the same time. If in fact we were required to use *or* in a strictly exclusive manner, it would be requisite to specify seven distinct alternatives in order to describe the negative of a combination of three single terms. The negatives of four or five terms would consist of fifteen or thirty-one alternatives. This consideration alone is sufficient to prove that the meaning of *or* cannot be always exclusive in common language.

Expressed symbolically, we may say that the negative of

$$A B C$$

is $\qquad$ not-A or not-B or not-C ;

that is, $\qquad a + b + c.$

Reciprocally the negative of

$$P + Q + R$$

is $\qquad pqr.$

Every disjunctive term, then, is the negative of a combined term, and *vice versâ*.

Apply this result to the combined term AAA, and its negative is

$$a + a + a.$$

Now since AAA is by the Law of Simplicity equivalent to A, so $a + a + a$ must be by the Law of Unity equivalent to a. Each law thus necessarily presupposes the other.

Symbolic expression of the Law of Duality.

We may now employ our symbol of alternation to express in a clear and formal manner the third Fundamental Law

of Thought, which I have called the Law of Duality.
Taking A to represent any class or object or quality, and
B any other class, object or quality, we may always
assert that A either agrees with B, or does not agree.
Thus we may say

$$A = AB + Ab.$$

This is a formula which will henceforth be constantly
employed, and it lies at the basis of reasoning.

The reader may perhaps wish to know why A is inserted
in both alternatives of the second member of the identity,
and why the law is not stated in the form

$$A = B + b.$$

But if he will consider the contents of the last section
(p. 87), he will see that the latter expression cannot be
correct, otherwise no term would have any negative.
For the negative of B + b is bB, or a self-contradictory
term : so that if A were identical with B + b, its negative a would be non-existent.   This result would generally
be an absurd one, and I see much reason to think that in
a strictly logical point of view it would always be absurd.
In all probability we ought to assume as a fundamental
logical axiom that *every term has its negative in thought.*
We cannot think at all without separating what we think
about from other different things, and these things necessarily form the negative notion[f].   If so, it follows that
any term of the form B + b is just as self-contradictory
as one of the form Bb.

It will be convenient to recapitulate in this place the
three great Laws of Thought in their symbolic form, thus

Law of Identity	$A = A.$
Law of Contradiction	$Aa = 0.$
Law of Duality	$A = AB + Ab.$

---

[f] 'Pure Logic,' p. 65. See also the criticism of this point by De
Morgan in the 'Athenæum,' No. 1892, 30th January, 1864 ; p. 155.

DISJUNCTIVE PROPOSITIONS. 89

Various Forms of the Disjunctive Proposition.

Disjunctive propositions may occur in a great variety of
forms, of which the old logicians took very insufficient
notice. There may be any number of alternatives each of
which may be a combination of any number of simple
terms. A proposition, again, may be disjunctive in one
or both members. The proposition

> Solids or liquids or gases are electrics or conductors of
> electricity

is an example of the doubly disjunctive form. The meaning of any such proposition is that whatever falls under
any one or more alternatives on one side must fall under
one or more alternatives on the other side. From what
has been said before, it is apparent that the proposition

$$A + B = C + D$$

will correspond to

$$ab = cd,$$

each member of the latter being the negative of a
member of the former proposition.

As an instance of a complex disjunctive proposition
I may give Senior's definition of wealth, namely 'Wealth
is what is transferable, limited in supply, and either
productive of pleasure or preventive of pain[g].'

> Let A = wealth
> B = transferable
> C = limited in supply
> D = productive of pleasure
> E = preventive of pain.

The definition takes the form

$$A = BC(D + E) :$$

but if we develop the alternatives by a method to be
afterwards more fully considered, it becomes

$$A = BCDE + BCDe + BCdE.$$

[g] Boole's 'Laws of Thought,' p. 106. Jevons' 'Pure Logic,' p. 69.

An example of a still more complex proposition may be found in De Morgan's writings[h], and is as follows:—
'He must have been rich, and if not absolutely mad was weakness itself, subjected either to bad advice or to most unfavourable circumstances.'

If we assign the letters of the alphabet in succession, thus,

A = he
B = rich
C = absolutely mad
D = weakness itself
E = subjected to bad advice
F = subjected to most unfavourable circumstances,

the proposition will take the form

$$A = AB\{C + D (E + F)\},$$

and if we develop the alternatives, expressing some of the different cases which may happen, we obtain

$$A = ABC + ABcDEF + ABcDEf + ABcDeF.$$

Inference by Disjunctive Propositions.

Before we can make a free use of disjunctive propositions in the processes of inference we must consider how disjunctive terms can be combined together or with simple terms. In the first place, to combine a simple term with a disjunctive one, we must combine it with every alternative of the disjunctive term. A vegetable, for instance, is either a herb, a shrub, or a tree. Hence an exogenous vegetable is either an exogenous herb, or an exogenous shrub, or an exogenous tree. Symbolically stated this process of combination is as follows—

$$A(B + C) = AB + AC.$$

Secondly, to combine two disjunctive terms with each other, combine each alternative of one separately with each

[h] 'On the Syllogism,' No. iii. p. 12. Camb. Phil. Trans., vol. x. part i.

alternative of the other. Since flowering plants are either exogens or endogens, and are at the same time either herbs, shrubs or trees, it follows that there are altogether six alternatives—namely, exogenous herbs, exogenous shrubs, exogenous trees, endogenous herbs, endogenous shrubs, endogenous trees. This process of combination is shown in the general form

$$(A + B)(C + D) = AC + AD + BC + BD.$$

It is hardly necessary to point out that, however numerous the terms combined, or the alternatives in those terms, we may effect the combination provided each alternative is combined with each alternative of the other terms, as in the algebraic process of multiplication.

Some processes of deduction may at once be exhibited. We may always, for instance, unite the same qualifying term to each side of an identity even though one or both members of the identity be disjunctive. Thus let

$$A = B + C.$$

Now it is self-evident that

$$AD = AD,$$

and in one side of this identity we may for A substitute its equivalent B + C obtaining

$$AD = BD + CD.$$

Since 'a gaseous element is either hydrogen, or oxygen, or nitrogen, or chlorine, or fluorine,' it follows that 'a free gaseous element is either free hydrogen, or free oxygen, or free nitrogen, or free chlorine, or free fluorine.'

This process of combination will lead to most useful inferences when the qualifying adjective combined with both sides of the proposition is a negative of one or more alternatives. Since chlorine is a coloured gas, we may infer that 'a colourless gaseous element is either (colourless) hydrogen, oxygen, nitrogen, or fluorine.' The alternative chlorine disappears because colourless chlorine does not exist. Again, since 'a tooth is either an incisor, canine, bicuspid, or molar,' it follows that 'a not-incisor tooth is either canine, bicuspid, or molar.' The general rule is that from the denial of any of the alternatives the affirmation of the remainder can be inferred. Now this result clearly follows from our process of substitution ; for if we have the proposition

$$A = B + C + D,$$

and insert this expression for A on one side of the self-evident identity

$$Ab = Ab,$$

we obtain $$Ab = ABb + AbC + AbD ;$$

and, as the first of the three alternatives is self-contradictory, we strike it out according to the law of contradiction : there remains

$$Ab = AbC + AbD.$$

Thus our system fully includes and explains that mood of the Disjunctive Syllogism technically called the *modus tollendo ponens.*

But the reader must carefully observe that the Disjunctive Syllogism of the mood *ponendo tollens*, which affirms one alternative, and thence infers the denial of the rest, cannot be held true in this system. If I say, indeed, that

Water is either salt or fresh water,

it seems evident that 'water which is salt is not fresh.' But this inference really proceeds from our knowledge that water cannot be at once salt and fresh. This inconsistency of the alternatives, as I have fully shown, will not always hold. Thus, if I say

Gems are either rare stones or beautiful stones,   (1)

it will obviously not follow that

A rare gem is not a beautiful stone,   (2)

nor that

A beautiful gem is not a rare stone.   (3)

Our symbolic method gives only true conclusions ; for if we take

$$A = \text{gem}$$
$$B = \text{rare stone}$$
$$C = \text{beautiful stone,}$$

the proposition (1) is of the form

$$A = B + C$$

hence $\quad\quad AB = B + BC$

and $\quad\quad AC = BC + C$;

but these inferences are not equivalent to the false ones (2) and (3).

We can readily represent such disjunctive reasoning, when it is valid, by expressing the inconsistency of the alternatives explicitly. Thus if we resort to our instance of

Water is either salt or fresh,

and take $\quad\quad A = \text{Water}$

$$B = \text{salt}$$
$$C = \text{fresh,}$$

then the premise is apparently of the form

$$A = AB + AC ;$$

but in reality there are the unexpressed conditions that 'what is salt is not fresh,' and 'what is fresh is not salt;' or, in letter-terms,

$$B = Bc$$
$$C = bC.$$

Now, if we substitute these descriptions in the original proposition, we obtain

$$A = ABc + AbC ;$$

uniting B to each side we infer

$$AB = ABc + ABbC$$

or $\quad\quad AB = ABc ;$

that is,

Water which is salt is water salt and not fresh.

I should weary the reader if I attempted to illustrate the multitude of forms which disjunctive reasoning may take; and as in the next chapter we shall be constantly treating the subject, I must here restrict myself to a single instance. A very common process of reasoning consists in the determination of the name of a thing by the successive exclusion of alternatives, a process called by the old name *abscissio infiniti.* Take the case :—

Red-coloured metal is either copper or gold	(1)
Copper is dissolved by nitric acid	(2)
This specimen is red-coloured metal	(3)
This specimen is not dissolved by nitric acid	(4)
Therefore this specimen consists of gold.	(5)

Assigning our letter-symbols thus —

$$A = \text{this specimen}$$
$$B = \text{red-coloured metal}$$
$$C = \text{copper}$$
$$D = \text{gold}$$
$$E = \text{dissolved by nitric acid,}$$

the premises may be stated in the form

$B = BCd + BcD$	(1)
$C = CE$	(2)
$A = AB$	(3)
$A = Ae.$	(4)

Substituting for C in (1) by means of (2) we get

$$B = BCdE + BcD.$$

From (3) and (4) we may infer likewise

$$A = ABe,$$

and if in this we substitute for B its equivalent just stated, it follows that

$$A = ABCdEc + ABcDe.$$

The first of the alternatives being contradictory, the result is
$$A = ABcDe$$
which contains a full description of 'this specimen,' as furnished in the premises, but by ellipsis indicates that it is gold. It will be observed that in the symbolic expression (1) I have explicitly stated what is certainly implied, that copper is not gold, and gold not copper, without which condition the inference would not hold good.

CHAPTER VI.

THE forms of deductive reasoning as yet considered, are mostly cases of Direct Deduction as distinguished from those which we are now about to treat. The method of Indirect Deduction may be described as that which points out what a thing is, by showing that it cannot be anything else. We can define a certain space upon a map, either by colouring that space, or by colouring all except the space ; the first mode is positive, the second negative. The difference, it will be readily seen, is exactly analogous to that between the direct and indirect proof in geometry. Euclid often shows that two lines are equal, by showing that they cannot be unequal, and the proof rests upon the known number of alternatives, greater, equal or less, which are alone conceivable. In other cases, as for instance in the seventh proposition of the first book, he shows that two lines must meet in a particular point, by showing that they cannot meet elsewhere.

In logic we can always define with certainty the utmost number of alternatives which are conceivable. The Law of Duality (pp. 6, 88) enables us always to assert that any quality or circumstance whatsoever is either present or absent in anything. Whatever may be the meaning and nature of the terms A and B it is certainly true that

$$A = AB + Ab$$
$$B = AB + aB.$$

These are universal though tacit premises which may be employed in the solution of every problem, and which are such invariable and necessary conditions of all thought, that they need not be specially laid down. The Law of Contradiction is a further condition of all thought and of all logical symbols; it enables, and in fact obliges, us to reject from further consideration all terms which imply the presence and absence of the same quality. Now, whenever we bring both these Laws of Thought into explicit action by the method of substitution, we employ the Indirect Method of Inference. It will be found that we can treat not only those arguments already exhibited according to the direct method, but we can also include an infinite multitude of other arguments which are incapable of solution by any other means.

Some philosophers, especially those of France, have held that the Indirect Method of Proof has a certain inferiority to a direct method, which should prevent our using it except when obliged. But there are an unlimited number of truths which we can prove only indirectly. We can prove that a number is a prime only by the purely indirect method of showing that it is not any of the numbers which have divisors, and the remarkable process known as Eratosthenes' Sieve is the only mode by which we can select the prime numbers[a]. It bears a strong analogy to the indirect method here to be described. We can also prove that the side and diameter of a square are incommensurable, but only in the negative or indirect manner, by showing that the contrary supposition constantly and inevitably leads to contradiction[b]. Many other demonstrations in various branches of the mathematical sciences rest upon a like method. Now if there is only one important truth which must be, and can only be

---

[a] See Horsley, 'Philosophical Transactions,' 1772; vol. lxii. p. 327. Montucla, 'Histoire des Mathematiques,' vol. i. p. 239. 'Penny Cyclopædia,' article *Eratosthenes*.

[b] Euclid, Book x. Prop. 117.

proved indirectly, we may say that the process is a
necessary and sufficient one, and the question of its comparative excellence or usefulness is not worth discussion.
As a matter of fact I believe that nearly half our logical
conclusions rest upon its employment.

Simple Illustrations.

In tracing out the powers and results of this method, we
will begin with the simplest possible instance. Let us take
a proposition of the very common form, $A = AB$, say,

A Metal is an Element,

and let us investigate its full meaning. Any person who
has had the least logical training, is aware that we can
draw from the above proposition an apparently different
one, namely,

A Not-element is a Not-metal.

While some logicians, as for instance De Morgan,[c] have
considered the relation of these two propositions to be
purely self-evident, and neither needing nor allowing
analysis, a great many more persons, as I have observed
while teaching logic, are at first unable to perceive the
close connection between them. I believe that a true and
complete system of logic will furnish a clear analysis of
this process which has been called *Contrapositive Conversion;* the full process is as follows :—

Firstly, by the Law of Duality we know that

Not-element is either Metal or Not-metal.

Now if it be metal, we know that it is by the premise
an element; we should thus be supposing that the very
same thing is an element and a not-element, which is
in opposition to the Law of Contradiction. According to
the only other alternative, then, the not-element must be
a not-metal.

c 'Philosophical Magazine,' December 1852, Fourth Series, vol. iv.
p. 435, 'On Indirect Demonstration.'

To represent this process of inference symbolically we take the premise in the form

$$A = AB. \qquad (1)$$

We observe that by the Law of Duality the term not-B is thus described

$$b = Ab + \dot{a}b. \qquad (2)$$

For A in this proposition we substitute its description as given in (1), obtaining

$$b = ABb + ab.$$

But according to the Law of Contradiction the term ABb must be excluded from thought or

$$ABb = 0.$$

Hence it results that b is either nothing at all, or it is ab; and the conclusion is

$$b = ab.$$

As it will often be necessary to refer to a conclusion of this kind I shall call it, as is usual, the *Contrapositive Proposition* of the original. The reader need hardly be cautioned to observe that from all A's are B's it does not follow that all not-A's are not-B's. For by the Law of Duality we have

$$a = aB + ab,$$

and it will not be found possible to make any substitution in this by our original premise $A = AB$. It still remains doubtful, therefore, whether not-metal is element or not-element.

The proof of the Contrapositive Proposition given above is exactly the same as that which Euclid applies in the case of geometrical notions. De Morgan describes Euclid's process as follows [d] :—'From every not-B is not-A he produces every A is B, thus—If it be possible, let this A be not-B, but every not-B is not-A, therefore this A is not-A, which is absurd: whence every A is B.' Now De Morgan thinks that this proof is entirely needless, because common

[d] 'Philosophical Magazine,' Dec. 1852; p. 437.

logic gives the inference without the use of any geometrical reasoning. I conceive however that logic gives
the inference only by an indirect process. De Morgan
claims 'to see identity in every A is B and every not-B
is not-A, by a process of thought prior to syllogism.'
But whether prior to syllogism or not, I claim that it
is not prior to the laws of thought and the process of
substitutive inference by which it may be undoubtedly
demonstrated.

Employment of the Contrapositive Proposition.

We can frequently employ the contrapositive form of
a proposition by the method of substitution ; and certain
moods of the ancient syllogism, which we have hitherto
passed over, may thus be satisfactorily comprehended
in our system. Take for instance the following syllogism
in the mood Camestres :—

    'Whales are not true fish : for they do not respire
        water, whereas true fish do respire water.'

Let us take

$$A = \text{whales},$$
$$B = \text{true fish},$$
$$C = \text{respiring water}.$$

The premises are of the form

$$A = Ac, \tag{1}$$
$$B = BC. \tag{2}$$

Now, by the process of contraposition we obtain from (2)

$$c = bc,$$

and we can substitute this expression for c in (1), obtaining

$$A = Abc,$$

or ' Whales are not true fish, not respiring water.'

The mood Cesare does not really differ from Camestres
except in the order of the premises, and it could be
exhibited in an exactly similar manner.

The mood Baroko gave much trouble to the old logicians who could not *reduce* it to the first figure in the same manner as the other moods, and were obliged to invent, specially for it and for Bokardo, a method of Indirect Reduction closely analogous to the Indirect proof of Euclid. Now these moods require no exceptional processes in this system. Let us take as an instance of Baroko, the argument

All heated solids give continuous spectra,   (1)

Some nebulæ do not give continuous spectra ;   (2)

Therefore some nebulæ are not heated solids.   (3)

Treating the little word *some* as an indeterminate adjective of selection, to which we assign a symbol like any other adjective, let

$$A = \text{some}$$
$$B = \text{nebulæ}$$
$$C = \text{giving continuous spectra}$$
$$D = \text{heated solid.}$$

The premises then become

$$D = DC \qquad\qquad (1)$$
$$AB = ABc. \qquad\qquad (2)$$

Now from (1) we obtain by the Indirect method the Contrapositive

$$c = cd,$$

and if we substitute this expression for c in (2) we have

$$AB = ABcd;$$

the full meaning of which is that 'some nebulæ do not give continuous spectra and are not solids.'

We might similarly apply the contrapositive in many other instances. Take the argument—'All fixed stars are self-luminous ; but some of the heavenly bodies are not self-luminous, and are therefore not fixed stars.' Taking our terms

$$A = \text{fixed stars}$$
$$B = \text{self-luminous}$$

$$C = \text{some}$$
$$D = \text{heavenly bodies,}$$

we have the premises

$$A = AB, \tag{1}$$
$$CD = bCD. \tag{2}$$

Now from (1) we can draw the Contrapositive

$$b = ab,$$

and substituting this expression for b in (2) we obtain

$$CD = abCD,$$

which expresses the conclusion of the argument that 'some heavenly bodies are not fixed stars.'

Contrapositive of a Simple Identity.

The reader should carefully note that when we apply the process of Indirect Inference to a simple identity of the form

$$A = B,$$

we may obtain further results. If we wish to know what is the term not-B, we have as before, by the Law of Duality,

$$b = Ab + ab,$$

and substituting for A we obtain

$$b = Bb + ab = ab.$$

But we may now also draw a second Contrapositive ; for we have

$$a = aB + ab,$$

and substituting for B its equivalent A we have

$$a = aA + ab = ab.$$

Hence from the single identity $A = B$ we can draw the two propositions

$$a = ab$$
$$b = ab,$$

and observing that these propositions have a common term we can make a new substitution, getting

$$a = b.$$

This result is in strict accordance with the fundamental principles of inference, and it may be a question whether it is not a self-evident result, independent of the steps of deduction by which we have reached it. For where two classes are coincident like A and B, whatever is true of the one is true of the other ; what is excluded from the one must be excluded from the other similarly. Now as a bears to A exactly the same relation that b bears to B, the identity of either pair follows from the identity of the other pair. In every identity, equality, or similarity, we may argue from the negative of the one side to the negative of the other. Thus at ordinary temperatures

Mercury = liquid-metal,

hence obviously

Not-mercury = not-liquid-metal ;

or since

Sirius = brightest fixed star,

it follows that whatever star is not the brightest is not Sirius, and *vice versâ*. Every correct definition is of the form A = B, and may often require to be applied in the equivalent negative form.

Let us take as an illustration of the mode of using this result the argument following :—

Vowels are letters which can be sounded alone, (1)
The letter w cannot be sounded alone ; (2)
Therefore the letter w is not a vowel. (3)

Here we have a definition (1), and a comparison of a thing with that definition (2), leading to exclusion of the thing from the class defined.

Taking the terms

A = vowel,
B = letter which can be sounded alone,
C = letter w, the premises are plainly of the form

$$A = B, \qquad (1)$$
$$C = bC. \qquad (2)$$

Now by the Indirect method we obtain from (1) the Contrapositive

$$b = a,$$

and inserting in (2) the equivalent for b we have

$$C = aC, \qquad (3)$$

or 'the letter w is not a vowel.'

Miscellaneous Examples of the Method.

We can apply the Indirect Method of Inference however many may be the terms involved or the premises containing those terms. As the working of the method is best learnt from examples, I will take a case of two premises forming the syllogism Barbara : thus

$$\text{Iron is a metal} \qquad (1)$$
$$\text{Metal is element.} \qquad (2)$$

If we want to ascertain what inference is possible concerning the term *Iron*, we develop the term by the Law of Duality. Iron must be either metal or not-metal; iron which is metal must be either element or not-element; and similarly iron which is not-metal must be either element or not-element. There are then altogether four alternatives among which the description of iron must be contained ; thus

$$\text{Iron, metal, element,} \qquad (\alpha)$$
$$\text{Iron, metal, not-element,} \qquad (\beta)$$
$$\text{Iron, not-metal, element,} \qquad (\gamma)$$
$$\text{Iron, not-metal, not-element.} \qquad (\delta)$$

Our first premise informs us that iron is a metal, and if we substitute this description in (γ) and (δ) we shall have self-contradictory combinations. Our second premise likewise informs us that metal is element, and applying this description to (β) we again have self-contradiction, so that there remains only (a) as a description of iron— our inference is

$$\text{Iron} = \text{iron, metal, element.}$$

To represent this process of reasoning in general symbols, let

$$A = \text{iron}$$
$$B = \text{metal}$$
$$C = \text{element.}$$

The premises of the problem take the form

$$A = AB \qquad\qquad (1)$$
$$B = BC. \qquad\qquad (2)$$

By the Law of Duality we have

$$A = AB + Ab \qquad\qquad (3)$$
$$A = AC + Ac. \qquad\qquad (4)$$

Now, if we insert for A in the second side of (3) its description in (4), we obtain what I shall call the *development* of A

$$A = ABC + ABc + AbC + Abc. \qquad\qquad (5)$$

Wherever the letters A or B appear in the second side of (5) substitute their equivalents given in (1) and (2) and the results at full length are

$$A = ABC + ABCc + ABbC + ABbCc.$$

The last three alternatives break the Law of Contradiction, so that

$$A = ABC + 0 + 0 + 0$$
$$A = ABC.$$

This conclusion is, indeed, no more than we could obtain by the direct process of substitution; it is the characteristic of the Indirect process that it gives all possible logical conclusions, both those which we have previously obtained, and an almost infinite number of others of which the ancient logic took little or no account. From the same premises, for instance, we can obtain a description of the class *not-element* or *c*. By the Law of Duality we can develop *c* into four alternatives, thus—

$$c = ABc + Abc + aBc + abc.$$

Now if we substitute for A and B as before, we get

$$c = ABCc + ABbc + aBCc + abc,$$

and striking out the terms which break the Law of Contradiction there remains

$$c = abc,$$

or what is not element is also not iron and not metal. This Indirect Method of Inference thus furnishes a complete solution of the following problem—*Given any number of logical premises or conditions, required the description of any class of objects, or any term, as governed by those conditions.*

The steps of the process of inference may thus be concisely stated :—

1. By the Law of Duality develop the utmost number of alternatives which may exist in the description of the required class or term as regards the terms involved in the premises.

2. For each term in these alternatives substitute its description as given in the premises.

3. Strike out every alternative which is then found to break the Law of Contradiction.

4. The remaining terms may be equated to the term in question as the desired description or inference.

Abbreviation of the Process.

Before proceeding to illustrations of the use of this method, I must point out how much its practical employment can be simplified, and how much more easy it is than would appear from the description. When we want to effect at all a complete solution of a logical

problem it is best to form, in the first place, a complete
series of all the combinations of terms involved in it.
If there be two terms A and B, the utmost variety of
combinations in which they can appear are

$$AB$$
$$Ab$$
$$aB$$
$$ab.$$

The term A appears in the first and second; B in the
first and third; a in the third and fourth; and b in the
second and fourth. Now if we have any premise, say

$$A = B,$$

we must ascertain which of these combinations would be
rendered self-contradictory by substitution; the second
and third would have to be struck out, and there would
remain

$$AB$$
$$ab.$$

Hence we draw the following inferences

$$A = AB, \quad B = AB, \quad a = ab, \quad b = ab.$$

Exactly the same method must be followed where a
question involves a greater number of terms. Thus by
the Law of Duality the three terms A, B, C, give rise to
eight conceivable combinations, namely

ABC	(a)
ABc	(β)
AbC	(γ)
Abc	(δ)
aBC	(ϵ)
aBc	(ζ)
abC	(η)
abc.	(θ)

The development of the term A is formed by the first four
of these; for B we must select (a), (β), (ϵ), (ζ); C consists
of (a), (γ), (ϵ) (η); b of (γ), (δ), (η), (θ), and so on.

Now if we want to investigate completely the meaning of the premises

$$A = AB \qquad (1)$$
$$B = BC, \qquad . \qquad (2)$$

we examine each of the eight combinations as regards each premise; (γ) and (δ) are contradicted by (1), and (β) and (ζ) by (2), so that there remain only

ABC	(a)
aBC	(ϵ)
abC	(η)
abc.	(θ)

To describe any term under the conditions of the premises (1) and (2), we have only to draw out the proper combinations from this list; thus—A is represented only by ABC or

$$A = ABC,$$

similarly $\qquad c = abc.$

For B we have two alternatives thus stated,

$$B = ABC + aBC;$$

and for b we have

$$b = abC + abc.$$

When we have a problem involving four distinct terms we need to double the number of combinations, and as we add each new term the combinations become twice as numerous. Thus

A, B	produce	four	combinations
A, B, C,	„	eight	„
A, B, C, D	„	sixteen	„
A, B, C, D, E	„	thirty-two	„
A, B, C, D, E, F	„	sixty-four	„

and so on.

I propose to call any such series of combinations the *Logical Abecedarium.* It holds in logical science a position of importance which cannot be exaggerated. As we proceed from logical to mathematical considerations it will

become apparent that there is a close connection between
these combinations and the most fundamental theorems of
mathematical science.   For the convenience of the reader
who may wish to employ the abecedarium in logical
questions, I have had printed on the next page a complete
series of the combinations up to those of six terms.   At
the very commencement in the first column is placed a
single letter X which might seem to be superfluous.   This
letter serves to denote that it is always some higher class
which is divided up.   Thus the combination AB really
means ABX, or that part of some larger class, say X,
which has the qualities of A and B present.   The letter
X is omitted in the greater part of the table merely for
the sake of brevity and clearness.   In a later chapter on
Combinations it will become apparent that the introduction of this unit class is requisite in order to complete the analogy with the Arithmetical Triangle there
described.

The reader ought to bear in mind that though the
abecedarium seems to give mere lists of combinations,
these combinations are intended in every case to constitute the development of a term of a proposition.
Thus the four combinations AB, Ab, aB, ab really mean
that any class X is described by the following proposition,

$$X = X (AB + Ab + aB + ab).$$

If we select the A's, we obtain the following proposition

$$A X = X (AB + Ab).$$

Thus whatever group of combinations we treat must be
conceived as part of a higher class, *summum genus* or
universe symbolised in the term X; but bearing this in
mind, it is needless to complicate our formulæ by always
introducing the letter.   All inference consists in passing
from propositions to propositions, and combinations *per se*
have no meaning.   They are consequently to be regarded
in all cases as forming parts of propositions.

The Logical Abecedarium.

I.	II.	III.	IV.	V.	VI.	VII.
X	AX	AB	ABC	ABCD	ABCDE	ABCDEF
	aX	Ab	ABc	ABCd	ABCDe	ABCDEf
		aB	AbC	ABcD	ABCdE	ABCDeF
		ab	Abc	ABcd	ABCde	ABCDef
			aBC	AbCD	ABcDE	ABCdEF
			aBc	AbCd	ABcDe	ABCdEf
			abC	AbcD	ABcdE	ABCdeF
			abc	Abcd	ABcde	ABCdef
				aBCD	AbCDE	ABcDEF
				aBCd	AbCDe	ABcDEf
				aBcD	AbCdE	ABcDeF
				aBcd	AbCde	ABcDef
				abCD	AbcDE	ABcdEF
				abCd	AbcDe	ABcdEf
				abcD	AbcdE	ABcdeF
				abcd	Abcde	ABcdef
					aBCDE	AbCDEF
					aBCDe	AbCDEf
					aBCdE	AbCDeF
					aBCde	AbCDef
					aBcDE	AbCdEF
					aBcDe	AbCdEf
					aBcdE	AbCdeF
					aBcde	AbCdef
					abCDE	AbcDEF
					abCDe	AbcDEf
					abCdE	AbcDeF
					abCde	AbcDef
					abcDE	AbcdEF
					abcDe	AbcdEf
					abcdE	AbcdeF
					abcde	Abcdef
						aBCDEF
						aBCDEf
						aBCDeF
						aBCDef
						aBCdEF
						aBCdEf
						aBCdeF
						aBCdef
						aBcDEF
						aBcDEf
						aBcDeF
						aBcDef
						aBcdEF
						aBcdEf
						aBcdeF
						aBcdef
						abCDEF
						abCDEf
						abCDeF
						abCDef
						abCdEF
						abCdEf
						abCdeF
						abCdef
						abcDEF
						abcDEf
						abcDeF
						abcDef
						abcdEF
						abcdEf
						abcdeF
						abcdef

In a theoretical point of view we may conceive that the abecedarium is always extended indefinitely. Every new quality or circumstance which can belong to an object, subdivides each combination or class, so that the number of such combinations when unrestricted by logical conditions is represented by an indefinitely high power of two. The extremely rapid increase in the number of subdivisions obliges us to confine our attention to a few circumstances at a time.

When contemplating the properties of this abecedarium, I am often inclined to think that Pythagoras perceived the deep logical importance of duality; for while unity was the symbol of identity and harmony, he described the number two as the origin of contrasts, or the symbol of diversity, division and separation. The number four or the *Tetractys* was also regarded by him as one of the chief elements of existence, for it represented the generating virtue whence come all combinations.

In one of the golden verses ascribed to Pythagoras, he conjures his pupil to be virtuous[e]:

> 'By him who stampt *The Four* upon the Mind,
> *The Four,* the fount of Nature's endless stream.'

Now four and the higher powers of duality do represent in this logical system the variety of combinations which can be generated in the absence of logical restrictions. The followers of Pythagoras may have shrouded their master's doctrines in mysterious and superstitious notions, but in many points these doctrines seem to have some basis in logical philosophy.

The Logical Slate.

To a person who has once comprehended the extreme significance and utility of the Logical Abecedarium, the

[e] Whewell, 'History of the Inductive Sciences,' vol. i. p. 222.

indirect process of inference becomes reduced to the repetition of a few uniform operations of classification, selection,
and elimination of contradictories. Logical deduction even
in the most complicated questions becomes a matter of
mere routine, and the amount of labour required is the
only impediment when once the meaning of the premises
is rendered clear. But the amount of labour is often
found to be considerable. The mere writing down of
sixty-four combinations of six letters each is no small
task, and, if we had a problem of five premises, each of
the sixty-four combinations would have to be examined
in connection with each premise. The requisite comparison is often of a very tedious character and considerable chance of errors thus arises.

I have given much attention therefore to reducing both
the manual and mental labour of the process, and I shall
describe several devices which may be adapted for saving
trouble and risk of mistake.

In the first place, as the same sets of combinations
occur over and over again in different problems, we may
avoid the labour of writing them out by having the sets
of letters ready printed upon small sheets of writing paper.
It has also been suggested by a correspondent that, if any
one series of combinations were marked upon the margin
of a sheet of paper, and a slit cut between each pair of
combinations, it would be easy to fold down any particular
combination, and thus strike it out of view. The combinations consistent with the premises would then remain
in a broken series. This method answers sufficiently well
for occasional use.

A more convenient mode, however, is to have the series
of letters shown on p. 109, engraved upon a common
school writing slate, of such a size, that the letters may
occupy only about a third of the space on the left hand
side of the slate. The conditions of the problem can then

be written down on the unoccupied part of the slate, and the proper series of combinations being chosen, the contradictory combinations can be struck out with the pencil. I have used a slate of this kind, which I call a *Logical Slate*, for more than ten years, and it has saved me much trouble. It is hardly possible to apply this process to problems of more than six terms, owing to the large number of combinations which would require examination ; thus seven terms would give 128 combinations, eight terms 256, nine terms 512, ten terms 1024, eleven terms 2048, twelve terms 4096, and so on in geometrical progression.

Abstraction of Indifferent Circumstances.

There is a simple but highly important process of inference which enables us to abstract, eliminate or disregard all circumstances indifferently present and absent. Thus if I were to state that 'a triangle is a figure of three sides, with or without equal angles,' the latter qualification would be superfluous, because by a law of thought I know that angles must be either equal or unequal. To add the qualification gives no new knowledge since the existence of the two alternatives will be understood in the absence of any information to the contrary. Accordingly, when two alternatives differ only as regards a single component term which is positive in one and negative in the other, we may always reduce them to one term by striking out their indifferent part. It is really a process of substitution which enables us to do this; for having any proposition of the form

$$A = ABC + ABc, \tag{1}$$

we know by the Law of Duality that

$$B = BC + Bc. \tag{2}$$

Hence
$$AB = ABC + ABc. \tag{3}$$

And as the second member of this is identical with the second member of (1) we may substitute, obtaining

$$A = AB.$$

This process of reducing useless alternatives, may be applied again and again ; for it is plain that

$$A = AB (CD + Cd + cD + cd)$$

communicates no more information than that A is B. This abstraction of indifferent terms is in fact the converse process to that of development described in p. 104 ; and it is one of the most important operations in the whole sphere of reasoning.

The reader should observe that in the proposition

$$AC = BC$$

we cannot abstract C and infer

$$A = B ;$$

but from

$$AC + Ac = BC + Bc$$

we may abstract all reference to the term C.

Illustrations of the Indirect Method.

An infinite variety of arguments and logical problems might be introduced here to show the comprehensive character and powers of the Indirect Method. We can treat either a single premise or a series of premises.

Take in the first place a simple definition, such as 'a triangle is a three-sided rectilinear figure.' Let

$$A = \text{triangle}$$
$$B = \text{three-sided}$$
$$C = \text{rectilinear figure,}$$

then the definition is of the form

$$A = BC.$$

If we take the series of eight combinations of three letters (see p. 106) and strike out those which are inconsistent with the definition, we have the following result :—

$$
\begin{array}{ll}
\mathrm{ABC} & \text{aBC} \\
\text{ABc} & a\mathrm{Bc}. \\
\text{AbC} & ab\mathrm{C} \\
\text{Abc} & abc.
\end{array}
$$

For the description of the class C we have
$$ \mathrm{C} = \mathrm{ABC} + ab\mathrm{C}, $$
that is, 'a rectilinear figure is either a triangle and three-sided, or not a triangle and not three-sided.'

For the class b we have
$$ b = ab\mathrm{C} + abc. $$

To the second side of this we may apply the process of simplification by abstraction described in the last section ; for by the Law of Duality
$$ ab = ab\mathrm{C} + abc ; $$
and as we have two propositions identical in the second side of each we may substitute, getting
$$ b = ab, $$
or what is not three-sided is not a triangle (whether it be rectilinear or not).

Let us treat by this method the following argument :—

> 'Blende is not an elementary substance ; elementary substances are those which are undecomposable; blende, therefore, is decomposable.'

Taking our letters thus—

$$ \mathrm{A} = \text{blende}, $$
$$ \mathrm{B} = \text{elementary substance}, $$
$$ \mathrm{C} = \text{undecomposable}, $$

the premises are of the form

$$ \mathrm{A} = \mathrm{A}b, \qquad (1) $$
$$ \mathrm{B} = \mathrm{C}. \qquad (2) $$

No immediate substitution can be made; but if we take the contrapositive of (2), namely

$$ b = c, \qquad (3) $$

we can substitute in (1) obtaining the conclusion
$$A = Ac.$$
But the same result may be obtained by taking the eight combinations of A, B, C, of the abecedarium ; it will be found that only three combinations, namely

$$Abc$$
$$aBC$$
$$abc,$$

are consistent with the premises, whence it results that
$$A = Abc,$$
or by the process of Ellipsis before described (p. 69)
$$A = Ac.$$

As a somewhat more complex example I take the argument thus stated, one which could not be thrown into the syllogistic form.

'All metals except gold and silver are opaque ; therefore what is not opaque is either gold or silver or is not-metal.'

There is more implied in this statement than is distinctly asserted, the full meaning being as follows :

All metals not gold or silver are opaque, (1)
Gold is not opaque but is a metal. (2)
Silver is not opaque but is a metal, (3)
Gold and silver are distinct substances. (4)

Taking our letters thus—

A = metal   C = silver
B = gold   D = opaque,

we may state the premises in the form

$$Abc = AbcD \quad (1)$$
$$B = ABd \quad (2)$$
$$C = ACd \quad (3)$$
$$B = Bc. \quad (4)$$

To obtain a complete solution of the question we take the sixteen combinations of A, B, C, D, and striking out those which are inconsistent with the premises, there remain only

$$ABcd$$
$$AbCd$$
$$AbcD$$
$$abcD$$
$$abcd.$$

The expression for not-opaque things consists of the three combinations containing d, thus

$$d = ABcd + AbCd + abcd,$$

or $\qquad d = Ad\,(Bc + bC) + abcd.$

In ordinary language, what is not-opaque is either metal which is gold, and then not-silver, or silver and then not gold, or else it is not-metal and neither gold nor silver.

A good example for the illustration of the Indirect Method is to be found in De Morgan's Formal Logic (p. 123), the premises being substantially as follows:—

From A follows B, and from C follows D; but B and D are inconsistent with each other; therefore A and C are inconsistent.

The meaning no doubt is that where A is, B will be found, or that every A is a B, and similarly every C is a D; but B and D cannot occur together. The premises therefore appear to be of the form

$$A = AB, \qquad (1)$$
$$C = CD, \qquad (2)$$
$$B = Bd. \qquad (3)$$

On examining the series of sixteen combinations, but five are found to be consistent with the above conditions, namely,

$$ABcd$$
$$aBcd$$
$$abCD$$
$$abcD$$
$$abcd.$$

In these combinations the only A which appears is joined to c, and similarly C is joined to a, or A is inconsistent with C.

A more complex argument, also given by De Morgan [f], contains five terms, and is as stated below, except that I have altered the letters.

> 'Every A is one only of the two B or C; D is both B and C, except when B is E, and then it is neither ; therefore no A is D.'

A little reflection will show that these premises are capable of expression in the following symbolic forms—

$$A = ABc + AbC, \qquad (1)$$
$$De = DeBC, \qquad (2)$$
$$DE = DEbc. \qquad (3)$$

As five letters, A, B, C, D, E, enter into these premises it is requisite to treat their thirty-two combinations, and it will be found that fourteen of them remain consistent with the premises, namely

ABcdE	aBCDe	abCdE
ABcde	aBCdE	abCde
AbCdE	aBCde	abcDE
AbCde .	aBcdE	abcdE
	aBcde	abcde.

Now if we examine the first four combinations, all of which contain A, we find that they none of them contain D ; or again if we select those which contain D, we have only two, thus—

$$D = aBCDe + abcDE.$$

Hence it is clear that no A is D, and *vice versâ* no D is A. We might also draw many other conclusions from the premises ; for instance—

$$DE = abcDE,$$

or D and E never meet but in the absence of A, B, and C.

---

[f] 'Formal Logic,' p. 124.

Fallacies analysed by the Indirect Method.

It has been sufficiently shown, perhaps, that we can by the Indirect Method of Inference extract the whole truth from any series of propositions, and exhibit it anew in any required form of conclusion. But it may also need to be shown by examples that so long as we follow correctly the almost mechanical rules of the method, we cannot fall into any of the common fallacies or paralogisms which are not seldom committed in ordinary discussion. Let us take the example of a fallacious argument, previously treated by the Method of Direct Inference (p. 75),

$$\text{Granite is not a sedimentary rock,} \qquad (1)$$
$$\text{Basalt is not a sedimentary rock,} \qquad (2)$$

and let us ascertain whether any precise conclusion can be drawn concerning the relation of granite and basalt. Taking as before

$$A = \text{granite,}$$
$$B = \text{sedimentary rock,}$$
$$C = \text{basalt,}$$

the premises become $A = Ab,$ (1)
$$C = Cb. \qquad (2)$$

Of the eight conceivable combinations of A, B, C, five agree with these conditions, namely

AbC	aBc
Abc	abC
	abc ;

the description of granite is found to be

$$A = AbC + Abc = Ab(C + c),$$

that is, granite is not a sedimentary rock but is either basalt or not-basalt. If we want a description of basalt the answer is of like form

$$C = AbC + abC = bC (A + a).$$

Basalt is a sedimentary rock, and either granite or not-granite. As it is already perfectly evident that basalt must be either granite or not, and *vice versâ*, the premises fail to give us any information on the point, that is to say the Method of Indirect Inference saves us from falling into any fallacious conclusions. This example sufficiently illustrates both the fallacy of Negative premises and that of Undistributed Middle of the old logic (pp. 75–77).

The fallacy called the Illicit Process of the Major Term is also incapable of commission in following the rules of the method. Our example was (p. 77)

$$\text{All planets are subject to gravity,} \qquad (1)$$
$$\text{Fixed stars are not planets.} \qquad (2)$$

The false conclusion is that 'fixed stars are not subject to gravity.' The terms are

$$A = \text{planet}$$
$$B = \text{fixed star}$$
$$C = \text{subject to gravity.}$$

And the premises are

$$A = AC, \qquad (1)$$
$$B = aB. \qquad (2)$$

The combinations which remain uncontradicted on comparison with these premises are

$$
\begin{array}{ll}
AbC & aBc \\
aBC & abC \\
& abc.
\end{array}
$$

For fixed star we have the description

$$B = aBC + aBc,$$

that is, 'a fixed star is not a planet, but is either subject or not, as the case may be, to gravity.'

The Logical Abacus.

The Indirect Method of Inference has now been sufficiently described, and a careful examination of its powers will show that it is capable of giving a full analysis and solution of every question involving any simply logical relations. The chief difficulty of the method consists in the great number of combinations which may have to be examined; not only may the requisite labour become formidable, but a considerable chance of mistake may arise. I have therefore given much attention to modes of facilitating the work, and have succeeded in reducing the method to an almost mechanical form. It soon appeared obvious that if the conceivable combinations of the abecedarium, for any number of letters, instead of being printed in fixed order on a piece of paper or slate, were marked upon light moveable pieces of wood, mechanical arrangements could readily be devised for selecting the combinations in any required order. The labour of comparison and rejection might thus be immensely reduced. This idea was first carried out in the Logical Abacus, which I have found useful in the lecture-room for exhibiting the complete solution of logical problems. A minute description of the construction and use of the abacus, together with figures of the parts, has already been given in my essay called *The Substitution of Similars*[g], and I will here give only a general description.

The abacus consists of a common school black-board placed in a sloping position and furnished with four horizontal and equi-distant ledges. The combinations of the letters shown in the first four columns of the abecedarium (see p. 109), are printed in somewhat large type, so that each letter is about an inch from the neighbouring one, but the letters are placed one above the other instead of being in horizontal lines as in p. 109. Each combination of letters is separately fixed to the surface of a thin slip of wood one inch broad and about one-eighth inch thick. Short steel pins are then driven in an inclined position into the wood. When a letter is a large capital representing a positive term, the pin is fixed in the upper part of its space; when the letter is a small italic representing a negative term, the pin is fixed in the lower part of the space. Now, if one of the series of combinations be ranged upon a ledge of the black-board, the sharp edge of a flat rule can be inserted beneath the pins belonging to any one letter—say A, so that all the combinations marked A can be lifted out and placed upon a separate ledge. Thus we have represented the act of thought which separates the class A from what is not-A. The operation can be repeated; out of the A's we can in like manner select those which are B, obtaining the AB's; and in like manner we might select any other class such as the aB's, the ab's or the abc's.

If now we take the series of eight combinations of the letters A, B, C, a, b, c, and wish to analyse the argument anciently called Barbara, having the premises

$$A = AB \qquad\qquad (1)$$
$$B = BC, \qquad\qquad (2)$$

we proceed as follows :—Firstly we raise the combinations marked a, leaving the A's behind; out of these A's we move to a lower ledge such as are not-B's, and to the remaining AB's we join the a's which have been raised. The result is that we have divided all the combinations into two classes, namely, the Ab's which are incapable of existing consistently with premise (1), and the combinations which are consistent with the premise. Turning now to the second premise, we raise out of those which agree with (1) the b's, then we lower the Bc's; lastly we join the b's to the BC's. We should now find our combinations arranged as below.

A B C				a B C		a b C	a b c
	A B c	A b C	A b c		a B c		

The lower line contains all the combinations which are inconsistent with either premise; we have carried out in a mechanical manner that exclusion of self-contradictories which was formerly done upon the slate or paper. Accordingly, from the remaining combinations in the upper line we can draw any inference which the premises yield. If we raise the A's we find only one, and that is C, so that A must be C. If we select the c's we again find only one which is a and also b, so that we prove that not-C is not-A and not-B.

When a disjunctive proposition occurs among the premises the requisite movements become rather more complicated. Take the disjunctive argument

A is either B or C or D,
A is not C and not D,
Therefore A is B.

The premises are represented accurately as follows :—

$$A = AB + AC + AD \qquad (1)$$
$$A = Ac \qquad (2)$$
$$A = Ad. \qquad (3)$$

As there are four terms we choose the series of sixteen combinations and place them on the highest ledge of the board but one. We raise the a's and lower the b's. But we are not to reject all the Ab's as contradictory, because by the first premise A's may be either B's or C's or D's. Accordingly out of the Ab's we must select the c's, and out of these again the d's, so that only Abcd will remain to be rejected finally. Joining all the other fifteen combinations together again we raise the a's and lower the AC's, and thus reject the combinations inconsistent with (2); similarly we reject the AD's which are inconsistent with (3). It will be found that there remain in addition to all the eight combinations containing a only one containing A, namely

ABcd,

whence it is apparent that A must be B, the true conclusion of the argument.

In my previous Essay[h] I have described the working of two other logical problems upon the abacus, which it would be tedious to repeat in this place.

The Logical Machine.

Although the Logical Abacus considerably reduced the labour of using the Indirect Method, it was not free from the possibility of error. I thought moreover that it would afford a conspicuous proof of the generality and power of the method if I could reduce it to a purely mechanical form. Logicians had long been accustomed to speak of Logic as an Organon or Instrument, and even Bacon, while he rejected the old syllogistic logic, had insisted, in the second aphorism of his ' New Instrument,' that the mind required some kind of systematic aid. In the kindred science of mathematics mechanical assistance of one kind or another had long been employed. Orreries, globes, mechanical clocks, and such like instruments, are really aids to calculation and are of considerable antiquity. The arithmetical machine of Pascal is more than two centuries old, having been constructed in 1642–45. M. Thomas of Colmar has recently manufactured an arithmetical machine on Pascal's principles which is extensively employed by engineers and others who need frequently to multiply or divide. To Babbage, however, was entirely due the merit of embodying the Calculus of Differences in a machine, which thus became capable of calculating the most complicated tables of figures. It seemed strange that in the more intricate science of quantity mechanism could be applicable, whereas in the simple science of

[h] 'Substitution of Similars,' pp. 56–59.

qualitative reasoning, the syllogism was only by analogy
or simile called an Instrument. Swift satirically described
the Professors of Laputa as in possession of a thinking
machine, and in 1851 Mr. Alfred Smee actually proposed
the construction of a Relational machine and a Differential
machine, the first of which would be a mechanical dictionary and the second a mode of comparing ideas ; but
with these exceptions I have not yet met with so much
as a suggestion of a reasoning machine. It may be added
that Mr. Smee's designs, though highly ingenious, appear
impracticable, and in any case do not attempt the performance of logical inference[i].

The Logical Abacus soon suggested the notion of a
Logical Machine, which, after two unsuccessful attempts,
I succeeded in constructing in a comparatively simple and
effective form. The details of the Logical Machine have
been fully described by the aid of plates in the Philosophical Transactions[k], and it would be both tedious
and needless to repeat the account of the somewhat
intricate movements of the machine in this place.

The general appearance of the machine is shown in a
plate facing the title-page of this volume. It somewhat
resembles a very small upright piano or organ, and has
a keyboard containing twenty-one keys. These keys are
of two kinds, sixteen of them representing the terms or
letters A, a, B, b, C, c, D, d, which have so often been
employed in our logical notation. When letters occur
on the left-hand side of a proposition, formerly called
the subject, each is represented by a key on the left-hand
half of the keyboard ; but when they occur on the right-

[i] See his work called 'The Process of Thought adapted to Words and
Language, together with a description of the Relational and Differential
Machines.' Also 'Philosophical Transactions,' [1870] vol. 160, p. 518.

[k] 'Philosophical Transactions,' [1870] vol. 160, p. 497. 'Proceedings
of the Royal Society,' vol. xviii. p. 166, Jan. 20, 1870. 'Nature,' vol. i.
p. 343.

hand side, or as it used to be called the predicate of the proposition, the letter keys on the right-hand side of the keyboard are the proper representatives. The five other keys may be called operation keys, to distinguish them from the letter or term keys. They stand for the stops, copula, and disjunctive conjunctions of a proposition. The middle key of all is the copula, to be pressed when the verb *is* or the sign = is met. The extreme right-hand key is called the Full Stop, because it should be pressed when a proposition is completed, in fact in the proper place of the full stop. The extreme left-hand key is used to terminate an argument or to restore the machine to its initial condition ; it is called the Finis key. The last key but one on the right and left complete the whole series, and represent the conjunction *or* in its un-exclusive meaning, or the sign + which I have employed, according as it occurs in the right or left hand side of the proposition. The whole keyboard is arranged as shown below—

Left-hand side of Proposition, or Subject.								Copula.	Right-hand side of Proposition, or Predicate.								Full Stop.
+ Or	*d*	D	*c*	C	*b*	B	*a*	A		A	*a*	B	*b*	C	*c*	D	*d*	+ Or

To work the machine it is only requisite to press the keys in succession as indicated by the letters and signs of a symbolical proposition. All the premises of an argument are supposed to be reduced to the simple notation which has been employed in the previous pages. Taking then such a simple proposition as

$$A = AB,$$

we press the keys A (subject), copula, A (predicate), B (predicate), and full stop.

If there be a second premise, for instance
$$B = BC,$$
we press in like manner the keys—

B (subj.), copula, B (pred.), C (pred.), full stop.

The process is exactly the same however numerous the premises may be. When they are completed the operator will see indicated on the face of the machine the exact combinations of letters which are consistent with the premises according to the principles of thought.

As shown in the figure opposite the title-page, the machine exhibits in front an abecedarium of sixteen combinations, exactly like that of the abacus, except that the letters of each combination are separated by a certain interval. After the above problem has been worked upon the machine the abecedarium will present the following appearance—

A	A						a	a		a	a	a	a
B	B					B	B		b	b	b	b
C	C				C	C		C	C	c	c
D	d				D	d		D	d	D	d

The operator will collect the various conclusions, as for instance that **A** is always C, that not-C is not-B and not-A ; that not-B is not-A but either **C** or not-C, as in the use of the Logical Slate or Abacus.

Disjunctive propositions are to be treated in an exactly similar manner. Thus, to work the premises

$$A = AB + AC$$
$$B + C = BD + CD,$$

it is only necessary to press in succession the keys

A (subj.), copula, A (pred.), B, + A,C, full stop.

B (subj.), + C, copula, B (pred.), D, +, C, D, full stop.

The combinations then remaining will be as follows

ABCD	aBCD
ABcD	aBcD
AbCD	abCD
	abcD
	abcd.

On pressing the subject key A, all the possible combinations which do not contain A will disappear, and the description of A may be gathered from what remains, namely that it is always D. The full-stop key restores all combinations consistent with the premises and any other selection may be made, as say not-D, which will be found to be always not-A, not-B, and not-C.

At the end of every problem, when no further questions need be addressed to the machine, it is desirable to press the Finis key, which has the effect of bringing into view the whole of the conceivable combinations of the abecedarium. This key in fact obliterates the conditions impressed upon the machine by moving back into their ordinary places those combinations which had been rejected as inconsistent with the premises. Before beginning any new problem it is requisite to observe that the whole sixteen combinations are visible. After the Finis key has been used the machine represents a mind endowed with powers of thought, but wholly devoid of knowledge. It would not in that condition give any answer but such as would consist in the primary laws of thought themselves. But when any proposition is worked upon the keys, the machine analyses or digests the meaning of it and becomes charged with the knowledge embodied in that proposition. Accordingly it is able to return as an answer any description of a term or class so far as furnished by that proposition in accordance with the Laws of Thought. The machine is thus the embodiment of a true logical system. The combinations are classified, selected or rejected just as they should be by a reasoning mind, so that at each step in a problem, the abecedarium represents the proper condition of a mind exempt from mistake. It cannot be asserted indeed that the machine entirely supersedes the agency of conscious thought; mental labour is required in interpreting the meaning of grammatical expressions and in correctly impressing that meaning on the machine; it is further required in gathering the conclusion from the remaining combinations. Nevertheless the true process of logical inference is actually accomplished in a purely mechanical manner.

It is worthy of remark that the machine can detect any self-contradiction existing between the premises presented to it, for it will then be found that one or more of the terms disappear entirely from the abecedarium. Thus if we worked the two propositions, A is B, and A is not-B, and then inquired for a description of A, the machine would refuse to give it by exhibiting no combination at all containing A. This result is in agreement with the law which I have explained that every term must have its negative (p. 88). Accordingly whenever any one of the letters A, B, C, D, a, b, c, d wholly disappears from the abecedarium, it may be safely inferred that some self-contradiction has been committed in the premises.

It ought to be carefully observed that the logical machine cannot receive a simple identity of the form

A = B except in the double form of A = AB and B = AB. To work the proposition A = B it is therefore necessary to press the keys—A (subj.), Copula, A (pred.), B (pred.), Full stop, B (subj.), Copula, A (pred.), B (pred.), Full stop. The same double operation will be necessary whenever the proposition is not of the kind called a partial identity (p. 47). Thus AB = CD, AB = AC, A = B + C, A + B = C + D, all require to be read from both ends separately. This is a remarkable fact which some persons may consider as militating against the equational form of proposition, but I do not think this is really the case.

Before leaving the subject I may remark that these mechanical devices are not likely to possess great practical utility. We do not require in common life to be constantly solving complex logical questions. Even in mathematical calculation the ordinary rules of arithmetic are generally sufficient, and a calculating machine could only be used with advantage in peculiar cases. But the machine and abacus have nevertheless two important uses.

1. I trust that the time is not very far distant when the predominance of the ancient Aristotelian Logic will be a matter of history, and the teaching of logic will be placed on a footing more worthy of its supreme importance. It will then be found that the solution of logical questions is an exercise of mind at least as valuable and necessary as mathematical calculation. I believe that these mechanical devices, or something of the same kind, will then become useful for exhibiting to a class of students a clear and visible analysis of logical problems of any degree of complexity, the nature of each step being rendered plain to the eye. For this purpose I have already often used the machine or abacus in my class lectures at the Owens College.

2. The more immediate importance of the machine seems to consist in the unquestionable proof which it affords that most comprehensive views of the principles of reasoning have now been attained, although they were almost wholly unknown to Aristotle and his followers. The time must come when the inevitable results of the admirable writings of the late Dr. Boole must be recognised at their true value, and the plain and palpable form in which the machine presents those results will, I hope, hasten the time. Undoubtedly his life marks an era in the high science of human reason. It may seem strange that it had remained for him first to set forth in its full extent the problem of logic, but I am not aware that any one before him had treated logic as a symbolic method for evolving from any premises the description of any class whatsoever as defined by those premises. His quasi-mathematical system indeed could not be regarded as a final and complete solution of the problem. Not only did it require the manipulation of mathematical symbols in a very intricate and perplexing manner, but the results when obtained were devoid of demonstrative force, because they turned upon the employment of unintelligible symbols, acquiring meaning only by analogy. I have also pointed out that he imported into his system a condition concerning the exclusive nature of alternatives (p. 83), which is not necessarily true of logical terms. I shall have to show in the next chapter that logic is really the basis of the whole science of mathematical reasoning, so that Boole completely inverted the true order of proof when he proposed to infer logical truths by algebraic processes. It is a wonderful evidence of his mental power that by methods fundamentally false he should have succeeded in reaching true conclusions and widening the sphere of reason.

The mechanical performance of logical inference affords a demonstration both of the truth of Boole's results and of the mistaken nature of his mode of deducing them. Conclusions which he could only obtain by pages of intricate calculation, are exhibited by the machine after one or two minutes of manipulation. And not only are those conclusions easily reached, but they are demonstratively true, because every step of the process involves nothing more obscure that the Laws of Thought.

The Order of Premises.

Before quitting the subject of deductive reasoning, I may remark that the order in which the premises of an argument, or any propositions whatsoever, are placed, is a matter of logical indifference. Much discussion has taken place at various times concerning the arrangement of the premises of a syllogism; and it has been generally held, in accordance with the opinion of Aristotle, that the so-called major premise, containing the major term, or the predicate of the conclusion, should stand first. This distinction however falls to the ground in our system, since the proposition is reduced to an identical form in which there is no distinction of subject and predicate. In a strictly logical and philosophic point of view the order of statement is wholly devoid of significance. The premises are simultaneously coexistent, and are not related to each other according to any of the properties of space or time. Just as the qualities of the same object are neither before nor after each other in nature (p. 40), and are only thought of in some one order owing to our limited capacity of mind, so the premises of an argument are neither before nor after each other, and are only thought of in succession because the mind cannot grasp many ideas at once. The logical combinations of the Abecedarium are exactly the same in whatever order the premises be treated on the logical slate or machine.

Some difference may doubtless exist as regards convenience to human memory. The mind may take in the results of an argument more easily in one mode of statement than another, although there is no real difference in the logical results. But in this point of view I think that Aristotle and the old logicians were clearly wrong. It is more easy to conclude that 'all A's are C's' from 'all A's are B's and all B's are C's,' than from the same propositions in inverted order, 'all B's are C's and all A's are B's.'

The Equivalency of Propositions.

One great advantage which arises from the study of this Indirect Method of Inference consists in the clear notion which we thus gain of the Equivalency of Propositions. The older logicians showed how from certain simple premises we might draw an inference, but they failed to point out whether that inference contained the whole, or only a part, of the information embodied in the premises. Now any one proposition or group of propositions may be classed with respect to another proposition or group of propositions, as

1. Equivalent,
2. Inferrible,
3. Consistent,
4. Contradictory.

Taking the proposition 'All men are mortals' as the original, 'All immortals are not men' is its equivalent; 'Some mortals are men' is inferrible, or capable of inference, but is not equivalent; 'All not men are not mortals' cannot be inferred, but is consistent, that is, may

be true at the same time ; ' All men are immortals ' is of
course contradictory.

One sufficient test of equivalency is the capability of
mutual inference.   Thus from

All electrics = all non-conductors,
I can infer

All non-electrics = all conductors,
and *vice versâ* from the latter I can pass back to the
former.   In short $A = B$ is equivalent to $a = b$.   Again,
from the union of the two propositions, $A = AB$ and
$B = AB$, I get $A = B$, and from this I might as easily
deduce the two with which I started.   In this case one
proposition is equivalent to two other propositions.   There
are indeed no less than four modes in which we may
express the identity of two classes A and B, namely,

FIRST MODE.	SECOND MODE.	THIRD MODE.	FOURTH MODE.
$A = B$	$a = b$	$\left.\begin{array}{l} A = AB \\ B = AB \end{array}\right\}$	$\left.\begin{array}{l} a = ab \\ b = ab \end{array}\right\}$

The Indirect Method of Inference furnishes an universal
and clear criterion as to the relationship of propositions.
The import of a statement is always to be measured by
the combinations of terms which it destroys.   Hence two
propositions are exactly equivalent when they remove
exactly the same combinations from the Abecedarium,
and neither more nor less.   A proposition is inferrible
but not equivalent to another when it removes some but
not all the combinations which the other removes.   Again,
propositions are consistent provided that they leave some
one combination containing each term, and the negative
of each term.   If after all the combinations inconsistent
with two propositions are struck out, there still appears
in the Abecedarium each of the letters A, *a*, B, *b*, C, *c*, D, *d*,
which were there before, then no inconsistency between
the propositions exists, although they may not be equivalent or even inferrible.   Finally, contradictory propositions

are those which altogether remove any one or more letter-terms from the Abecedarium.

What is true of single propositions applies also to groups of propositions, however large or complicated; that is to say, one group may be equivalent, inferrible, consistent, or contradictory as regards another, and we may similarly compare one proposition with a group of propositions.

To give in this place illustrations of all the four kinds of relation would require much space : as the examples given in previous sections or chapters may serve more or less to explain the relations of inference, consistency, and contradiction, I will only add a few instances of equivalent propositions or groups.

In the following list each proposition or group of propositions is exactly equivalent in meaning to the corresponding one in the other column, and the truth of this statement may be tested by working out the combinations of the Abecedarium, which ought to be found exactly the same in the case of each pair of equivalents.

$$A = Ab \qquad\qquad B = aB$$

$$A = b \qquad\qquad a = B$$

$$A = BC \qquad\qquad a = b \dagger c$$

$$A = AB \dagger AC \qquad\qquad b = ab \dagger AbC$$

$$A \dagger B = C \dagger D \qquad\qquad ab = cd$$

$$A \dagger c = B \dagger d \qquad\qquad aC = bD$$

$$A = ABc \dagger AbC \qquad\qquad \left.\begin{array}{l} A = AB \dagger AC \\ AB = ABc \end{array}\right\}$$

$$\left.\begin{array}{l} A = B \\ B = C \end{array}\right\} \qquad\qquad \left.\begin{array}{l} A = B \\ A = C \end{array}\right\}$$

$$\left.\begin{array}{l} A = AB \\ B = BC \end{array}\right\} \qquad\qquad \left.\begin{array}{l} A = AC \\ B = A \dagger aBC \end{array}\right\}$$

$$\left.\begin{array}{l} A = AB \\ A = AC \\ A = AD \end{array}\right\} \qquad\qquad A = ABCD.$$

Although in these and many other cases the equivalents of certain propositions can readily be given, yet I believe that no uniform and infallible process can be pointed out by which the exact equivalents of premises can be ascertained. Ordinary deductive inference usually gives us only a portion of the contained information. It is true that the combinations consistent with a set of propositions are logically equivalent to them, but the difficulty consists in passing back from the combinations to a new set of propositions. The task is here of a different character from any which we have yet attempted. It is in reality an inverse process, and is just as much more troublesome and uncertain than the direct process, as seeking is compared with hiding. Not only may several different answers equally apply, but there is no method of discovering any of those answers except by repeated trial. The problem which we have here met is really that of induction, the inverse of deduction ; and, as I shall soon show, induction is always tentative, and unless conducted with peculiar skill and insight must be exceedingly laborious in cases of any considerable complexity.

The late Professor de Morgan was unfortunately led by this equivalency of propositions into the most serious error of his ingenious system of Logic. He held that because the proposition 'All A's are all B's,' was but another expression for the two propositions 'All A's are B's' and 'All B's are A's,' it must be a composite and not really an elementary form of proposition[1]. But on taking a general view of the equivalency of propositions such an objection seems to have no weight. Logicians have, with few exceptions, persistently upheld the original error of Aristotle in rejecting from their science the one simple

[1] 'Syllabus of a proposed system of Logic,' §§ 57, 121, &c. 'Formal Logic,' p. 66.

relation of identity on which all more complex logical relations must really rest.

The Nature of Inference.

The question, What is Inference? is involved, even to the present day, in as much uncertainty as that ancient question, What is Truth? I shall in more than one part of this work endeavour to show that inference never does more than explicate, unfold, or develop the information contained in certain premises or facts. Neither in deductive nor inductive reasoning can we add a tittle to our implicit knowledge, which is like that contained in an unread book or a sealed letter. Sir W. Hamilton has well said, 'Reasoning is the showing out explicitly that a proposition not granted or supposed, is implicitly contained in something different which is granted or supposed [m].'

Professor Bowen has explained [n] with much clearness that the conclusion of an argument states explicitly what is virtually or implicitly thought. ' The process of reasoning is not so much a mode of evolving a new truth, as it is of establishing or proving an old one, by showing how much was admitted in the concession of the two premises taken together.' It is true that the whole meaning of these statements rests upon that of such words as 'explicit,' 'implicit,' 'virtual.' That is implicit which is wrapped up, and we render it explicit when we unfold it. Just as the conception of a circle involves a hundred important geometrical properties, all following from what we know, if we have acuteness to unfold the results, so every fact and statement involves more meaning than seems at first sight. Reasoning explicates or brings to conscious possession what was before unconscious. It does not create, nor

m Lectures on Metaphysics, vol. iv. p. 369.
n Bowen, ' Treatise on Logic,' Cambridge, U. S., 1866; p. 362.

does it destroy, but it transmutes and throws the same matter into a new form.

The difficult question still remains, Where does novelty of form begin? Is it a case of inference when we pass from 'Sincerity is the parent of truth' to 'The parent of truth is sincerity?' The old logicians would have called this change conversion, one case of immediate inference. But as all identity is necessarily reciprocal, and the very meaning of such a proposition is that the two terms are identical in their signification, I fail to see any difference between the statements whatever. As well might we say that $a = b$ and $b = a$ are different equations.

Another point of difficulty is to decide when a change is merely grammatical and when it involves a real logical transformation. Between a *table of wood* and a *wooden table* there is no logical difference (p. 37), the adjective being merely a convenient substitute for the prepositional phrase. But it is uncertain to my mind whether the change from 'All men are mortal' to 'No men are not mortal' is purely grammatical. Logical change may perhaps be best described as consisting in the determination of a relation between certain classes of objects from a relation between certain other classes. Thus I consider it a truly logical inference when we pass from 'All men are mortal' to 'All immortals are not-men,' because the classes *immortals* and *not-men* are different from *mortals* and *men*, and yet the propositions contain at the bottom the very same truth, as shown in the combinations of the Abecedarium.

From logical inference we must discriminate the passage from the qualitative to the quantitative form of a proposition. We state the same truth when we say that 'mortality belongs to all men,' as when we assert that 'all men are mortals.' Here we do not pass from class to class, but from one kind of term, the abstract, to another kind, the concrete. But inference probably enters when we pass from either of the above propositions to the assertion that the class of immortal men is zero, or contains no objects.

It is really a question of words to what processes we shall or shall not apply the name 'inference,' and I have no wish to continue the trifling discussions which have already taken place upon the subject. We shall not commit any serious error, provided that we always bear in mind that two propositions may be connected together in four different ways. They may be—

1. *Tautologous* or *identical*, involving the same relation between the same terms and classes, and only differing in the order of statement; thus ' Victoria is the Queen of England' is tautologous with 'The Queen of England is Victoria.'

2. *Grammatically equivalent*, in which the classes or objects are the same and similarly related, and the only difference is in the words; thus ' Victoria is the Queen of England' is grammatically equivalent to 'Victoria is England's Queen.'

3. Equivalent in qualitative and quantitative form, the classes being the same, but viewed in a different manner.

4. *Logically equivalent*, when the classes and relations are different, but involve the same knowledge of the possible combinations.

CHAPTER VII

WE enter in this chapter upon the second great department of logical method, that of Induction or the Inference of general from particular truths. It cannot be said that the Inductive process is of greater importance than the Deductive process already considered, because the latter process is absolutely essential to the existence of the former. Each is the complement and counterpart of the other. The principles of thought and existence which underlie them are at the bottom the same, just as subtraction of numbers necessarily rests upon the same principles as addition. Induction is, in fact, the inverse operation to deduction, and cannot be conceived to exist without the corresponding operation, so that the question of relative importance cannot arise. Who thinks of asking whether addition or subtraction is the more important process in arithmetic? But at the same time much difference in difficulty may exist between a direct and inverse operation ; the integral calculus, for instance, is almost infinitely more difficult than the differential calculus of which it is the inverse. It must be allowed that in logic inductive investigations are of a far higher degree of difficulty, variety, and complexity than any questions of deduction ; and it is this fact no doubt which has led some logicians to erroneous opinions concerning the exclusive importance of induction.

Hitherto we have been engaged in considering how

from certain conditions, laws, or identities governing the combinations of qualities, we may deduce the nature of the combinations agreeing with those conditions. Our work has been to unfold the results of what is contained in any statements, and the process has been one of *Synthesis*. The terms or combinations of which the character has been determined have usually, though by no means always, involved more qualities, and therefore, by the relation of extension and intension, fewer objects than the terms in which they were described. The truths inferred were thus usually less general than the truths from which they were inferred.

In induction all is inverted. The truths to be ascertained are more general than the data from which they are drawn. The process by which they are reached is *analytical*, and consists in separating the complex combinations in which natural phenomena are presented to us, and determining the relations of separate qualities. Given events obeying certain unknown laws, we have to discover the laws obeyed. Instead of the comparatively easy task of finding what effects will follow from a given law, the effects are now given and the law is required. We have to interpret the will by which the conditions of creation were laid down.

Induction an Inverse Operation.

I have already asserted that induction is the inverse operation of deduction, but the difference is one of such great importance that I must dwell upon it. There are many cases where we can easily and infallibly do a certain thing but may have much trouble in undoing it. A person may walk into the most complicated labyrinth or the most extensive catacombs, and turn hither and thither at his will; it is when he wishes to return that doubt and difficulty commence. In entering, any path served him; in leaving, he must select certain definite paths, and in this selection he must either trust to memory of the way he entered or else make an exhaustive trial of all possible ways. The explorer entering a new country makes sure his line of return by barking the trees.

The same difficulty arises in many scientific processes. Given any two numbers, we may by a simple and infallible process obtain their product, but it is quite another matter when a large number is given to determine its factors. Can the reader say what two numbers multiplied together will produce the number 8,616,460,799 ? I think it unlikely that any one but myself will ever know; for they are two large prime numbers, and can only be rediscovered by trying in succession a long series of prime divisors until the right one be fallen upon. The work would probably occupy a good computer for many weeks, but it did not occupy me many minutes to multiply the two factors together. Similarly there is no direct process for discovering whether any number is a prime or not; it is only by exhaustingly trying all inferior numbers which could be divisors, that we can show there is none, and the labour of the process would be intolerable were it not performed systematically once for all in the process known as the Sieve of Eratosthenes, the results being registered in tables of prime numbers.

The immense difficulties which are encountered in the solution of algebraic equations are another illustration. Given any algebraic factors, we can easily and infallibly arrive at the product, but given a product it is a matter of infinite difficulty to resolve it into factors. Given any series of quantities however numerous, there is very little trouble in making an equation which shall have those quantities as roots. Let a, b, c, d, &c., be the quantities; then

$$(x-a)\,(x-b)\,(x-c)\,(x-d)\ldots\ldots = 0$$

is the equation required, and we only need to multiply out the expression on the left hand by ordinary rules. But having given a complex algebraic expression equated to zero, it is a matter of exceeding difficulty to discover all the roots. Mathematicians have exhausted their highest powers in carrying the complete solution up to the fourth degree. In every other mathematical operation the inverse process is far more difficult than the direct process, subtraction than addition, division than multiplication, evolution than involution; but the difficulty increases vastly as the process becomes more complex. The differentiation, the direct process, is always capable of performance by certain fixed rules, but as these produce considerable variety of results, the inverse process of integration presents immense difficulties, and in an infinite majority of cases surpasses the present resources of mathematicians. There are no infallible and general rules for its accomplishment; it must be done by trial, by guesswork, by remembering the results of differentiation, and using them as a guide.

Coming more nearly to our own immediate subject, exactly the same difficulty exists in determining the law which certain numbers obey. Given a general mathematical expression, we can infallibly ascertain its value for any required value of the variable. But I am not aware that mathematicians have ever attempted to lay down the rules of a process by which, having given certain numbers, one might discover a rational or precise formula from which they proceed. The problem is always indeterminate, because an infinite number of formulæ agreeing with certain numbers, might always be discovered with sufficient trouble.

The reader may test his power of detecting a law, by contemplation of its results, if he, not being a mathematician, will attempt to point out the law obeyed by the following numbers:

$$-\frac{1}{2}, \frac{1}{6}, -\frac{1}{30}, \frac{1}{42}, -\frac{1}{30}, \frac{5}{66}, \frac{691}{2730}, \frac{7}{6}, \frac{3617}{510}, \text{ etc.}$$

These numbers are sometimes negative, more often positive ; sometimes in low terms, but unexpectedly springing up to high terms ; in absolute magnitude they are very variable. They seem to set all regularity and method at defiance, and it is hardly to be supposed that any one could, from contemplation of the numbers, have detected the relation between them. Yet they are derived from the most regular and symmetrical laws of relation, and are of the highest importance in mathematical analysis, being known as the numbers of Bernouilli.

Compare again the difficulty of decyphering with that of cyphering. Any one can invent a secret language, and with a little steady labour can translate the longest letter into the character. But to decypher the letter having no key to the signs adopted, is a wholly different matter. As the possible modes of secret writing are infinite in number and exceedingly various in kind, there is no direct mode of discovery whatever. Repeated trial, guided more or less by knowledge of the customary form of cypher, and resting entirely on the principles of probability, is the only resource. A peculiar tact or skill is requisite for the process, and a few men, such as Wallis or Mr. Wheatstone, have attained great success.

Induction is the decyphering of the hidden meaning of natural phenomena. Given events which happen in certain definite combinations, we are required to point out the laws which have governed those combinations. Any laws being supposed, we can, with ease and certainty, decide whether the phenomena obey those laws. But the laws which may exist are infinite in variety, so that the chances are immensely against mere random guessing. The difficulty is much increased by the fact that several laws will usually be in operation at the same time, the effects of which are complicated together. The only modes of discovery consist either in exhaustively trying a great number of supposed laws, a process which is exhaustive in more senses than one, or else by carefully contemplating the effects, endeavouring to remember cases in which like effects followed from known laws. However we accomplish the discovery, it must be done by the more or less apparent application of the direct process of deduction.

The Logical Abecedarium illustrates induction as well as it does deduction. In the Indirect process of Inference we found that from certain propositions we could infallibly determine the combinations of terms agreeing with those premises. The inductive problem is just the inverse. Having given certain combinations of terms, we need to ascertain the propositions with which they are consistent, and from which they may have proceeded. Now if the reader contemplates the following combinations

$$ABC \qquad abC$$
$$aBC \qquad abc,$$

he will probably remember at once that they belong to the premises $A = AB$, $B = BC$. If not, he will require a few trials before he meets with the right answer, and every trial will consist in assuming certain laws and observing whether the deduced results agree with the data. To test the facility with which he can solve this inductive problem, let him casually strike out any of the combinations, say of the fourth column of the Abecedarium (p. 109), and say what laws the remaining combinations obey, observing that every one of the letter-terms and their negatives ought to appear in order to avoid self-contradiction in the premises (pp. 88, 128). Let him say, for instance, what laws are embodied in the combinations

$$ABC \qquad aBC$$
$$Abc \qquad abC.$$

The difficulty becomes much greater when more terms
enter into the combinations. It would be no easy matter
to point out the complete conditions fulfilled in the combinations

$$ACe$$
$$aBCe$$
$$aBcdE$$
$$abCe$$
$$abcE.$$

After some trouble the reader may discover that the
principal laws are $C = e$, and $A = Ae$; but he would hardly
discover the remaining law, namely that $BD = BDe$.

The difficulties encountered in the inductive investigations of nature, are of an exactly similar kind.

We seldom observe any great law in uninterrupted and
undisguised operation. The acuteness of Aristotle and
the ancient Greeks, did not enable them to detect that all
terrestrial bodies tend to fall towards the centre of the
earth. A very few nights of observation would have convinced an astronomer viewing the solar system from its
centre, that the planets travelled round the sun; but the
fact that our place of observation is one of the travelling
planets, so complicates the apparent motions of the other
bodies, that it required all the industry and sagacity of
Copernicus to prove the real simplicity of the planetary
system. It is the same throughout nature; the laws may
be simple, but their combined effects are not simple, and
we have no clue to guide us through their intricacies. ' It
is the glory of God,' said Solomon, ' to conceal a thing, but
the glory of a king to search it out.' The laws of nature
are the invaluable secrets which God has hidden, and it is
the kingly prerogative of the philosopher to search them
out by industry and sagacity.

Induction of Simple Identities.

Many of the most important laws of nature are expressible in the form of simple identities, and I can at once adduce them as examples to illustrate what I have said of the difficulty of the inverse process of induction. There are many cases in which two phenomena are usually conjoined. Thus all gravitating matter is exactly coincident with all matter possessing inertia ; where one property appears, the other likewise appears. All crystals of the cubical system, are all the crystals which do not doubly refract light. All exogenous plants are, with some exceptions, those which have two cotyledons or seed-leaves.

A little reflection will show that there is no direct and infallible process by which such complete coincidences may be discovered. Natural objects are aggregates of many qualities, and any one of those qualities may prove to be in close connection with some others. If each of a numerous group of objects is endowed with a hundred distinct physical or chemical qualities, there will be no less than $\frac{1}{2}$ (100 × 99) or 4950 pairs of qualities, which may be connected, and it will evidently be a matter of great intricacy and labour to ascertain exactly which qualities are connected by any simple law.

One principal source of difficulty is that the finite powers of the human mind are not sufficient to compare by a single act any large group of objects with another large group. We cannot hold in the conscious possession of the mind at any one moment more than five or six different ideas. Hence we must treat any more complex group by successive acts of attention. The reader will perceive by an almost individual act of comparison that the words *Roma* and *Mora* contain the same letters. He may perhaps see at a glance whether the same is true of *Causal* and *Casual,* and of *Logica* and *Caligo.* To assure himself that the letters in *Astronomers* make *No more stars*, that *Serpens in aculeo* is an anagram of *Joannes Keplerus*, or *Great gun do us a sum* an anagram of *Augustus de Morgan*, it will certainly be necessary to break up the act of comparison into several successive acts. The process will acquire a double character, and will consist in ascertaining that each letter of the first group is among the letters of the second group, and *vice versâ*, that each letter of the second is among those of the first group. In the same way we can only prove that two long lists of names are identical, by showing that each name in one list occurs in the other, and *vice versâ*.

This process of comparison really consists in establishing two partial identities, which are, as already shown (p. 133), equivalent in conjunction to one simple identity. We first ascertain the truth of the two propositions $A = AB$, $B = AB$, and we then rise by substitution to the single law $A = B$.

There is another process, it is true, by which we may get to exactly the same result, for the two propositions $A = AB$, $a = ab$ are also equivalent to the simple identity $A = B$ (p. 133). If then we can show that all objects included under A are included under B, and also that all objects not included under A are not included under B, our purpose is effected. By this process we should usually compare two lists if we are allowed to mark them. For each name in the first list we should strike off one in the second, and if, when the first list is exhausted the second list is also exhausted, it follows that all names absent from the first must be absent from the second, and the coincidence must be complete.

The two modes of proving a simple identity are so closely allied that it is doubtful how far we can detect any difference in their powers and instances of application. The first method is perhaps more convenient where the phenomena to be compared are rare. Thus we prove that all the musical concords coincide with all the more simple numerical ratios, by showing that each concord arises from a simple ratio of undulations, and then showing that each simple ratio gives rise to one of the concords. To examine all the possible cases of discord or complex ratio of undulation would be impossible. By a happy stroke of induction Sir John Herschel discovered that all crystals of quartz which rotate the plane of polarization of light are precisely those crystals which have plagihedral faces, that is, oblique faces on the corners of the prism unsymmetrical with the ordinary faces. This singular relation would be proved by observing that all plagihedral crystals possessed the power of rotation, and *vice versâ* all crystals possessing this power were plagihedral. But it might at the same time be noticed that all ordinary crystals were devoid of the power. There is no reason why we should not observe any of the four propositions $A = AB$, $B = AB$, $a = ab$, $b = ab$, all of which follow from $A = B$ (see p. 133).

Sometimes the terms of the identity may be singular objects; thus we observe that diamond is a combustible gem, and being unable to discover any other that is, we affirm

Diamond = combustible gem.

In a similar manner we ascertain that

Mercury = metal liquid at ordinary temperatures,

Substance of least density = substance of least atomic weight.

Two or three objects may occasionally enter into the induction, as when we learn that

Sodium + potassium = metal of less density than water,

Venus + Mercury + Mars = major planet devoid of satellites.

Induction of Partial Identities.

We found in the last section that the simple identity of two classes is almost always discovered not by direct observation of the fact, but by first establishing two partial identities. There are also a great multitude of cases in which the partial identity of one class with another is the only relation to be discovered. Thus the most common of all inductive inferences consists in establishing the fact that all objects having the properties of A have also those of B, or that $A = AB$. To ascertain the truth of a proposition of this kind it is merely necessary to assemble together, mentally or physically, all the objects included under A, and then observe whether B is present in each of them, or, which is the same, whether it would be impossible to select from among them any not-B. Thus, if we mentally assemble together all the heavenly bodies which move with apparent rapidity, that is to say the planets, we find that they all possess the property of not scintillating. We cannot analyse any vegetable substance without discovering that it contains carbon and hydrogen, but it is not true that all substances containing carbon and hydrogen are vegetable substances.

The great mass of scientific truths consists of propositions of this form $A = AB$. Thus in astronomy we learn that all the planets are spheroidal bodies; that they all revolve in one direction round the sun; that they all shine by reflected light; that they all obey the law of gravitation. But of course it is not to be asserted that all bodies obeying the law of gravitation, or shining by reflected light, or revolving in a particular direction, or being spheroidal in form, are planets. In other sciences we have immense numbers of propositions of the same form, as for instance that all substances in becoming gaseous absorb heat; that all metals are elements; that they are all good conductors of heat and electricity; that all the alkaline metals are monad elements; that all foraminifera are marine organisms; that all parasitic animals are non-mammalian; that lightning never issues from stratous clouds[a]; that pumice never occurs where only Labrador felspar is present[b]: and scientific importance may attach even to such apparently trifling observations as that ‘white cats having blue eyes are deaf[c].’

The process of inference by which all such truths are obtained may readily be exhibited in a precise symbolic form. We must have one premise specifying in a disjunctive form all the possible individuals which belong to a class; we resolve the class, in short, into its constituents. We then need a number of propositions each of which affirms that one of the individuals possesses a certain property. Thus the premises must be of the form

$$A = B + C + D + \ldots\ldots + P + Q$$
$$B = BX$$
$$C = CX$$
$$\ldots \quad \ldots$$
$$\ldots \quad \ldots$$
$$Q = QX.$$

Now if we substitute for each alternative of the first premise its description as found among the succeeding premises we obtain

$$A = BX + CX + \ldots\ldots + PX + QX$$

or

$$A = (B + C + \ldots\ldots + Q)X.$$

[a] Arago's Meteorological Essays, p. 10.
[b] Lyell's Elements of Geology, Fourth ed. p. 373.
[c] Darwin's Variation of Animals, &c.

But for the aggregate of alternatives we may now substitute their equivalent as given in the first premise, namely A, so that we get the required result

$$A = AX.$$

It may be remarked that we should have reached the same final result if our original premise had been of the form

$$A = AB + AC + \ldots\ldots + AQ.$$

The difference of meaning is that all B's need not now be A's, nor all C's, &c. But we should still have

$$A = ABX + ACX + \ldots\ldots + AQX = AX.$$

We can always prove a proposition, if we find it more convenient, by proving its equivalent. To assert that all not-B's are not-A's, is exactly the same as to assert that all A's are B's. Accordingly we may ascertain that $A = AB$ by first ascertaining that $b = ab$. If we observe, for instance, that all substances which are not solids are also not capable of double refraction, it follows necessarily that all double refracting substances are solids. We may convince ourselves that all electric substances are nonconductors of electricity, by reflecting that all good conductors do not, and in fact cannot, retain electric excitation. When we come to questions of probability it will be found desirable to prove, as far as possible, both the original proposition and its equivalent, as there is then an increased area of observation.

The number of alternatives which may arise in the division of a class varies greatly, and may be any number from two upwards. Thus it is probable that every substance is either magnetic or diamagnetic, and no substance can be both at the same time. The division then must be made in the form

$$A = ABc + AbC.$$

If now we can prove that all magnetic substances are capable of polarity, say $B = BC$, and also that all diamagnetic substances are capable of polarity $C = CD$, it follows by substitution that all substances are capable of polarity, or $A = AD$. We may divide the class substance again into the three subclasses, solid, liquid, and gas; and if we can show that in each of these forms it obeys Carnot's thermodynamic law, it follows that all substances obey that law. Similarly we may show that all verte-brate animals possess red blood, if we can show separately that fish, reptiles, birds, marsupials, and mammals possess red blood, there being, as far as is known, only five principal subclasses of vertebrata.

Our inductions will often be embarrassed by exceptions, real or apparent. We might affirm that all gems are incombustible were not diamond undoubtedly combustible. Nothing seems more evident than that all the metals are opaque until we examine them in fine films, when gold and silver are found to be transparent. All plants absorb carbonic acid except certain fungi; all the bodies of the planetary system have a progressive motion from west to east, except the satellites of Uranus and Neptune. Even some of the profoundest laws of matter are not quite universal; all solids expand by heat except india-rubber, and possibly a few other substances; all liquids which have been tested expand by heat except water below 4°C and fused bismuth; all gases have a coefficient of expansion increasing with the temperature except hydrogen. In a later chapter I shall consider how such anomalous cases may be regarded and classified; here we have only to express them in a consistent manner in our nota-tion.

Let us take the case of the transparency of metals, and assign the terms thus

A = metal	D = iron
B = gold	E, F &c. = copper, lead, &c.
C = silver	X = opaque.

Our premises will be

$$A = B + C + D + E, \&c.$$
$$B = Bx$$
$$C = Cx$$
$$D = DX$$
$$E = EX,$$

and so on for the rest of the metals. Now evidently

$$Abc = (D + E + F + \ldots\ldots)bc,$$

and by substitution as before we shall obtain

$$Abc = AbcX,$$

or in words, 'All metals not gold nor silver are opaque;' at the same time we have

$$A(B + C) = AB + AC = ABx + ACx = A(B + C)x,$$

or 'Metals which are either gold or silver are not opaque.'

In some cases the problem of induction assumes a much higher degree of complexity. If we examine the properties of crystallized substances we may find some properties which are common to all, as cleavage or fracture in definite planes; but it would soon become requisite to break up the class into several minor ones. We should divide crystals according to the seven accepted systems—and we should then find that crystals of each system possess many common properties. Thus crystals of the Regular or Cubical system expand equally by heat, conduct heat and electricity with uniform rapidity, and are of like elasticity in all directions; they have but one index of refraction for light; and every facet is repeated in like relation to each of the three axes. Crystals of the system which possess one principal axis will be found to possess the various physical powers of conduction, refraction, elasticity, &c., uniformly in directions perpendicular to the principal axis, but in other directions their properties vary according to complicated laws. The remaining systems in which the crystals possess three unequal axes, or have inclined axes, exhibit still more complicated results, the

effects of the crystal upon light, heat, electricity, &c.,
varying in all directions. But when we pursue induction
into the intricacies of its application to Nature we really
enter upon the subject of classification which we must
take up again in a later part of this work.

Complete Solution of the Inverse or Inductive Logical Problem.

It is now plain that Induction consists in passing back
from a series of combinations to the laws by which such
combinations are governed. The natural law that all
metals are conductors of electricity really means that in
nature we find three classes of objects, namely—

1. Metals, conductors ;
2. Not-metals, conductors ;
3. Not-metals, not-conductors.

It comes to the same thing if we say that it excludes the
existence of the class, 'metals not-conductors.' In the
same way every other law or group of laws will really
mean the exclusion from existence of certain combinations
of the things, circumstances or phenomena governed by
those laws. Now in logic we treat not the phenomena
and laws but, strictly speaking, the general forms of the
laws ; and a little consideration will show that for a finite
number of things the possible number of forms or kinds
of law governing them must also be finite. Using general
terms we know that A and B can be present or absent in
four ways and no more—thus

$$AB, Ab, aB, ab;$$

therefore every possible law which can exist concerning
the relation of A and B must be marked by the exclusion
of one or more of the above combinations. The number
of possible laws then cannot exceed the number of selections which we can make from these four combinations,

and we arrive at this utmost number of cases by omitting any one or more of the four. The number of cases to be considered is therefore 2 × 2 × 2 × 2 or sixteen, since each may be present or absent; and these cases are all shown in the following table, in which the sign o indicates absence or non-existence of the combination shown at the left-hand column in the same line, and the mark I its presence :—

	1	2	3	4	5	6	7	8	9	10	11	12	13	14	15	16
AB	o	o	o	o	o	o	o	o	I	I	I	I	I	I	I	I
Ab	o	o	o	o	I	I	I	I	o	o	o	o	I	I	I	I
aB	o	o	I	I	o	o	I	I	o	o	I	I	o	o	I	I
ab	o	I	o	I	o	I	o	I	o	I	o	I	o	I	o	I

Thus in column sixteen we find that all the conceivable combinations are present, which means that there are no special laws in existence in such a case, and that the combinations are governed only by the universal Laws of Identity and Difference. The example of metals and conductors of electricity would be represented by the twelfth column; and every other mode in which two things or qualities might present themselves is shown in one or other of the columns. More than half the cases may indeed be at once rejected, because they involve the entire absence of a term or its negative. It has been shown to be a necessary logical principle that every term must have its negative (p. 88), and where this is not the case some inconsistency between the laws or conditions of combinations must exist. Thus if we laid down the two following propositions, 'Graphite conducts electricity,' and 'Graphite does not conduct electricity,' it would amount to asserting the impossibility of graphite existing at all; or in general terms, A is B and A is not B result in destroying altogether the combinations containing A.

We therefore restrict our attention to those cases which may be represented in natural phenomena where at least two combinations are present, and which correspond to those columns of the table in which each of A, a, B, b appears. These cases are shown in the columns marked with an asterisk.

We find that seven cases remain for examination, thus characterised—

Four cases exhibiting three combinations,
Two cases exhibiting two combinations,
One case exhibiting four combinations.

It has already been pointed out that a proposition of the form A = AB destroys one combination Ab, so that this is the form of law applying to the twelfth case. But by changing one or more of the terms in A = AB into its negative, or by interchanging A and B, a and b, we obtain no less than eight different varieties of the one form; thus—

12th case.	8th case.	15th case.	14th case.
A = AB	A = Ab	$a = a$B	$a = ab$
$b = ab$	B = aB	$b = Ab$	B = AB.

But the reader of the preceding sections will at once see that each proposition in the lower line is logically equivalent to, and is in fact the contrapositive of, that above it (p. 98). Thus the propositions A = Ab and B = aB both give the same combinations, shown in the eighth column of the table, and trial shows that the twelfth, eighth, fifteenth and fourteenth cases are thus fully accounted for. We come to this conclusion then—*The general form of proposition* A = AB *admits of four logically distinct varieties, each capable of expression in two different modes.*

In two columns of the table, namely the seventh and tenth, we observe that two combinations are missing. Now a simple identity A = B renders impossible both Ab and aB, accounting for the tenth case ; and if we change B into b the identity $A = b$ accounts for the seventh case. There may indeed be two other varieties of the simple identity, namely $a = b$ and $a = B$; but it has already been shown repeatedly that these are equivalent respectively to $A = B$ and $A = b$ (pp. 133, 134). As the sixteenth column has already been accounted for as governed by no special conditions, we come to the following general conclusion :— The laws governing the combinations of two terms must be capable of expression either in a partial identity $(A = AB)$, or a simple identity $(A = B)$; the partial identity is capable of only four logically distinct varieties, and the simple identity of two. Every logical relation between two terms must be expressed in one of these six laws, or must be logically equivalent to one of them.

In short, we may conclude that in treating of partial and complete identity, we have exhaustively treated the modes in which two terms or classes of objects can be related. Of any two classes it may be said that one must either be included in the other, or must be identical with it, or some similar relation must exist between one class and the negative of the other. We have thus completely solved the inverse logical problem concerning two terms[d].

The Inverse Logical Problem involving Three Terms.

No sooner do we introduce into the problem a third term C, than the investigation assumes a far more complex character, so that some readers may prefer to pass over this section. Three terms and their negatives may be combined, as we have frequently seen, in eight different

---

[d] The contents of this and the following section nearly correspond with those of a paper read before the Manchester Literary and Philosophical Society on December 26th, 1871. See Proceedings of the Society, vol. xi. pp. 65–68, and Memoirs, Third Series, vol. v. pp. 119–130.

combinations, and the effect of laws or logical conditions is to destroy any one or more of these combinations. Now we may make selections from eight things in 2^8 or 256 ways; so that we have no less than 256 different cases to treat, and the complete solution is at least fifty times as troublesome as with two terms. Many series of combinations, indeed, are contradictory, as in the simpler problem, and may be passed over. The test of consistency is that each of the letters A, B, C, a, b, c shall appear somewhere in the series of combinations; but I have not been able to discover any mode of calculating the number of cases in which inconsistency would happen. The logical complexity of the problem is so great that the ordinary modes of calculating numbers of combinations in mathematical science fail to give any aid, and exhaustive examination of the combinations in detail is the only method applicable.

My mode of solving the problem was as follows:—Having written out the whole of the 256 series of combinations, I examined them separately and struck out such as did not fulfil the test of consistency. I then chose some common form of proposition involving two or three terms, and varied it in every possible manner, both by the circular interchange of letters (A, B, C into B, C, A and then C, A, B), and by the substitution for any one or more of the terms of the corresponding negative terms. For instance, the proposition AB = ABC can be first varied by circular interchange, so as to give BC = BCA and then CA = CAB. Each of these three can then be thrown into eight varieties by negative change. Thus AB = ABC gives aB = aBC, Ab = AbC, AB = ABc, ab = abC, and so on. Thus there may possibly exist no less than twenty-four varieties of the law having the general form AB = ABC, meaning that whatever has the properties of A and B has those also of C. It by no means follows that some of the varieties may not be equivalent to others; and trial shows, in fact, that $AB = ABC$ is exactly the same in meaning as $Ac = Abc$ or $Bc = Bca$. Thus the law in question has but eight varieties of distinct logical meaning. I now ascertain by actual deductive reasoning which of the 256 series of combinations result from each of these distinct laws, and mark them off as soon as found. I now proceed to some other form of law, for instance $A = ABC$, meaning that whatever has the qualities of A has those also of B and C. I find that it admits of twenty-four variations, all of which are found to be logically distinct ; the combinations being worked out, I am able to mark off twenty-four more of the list of 256 series. I proceed in this way to work out the results of every form of law which I can find or invent. If in the course of this work I obtain any series of combinations which had been previously marked off, I learn at once that the law is logically equivalent to some law previously treated. It may be safely inferred that every variety of the apparently new law will coincide in meaning with some variety of the former expression of the same law. I have sufficiently verified this assumption in some cases and have never found it lead to error. Thus just as $AB = ABC$ is equivalent to $Ac = Abc$, so we find that $ab = abC$ is equivalent to $ac = acB$.

Among the laws treated were the two $A = AB$ and $A = B$ which involve only two terms, because it may of course happen that among three things two only are in special logical relation, and the third independent ; and the series of combinations representing such cases of relation are sure to occur in the complete enumeration. All single propositions which I could invent having been treated, pairs of propositions were next investigated. Thus we have the relations, 'All A's are B's and all B's are C's,' of which the old logical syllogism is the development. We may also have 'all A's are all B's, and all B's are C's,' or even 'all A's are all B's, and all B's are all C's.' All such premises admit of variations, greater or less in number, the logical distinctness of which can only be determined by trial in detail. Disjunctive propositions either singly or in pairs were also treated, but were often found to be equivalent to other propositions of a simpler form; thus $A = ABC + Abc$ is exactly the same in meaning as $AB = AC$.

This mode of exhaustive trial bears some analogy to that ancient mathematical process called the Sieve of Eratosthenes. Having taken a long series of the natural numbers, Eratosthenes is said to have calculated out in succession all the multiples of every number, and to have marked them off, so that at last the prime numbers alone remained, and the factors of every number were exhaustively discovered. My problem of 256 series of combinations is the logical analogue, the chief points of difference being that there is a limit to the number of cases, and that prime numbers have no analogue in logic, since every series of combinations corresponds to a law or group of conditions. But the analogy is perfect in the point that they are both inverse processes. There is no mode of ascertaining that a number is prime but by showing that it is not the product of any assignable factors. So there is no mode of ascertaining what laws are embodied in any series of combinations but trying exhaustively the laws which would give them. Just as the results of Eratosthenes' method have been worked out to a great extent and registered in tables for the convenience of other mathematicians, I have endeavoured to work out the inverse logical problem to the utmost extent which is at present practicable or useful.

I have thus found that there are altogether fifteen conditions or series of conditions which may govern the combinations of three terms, forming the premises of fifteen essentially different kinds of arguments. The following table contains a statement of these conditions, together with the number of combinations which are contradicted or destroyed by each, and the number of logically distinct variations of which the law is capable. There might be also added, as a sixteenth case, that case where no special logical condition exists, so that all the eight combinations remain.

Reference Number.	Propositions expressing the general type of the logical conditions.	Number of distinct logical variations.	Number of combinations contradicted by each.
I.	$A = B$	6	4
II.	$A = AB$	12	2
III.	$A = B, B = C$	4	6
IV.	$A = B, B = BC$	24	5
V.	$A = AB, B = BC$	24	4
VI.	$A = BC$	24	4
VII.	$A = ABC$	24	3
VIII.	$AB = ABC$	8	1
IX.	$A = AB, aB = aBc$	24	3
X.	$A = ABC, ab = abC$	8	4
XI.	$AB = ABC, ab = abc$	4	2
XII.	$AB = AC$	12	2
XIII.	$A = BC + Abc$	8	3
XIV.	$A = BC + bc$	2	4
XV.	$A = ABC, a = Bc + bC$	8	5
		192	

There are sixty-three series of combinations derived from self-contradictory premises, which with the above 192 series and the one case where there are no conditions or laws at all, make up the whole conceivable number of 256 series.

We learn from this table, for instance, that two propositions of the form $A = AB, B = BC$, which are such as constitute the premises of the old syllogism Barbara, negative or render impossible four of the eight combinations in which three terms may be united, and that these propositions are capable of taking twenty-four variations by transpositions of the terms or the introduction

M

of negatives. This table then presents the results of
a complete analysis of all the possible logical relations
arising in the case of three terms, and the old syllogism
forms but one out of fifteen typical forms. Generally
speaking every form can be converted into apparently
different propositions; thus the fourth type $A = B, B = BC$
may appear in the form $A = ABC, a = ab$, or again in the
form of three propositions $A = AB, B = BC, aB = aBc$; but
all these sets of premises yield identically the same series
of combinations, and are therefore of exactly equivalent
logical meaning. The fifth type, or Barbara, can also be
thrown into the equivalent forms $A = ABC, aB = aBC$ and
$A = AC, B = A + aBC$. In other cases I have obtained the
very same logical conditions in four modes of statement.
As regards mere appearance and mode of statement, the
number of possible premises would be almost unlimited.

The most remarkable of all the types of logical condition
is the fourteenth, namely $A = BC + bc$. It is that which
expresses the division of a genus into two doubly marked
species, and might be illustrated by the example—'Component of the physical universe = matter, gravitating, or
not-matter (ether), not-gravitating.'

It is capable of only two distinct logical variations,
namely, $A = BC + bc$ and $A = Bc + bC$. By transposition
or negative change of the letters we can indeed obtain
six different expressions of each of these propositions;
but when their meanings are analysed, by working out
the combinations, they are found to be logically equivalent to one or other of the above two. Thus the proposition $A = BC + bc$ can be written in any of the following
five other modes,

$$a = bC + Bc, \qquad B = CA + ca, \qquad b = cA + Ca,$$
$$C = AB + ab, \qquad c = aB + Ab.$$

I do not think it needful at present to publish the
complete table of 193 series of combinations and the

premises corresponding to each. Such a table enables us by mere inspection to learn the laws obeyed by any set of combinations of three things, and is to logic what a table of factors and prime numbers is to the theory of numbers, or a table of integrals to the higher mathematics. The table already given (p. 161) would enable a person with but little labour to discover the law of any combinations. If there be seven combinations (one contradicted) the law must be of the eighth type, and the proper variety will be apparent. If there be six combinations (two contradicted), either the second, eleventh, or twelfth type applies, and a certain number of trials will disclose the proper type and variety. If there be but two combinations the law must be of the third type, and so on.

The above investigations are complete as regards the possible logical relations of two or three terms. But when we attempt to apply the same kind of method to the relations of four or more terms, the labour becomes impracticably great. Four terms give sixteen combinations compatible with the laws of thought, and the number of possible selections of combinations is no less than 2^{16} or 65,536. The following table shows the extraordinary manner in which the number of possible logical relations increases with the number of terms involved.

Number of terms.	Number of possible combinations.	Number of possible selections of combinations corresponding to consistent or inconsistent logical relations.
2	4	16
3	8	256
4	16	65,536
5	32	4,294,967,296
6	64	18,446,744,073,709,551,616

Some years of continuous labour would be required to ascertain the precise number of types of laws which may govern the combinations of only four things, and but a small part of such laws would be exemplified or capable of practical application in science. The purely logical inverse problem, whereby we pass from combinations to their laws, is solved in the preceding pages, as far as it is likely to be for a long time to come ; and it is almost impossible that it should ever be carried more than a single step further.

Distinction between Perfect and Imperfect Induction.

We cannot proceed further with advantage, before noticing the extreme difference which exists between cases of perfect and those of imperfect induction. We call an induction perfect when all the objects or combinations of events which can possibly come under the class treated have been examined. But in the majority of cases it is impossible to collect together, or in any way to investigate, the properties of all portions of a substance or of all the individuals of a race. The number of objects would often be practically infinite, and the greater part of them might be beyond our reach, in the interior of the earth, or in the most distant parts of the Universe. In all such cases induction is said to be imperfect, and affected by more or less uncertainty. As some writers have fallen into much error concerning the functions and relative importance of these two branches of reasoning, I shall have to point out that—

1. Perfect Induction is a process absolutely requisite, both in the performance of imperfect induction and in the treatment of large bodies of facts of which our knowledge is complete.

2. Imperfect Induction is founded on Perfect Induction, but involves another process of inference of a widely different character.

It is certain that if I can draw any inference at all concerning objects not examined, it must be done on the data afforded by the objects which have been examined. If I judge that a distant star obeys the law of gravity, it must be because all other material objects sufficiently known to me obey that law. If I venture to assert that all ruminant animals have cloven hoofs, it is because all ruminant animals which have come to my notice have cloven hoofs. On the other hand I cannot safely say that all cryptogamous plants possess a purely cellular structure, because some such plants have a partially vascular structure. The probability that a new cryptogam will be cellular only can be estimated, if at all, on the ground of the comparative numbers of known cryptogams which are and are not cellular. Thus the first step in every induction will consist in accurately summing up the number of instances of a particular object or phenomenon which have fallen under our observation. Adams and Leverrier, for instance, must have inferred that the undiscovered planet Neptune would obey Bode's law, because *all the planets known at that time obeyed it.* On what principles and on what circumstances the passage from the known to the apparently unknown is warranted, must be carefully discussed in the next section, and in various parts of this work.

It would be a great mistake, however, to suppose that Perfect Induction is in itself useless. Even when the enumeration of objects belonging to any class is complete, and admits of no inference to unexamined objects, the enumeration of our knowledge in a general proposition is a process of so much importance that we may consider it practically necessary. In many cases we may render our investigations exhaustive; all the teeth or bones of an animal; all the cells in a minute vegetable organ; all the caves in a mountain side; all the strata in a geological section; all the coins in a newly found hoard, may be so completely scrutinized that we may make some general

assertion concerning them without fear of mistake. Every
bone might be proved to consist of phosphate of lime ;
every cell to enclose a nucleus ; every cave to contain
remains of extinct animals ; every stratum to exhibit signs
of marine origin ; every coin to be of Roman manufacture.
These are cases where our investigation is limited to a
definite portion of matter, or a definite area on the earth's
surface.

There is another class of cases where induction is
naturally and necessarily limited to a definite number of
alternatives. Of the regular solids we can say without
the least doubt that no one has more than twenty faces,
thirty edges, and twenty corners ; for by the principles
of geometry we learn that there cannot exist more than
five regular solids, of each of which we easily observe
that the above statements are true. In the theory of
numbers, an endless variety of perfect inductions might
be made ; we can show that no number less than sixty
possesses so many divisors, and the like is true of 360[e],
for it does not require any very great amount of labour to
ascertain and count all the divisors of numbers up to sixty
or 360. Similarly I can assert that between 60,041 and
60,077 no prime number occurs, because the exhaustive
examination of those who have constructed tables of prime
numbers proves it to be so.

In matters of human appointment or history, we can
frequently have a complete limitation to the numbers of
instances to be included in an induction. We might show
that none of the other kings of England reigned so long as
George III ; that Magna Charta has not been repealed by
any subsequent statute ; that the propositions of the third
book of Euclid treat only of circles ; that no part of the
works of Galen mentions the fourth figure of the syl-

---

[e] Wallis's 'Treatise of Algebra' (1685), p. 22.

logism ; that the price of corn in England has never been
so high since 1847 as it was in that year ; that the price
of the English funds has never been lower than it was on
the 23rd of January, 1798, when it fell to 47¼.

It has been urged against this process of Perfect Induction that it gives no new information, and is merely a
summing up in a brief form of a multitude of particulars.
But mere abbreviation of mental labour is one of the
most important aids we can enjoy in the acquisition of
knowledge. The powers of the human mind are so limited
that multiplicity of detail is alone sufficient to prevent its
progress in many directions. Thought would be practically impossible if every separate fact had to be separately
thought and treated. Economy of mental power may be
considered one of the main conditions on which our ele-
vated intellectual position depends. Most mathematical
processes are but abbreviations of the simpler acts of
addition and subtraction. The invention of logarithms
was one of the most striking additions ever made to
human power : yet it was a mere abbreviation of operations which could have been done before had a sufficient
amount of labour been available. Similar additions to
our power will, it is hoped, be made from time to time,
for the number of mathematical problems hitherto solved
is but an indefinitely small portion of those which await
solution, because the labour they have hitherto demanded
renders them impracticable. So it is really throughout
all regions of thought. The amount of our knowledge
depends upon our powers of bringing it within practicable compass. Unless we arrange and classify facts,
and condense them into general truths, they soon surpass our powers of memory, and serve but to confuse.
Hence Perfect Induction, even as a process of abbreviation, is absolutely essential to any high degree of
mental achievement.

Transition from Perfect to Imperfect Induction.

It is a question of profound difficulty on what grounds we are warranted in inferring the future from the present, or the nature of undiscovered objects from those which we have examined with our senses. We pass from Perfect to Imperfect Induction when once we allow our conclusion to pass, at all events apparently, beyond the data on which it was founded. In making such a step we seem to gain a nett addition to our knowledge; for we learn the nature of what was unknown. We reap where we have never sown. We appear to possess the divine power of creating knowledge, and reaching with our mental arms far beyond the sphere of our own observation. I shall, indeed, have to point out certain methods of reasoning in which we do pass altogether beyond the sphere of the senses, and acquire accurate knowledge which observation could never have given; but it is not imperfect induction that accomplishes such a task. Of imperfect induction itself, I venture to assert that it never makes any real addition to our knowledge, in the meaning of the expression sometimes accepted. As in other cases of inference it merely unfolds the information contained in past observations or events; it merely renders explicit what was implicit in previous experience. It transmutes knowledge, but certainly does not create knowledge.

There is no fact which I shall more constantly keep before the reader's mind in the following pages than that the results of imperfect induction, however well authenticated and verified, are never more than probable. We never can be sure that the future will be as the present. We hang ever upon the Will of the Creator: and it is only so far as He has created two things alike, or maintains the framework of the world unchanged from moment to moment, that our most careful inferences can be fulfilled. All predictions, all inferences which reach beyond their data, are purely hypothetical, and proceed on the assumption that new events will conform to the conditions detected in our observation of past events. No experience of finite duration can be expected to give an exhaustive knowledge of all the forces which are in operation. There is thus a double uncertainty; even supposing the Universe as a whole to proceed unchanged, we do not really know the Universe as a whole. Comparatively speaking we know only a point in its infinite extent, and a moment in its infinite duration. We cannot be sure, then, that our observations have not escaped some fact, which will cause the future to be apparently different from the past; nor can we be sure that the future really will be the outcome of the past. We proceed then in all our inferences to unexamined objects and times on the assumptions—

1. That our past observation gives us a complete knowledge of what exists.
2. That the conditions of things which did exist will continue to be the conditions of things which will exist.

We shall often need to illustrate the character of our knowledge of nature by the simile of a ballot-box, so often employed by mathematical writers in the theory of probability. Nature is to us like an infinite ballot-box, the contents of which are being continually drawn, ball after ball, and exhibited to us. Science is but the careful observation of the succession in which balls of various character usually present themselves; we register the combinations, notice those which seem to be excluded from occurrence, and from the proportional frequency of those which usually appear we infer the probable character of future drawings. But under such circumstances certainty of prediction depends on two conditions :—

1. That we acquire a perfect knowledge of the comparative numbers of balls of each kind within the box.

2. That the contents of the ballot-box remain unchanged.

Of the latter assumption, or rather that concerning the
constitution of the world which it illustrates, the logician
or physicist can have nothing to say. As the Creation of
the Universe is necessarily an act passing all experience
and all conception, so any change in that Creation, or, it
may be, a termination of it, must likewise be infinitely beyond the bounds of our mental faculties. No science, no
reasoning upon the subject, can have any validity; for
without experience we are without the basis and materials
of knowledge. It is the fundamental postulate accordingly
of all inference concerning the future, that there shall be
no arbitrary change in the subject of inference ; of the probability or improbability of such a change I conceive that
our faculties can give no estimate.

The other condition of inductive inference—that we
acquire an approximately complete knowledge of the
combinations in which events do occur, is at least in some
degree within the bounds of our perceptive and mental
powers. There are many branches of science in which
phenomena seem to be governed by conditions of a most
fixed and general character. We have much ground in
such cases for believing that the future occurrence of
such phenomena may be calculated and predicted. But
the whole question now becomes one of probability and
improbability. We leave the region of pure logic to enter
one in which the number of events is the ground of
inference. We do not leave the region of logic; we only
leave that where certainty, affirmative or negative, is the
result, and the agreement or disagreement of qualities the
means of inference. For the future, number and quantity
will enter into most of our processes of reasoning ; but then
I hold that number and quantity are but portions of the

great logical domain. I venture to assert that number is wholly logical, both in its fundamental nature and in all its complex developments. Quantity in all its forms is but a development of number. That which is mathematical is not the less logical; if anything it is the more logical, in the sense that it presents logical results in the highest degree of complexity and variety.

Before proceeding then from Perfect to Imperfect Induction I break off in some degree the course of the work, to treat of the logical conditions of number. I shall then employ number to estimate the variety of combinations in which natural phenomena may present themselves, and the probability or improbability of their occurrence under definite circumstances. It is in later parts of the work that I must endeavour to establish, in a complete manner, the notions which I have set forth upon the subject of Imperfect Induction, as applied in the investigation of Nature, which notions may be thus briefly stated:—

1. Imperfect Induction entirely rests upon Perfect Induction for its materials.

2. The logical process by which we seem to pass directly from examined to unexamined cases consists in an inverse and complex application of deductive inference, so that all reasoning may be said to be either directly or inversely deductive.

3. The result is always of a hypothetical character, and is never more than probable.

4. No nett addition is ever made to our knowledge by reasoning; what we know of future events or unexamined objects is only the unfolded contents of our previous knowledge, and it becomes less and less probable as it is more boldly extended to remote cases.

BOOK II.

CHAPTER VIII.

PRINCIPLES OF NUMBER.

NOT without much reason did Pythagoras represent the world as ruled by number. Into almost all our acts of clear thought number enters, and in proportion as we can define numerically we enjoy exact and useful knowledge of the Universe. The science of numbers, too, the study of the principles and methods of reasoning in number, has hitherto presented the widest and most practicable training in logic. So free and energetic has been the study of mathematical forms, compared with the forms and laws of logic, that mathematicians have passed far in advance of any pure logicians. Occasionally, in recent times, they have condescended to apply their great algebraic instruments to a reflex advancement of the primary logical science. It is thus that we chiefly owe to profound mathematicians, such as Sir John Herschel, Dr. Whewell, Professor De Morgan or Dr. Boole, the regeneration of logic in the present century, and I entertain no doubt that it is in maintaining a close alliance with the extensive branches of quantitative reasoning that we must look for still further progress in our comprehension of qualitative inference.

I cannot assent, indeed, to the common notion that

certainty begins and ends with numerical determination. Nothing is more certain and accurate than logical truth. The laws of identity and difference are the tests of all that is true and certain throughout the range of thought, and mathematical reasoning is cogent only when it conforms to these conditions, of which logic is the first development. And if it be erroneous to suppose that all certainty is mathematical, it is equally an error to imagine that all which is mathematical is certain. Many processes of mathematical reasoning are of most doubtful validity. There are many points of mathematical doctrine which are and must long remain matter of opinion ; for instance, the best form of the definition and axiom concerning parallel lines, or the true nature of a limit or a ratio of infinitesimal quantities. In the use of symbolic reasoning questions occur at every point on which the best mathematicians may differ, as Bernouilli and Leibnitz differed irreconcileably concerning the existence of the logarithms of negative quantities[a]. In fact we no sooner leave the simple logical conditions of number, than we find ourselves involved in a mazy and mysterious science of symbols.

Mathematical science enjoys no monopoly, and not even a supremacy in certainty of results. It is the boundless extent and variety of quantitative questions that surprises and delights the mathematical student. When simple logic can give but a bare answer Yes or No, the algebraist raises a score of subtle questions, and brings out a score of curious results. The flower and the fruit, all that is attractive and delightful, fall to the share of the mathematician, who too often despises the pure but necessary stem from which all has arisen. But in no part of human thought can a reasoner cast himself free from the prior conditions of logical correctness. The mathematician is

---

[a] Montucla, 'Histoire des Mathématiques,' vol. iii. p. 373.

only strong and true as long as he is logical, and if numbers rule the world, it is the laws of logic which rule number.

Nearly all writers have hitherto been strangely content to look upon numerical reasoning as something wholly apart from logical inference. A long divorce has existed between quality and quantity, and it has not been uncommon to treat them as contrasted in nature and restricted to independent branches of human thought. For my own part, I have a profound belief that all the sciences meet somewhere upon common ground. No part of knowledge can stand wholly disconnected from other parts of the great universe of thought; it is incredible, above all, that the two great branches of abstract science, interlacing and co-operating in every discourse, should rest upon totally distinct foundations. I assume that a connection exists, and care only to inquire, What is its nature? Does the science of quantity rest upon that of quality; or, *vice versâ*, does the science of quality rest upon that of quantity? There might conceivably be a third view, that they both rest upon some still deeper set of principles yet undiscovered, but there is an absence of any suggestions to this effect. The late Dr. Boole adopted the second view, and treated logic as a kind of algebra,—a special case of analytical reasoning which admits but the two quantities—unity and zero. He proved beyond doubt that a deep analogy does exist between the forms of algebraic and logical deduction; and could this analogy receive no other explanation we must have accepted his opinion, however strange. But I shall attempt to show that just the reverse explanation is the true one.

I hold that algebra is a highly developed logic, and number but logical discrimination. Logic resembles algebra, as the mould resembles that which is cast in it. Logic has imposed its own laws upon every branch of mathematical science, and it is no wonder that we ever meet with the traces of those laws from the domain of which we can never emerge.

The Nature of Number.

Number is but another name for *diversity.* Exact identity is unity, and with difference arises plurality. An abstract notion, as was pointed out (p. 33), possesses a certain *oneness.* The quality of *justice,* for instance, is one and the same in whatever just acts it be manifested. In justice itself there are no marks of difference by which to discriminate justice from justice. But one just act can be discriminated from another just act by many circumstances of time and place, and we can count and number many acts each thus discriminated from every other. In like manner pure gold is simply pure gold, and is so far one and the same throughout. But besides its intrinsic and invariable qualities, gold occupies space and must have shape or size. Portions of gold are always mutually exclusive and capable of discrimination, at least in respect that they must be each without the other. Hence they may be numbered.

Plurality arises when and only when we detect difference. For instance, in counting a number of gold coins I must count each coin once, and not more than once. Let C denote a coin, and the mark above it the position in the order of counting. Then I must count the coins

$$C' + C'' + C''' + C'''' + \ldots \ldots$$

If I were to make them as follows

$$C' + C'' + C''' + C''' + C'''' + \ldots \ldots,$$

I should make the third coin into two, and should imply the existence of difference where there is not difference[b]. C''' and C''' are but the names of one coin named twice

[b] 'Pure Logic,' Appendix, p. 82, § 192.

over. But according to one of the conditions of logical symbols, which I have called the Law of Unity (p. 86), the same name repeated has no effect, and

$$A \dagger A = A.$$

We must apply the Law of Unity, and must reduce all identical alternatives before we can count with certainty and use the processes of numerical calculation. Identical alternatives are harmless in logic, but produce deadly error in number. Thus logical science ascertains the nature of the mathematical unit, and the definition may be given in these terms—*A unit is any object of thought which can be discriminated from every other object treated as a unit in the same problem.*

It has often been said that units are units in respect of being perfectly similar to each other; but though they may be perfectly similar in some respects, they must be different in at least one point, otherwise they would be incapable of plurality. If three coins were so similar that they occupied the same space at the same time, they would not be three coins, but one. It is a property of space that every point is discriminable from every other point, and in time every moment is necessarily distinct from any other moment before or after. Hence we frequently count in space or time, and Locke, with some other philosophers, has even held that number arises from repetition in time. Beats of a pendulum might be so perfectly similar that we could discover no difference except that one beat is before and another after. Time alone is here the ground of difference and is a sufficient foundation for the discrimination of plurality; but it is by no means the only foundation. Three coins are three coins, whether we count them successively or regard them all simultaneously. In many cases neither time nor space is the ground of difference, but pure quality alone enters. We can discriminate for instance the weight, inertia, and hardness of gold as three qualities, though none of these is before or after the other, either in space or time. Every means of discrimination may be a source of plurality.

Our logical notation may be used to express the rise of number. The symbol A stands for one thing or one class, and in itself must be regarded as a unit, because no difference is specified. But the combinations AB and A*b* are necessarily *two*, because they cannot logically coalesce, and there is a mark B which distinguishes one from the other. A logical definition of the number *four* is given in the combinations ABC, AB*c*, A*b*C, A*bc*, where there is a double difference, and as Puck says—

> 'Yet but three? Come one more;
> Two of both kinds makes up four.'

I conceive that all numbers might be represented as arising out of the combinations of the Abecedarium, more or less of each series being struck out by various logical conditions. The number three, for instance, arises from the condition that A must be either B or C, so that the combinations are ABC, AB*c*, A*b*C.

Of Numerical Abstraction.

There will now be little difficulty in forming a clear notion of the nature of numerical abstraction. It consists in abstracting the character of the difference from which plurality arises, retaining merely the fact. When I speak of *three men* I need not at once specify the marks by which each may be known from each. Those marks must exist if they are really three men and not one and the same, and in speaking of them as many I imply the existence of the requisite differences. Abstract number, then, is *the empty form of difference*; the abstract

N

number *three* asserts the existence of marks without specifying their kind.

Numerical abstraction is then a totally different process from logical abstraction (see p. 33), for in the latter process we drop out of notice the very existence of difference and plurality. In forming the abstract notion *hardness*, for instance, I drop out of notice altogether the diverse circumstances in which the quality may appear. It is the concrete notion *three hard objects*, which asserts the existence of hardness along with sufficient other undefined qualities, to mark out *three* such objects. Numerical thought is indeed closely interwoven with logical thought. We cannot use a concrete term in the plural, as *men*, without implying that there are marks of difference. Only when we use a term in the singular and abstract sense *man*, do we deal with unity, unbroken by difference.

The origin of the great generality of number is now apparent. Three sounds differ from three colours, or three riders from three horses ; but they agree in respect of the variety of marks by which they can be discriminated. The symbols $1 + 1 + 1$ are thus the empty marks asserting the fact of discrimination which may apply to objects wholly independently of their peculiar nature.

Concrete and Abstract Numbers.

The common distinction between concrete and abstract numbers can now be easily stated. In proportion as we specify the logical character of the things numbered, we render them concrete. In the abstract number *three* there is no statement of the points in which the three objects agree ; but in three coins, three men, or three horses, not only are the variety of objects defined, but their nature is restricted. Concrete number thus implies the same consciousness of difference as abstract number, but it is mingled with a groundwork of similarity expressed in the logical terms. There is similarity or identity so far as logical terms enter ; difference so far as the terms are merely numerical.

The reason of the important Law of Homogeneity will now be apparent. This law asserts that in every arithmetical calculation the logical nature of the things numbered must remain unaltered. The specified logical agreement of the things numbered must not be affected by the unspecified numerical differences. A calculation would be palpably absurd which, after commencing with length, gave a result in hours. It is in reality equally absurd in a purely arithmetical point of view to deduce areas from the calculation of lengths, masses from the combination of volume and density, or momenta from mass and velocity. It must remain for subsequent consideration in what sense we may truly say that two linear feet multiplied by two linear feet give four superficial feet, but arithmetically it is absurd, because there is a change of unit.

As a general rule we treat in each calculation only objects of one nature. We do not, and cannot properly add, in the same sum yards of cloth and pounds of sugar. We cannot even conceive the result of adding area to velocity, or length to density, or weight to value. The unit numbered and added must have a basis of homogeneity, or must be reducible to some common denominator. Nevertheless it is quite possible, and in fact common, to treat in one complex calculation the most heterogeneous quantities, on the condition that each kind of object is kept distinct, and treated numerically only in conjunction with its own kind. Different units, so far as their logical differences are specified, must never be substituted one for the other. Chemists continually use equations which assert the equivalence of groups of atoms. Ordinary fermentation is represented by the formula

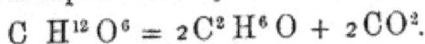

$$C\ H^{12}O^6 = 2C^2H^6O + 2CO^2.$$

Three kinds of units, the atoms respectively of Carbon, Hydrogen, and Oxygen, are here intermingled, but there is really a separate equation in regard to each kind. Mathematicians also employ compound equations of the same kind; for in $a + b\sqrt{-1} = c + d\sqrt{-1}$, it is impossible by ordinary addition to add a to $b\sqrt{-1}$. Hence we really have the separate equations $a = c$, and $b = d^c$. Similarly an equation between two quaternions is equivalent to four equations between ordinary quantities, whence indeed the origin of the name *quaternion*.

Analogy of Logical and Numerical Terms.

If my assertion is correct that number arises out of logical conditions, we ought to find number obeying all the laws and conditions of logic. It is almost superfluous to point out that this is the case with the fundamental laws of identity and difference, and it only remains for me to show that mathematical symbols do really obey the special conditions of logical symbols which were formerly pointed out (p. 39). Thus the Law of Commutativeness, is equally true of quality and quantity. As in logic we have

$$AB = BA,$$

so in mathematics it is familiarly known that

$$2 \times 3 = 3 \times 2, \quad \text{or} \quad x \times y = y \times x.$$

The properties of space, in short, are as indifferent in pure multiplication as we found them in pure logical thought.

<hr>

c De Morgan's 'Trigonometry and Double Algebra,' p. 126.

Similarly, just as in logic

$$\text{triangle or square} = \text{square or triangle},$$

or generally

$$A + B = B + A,$$

so in quantity

$$2 + 3 = 3 + 2,$$

or generally

$$x + y = y + x.$$

The symbol $+$ is not identical with $+$, but it is so far analogous.

How far, now, is it true that mathematical symbols obey the law of simplicity expressed in the form

$$AA = A,$$

or the example

$$\text{Round round} = \text{round ?}$$

Apparently there are but two numbers which obey this law; for it is certain that

$$x \times x = x$$

is true only in the two cases when $x = 1$ or 0.

In reality all numbers obey the law, for $2 \times 2 = 2$ is not really analogous to $AA = A$. According to the definition of a unit already given, each unit is discriminated from each other in the same problem, so that in $2' \times 2'$, the first *two* involves a different discrimination from the second *two*. I get four kinds of things, for instance, if I first discriminate ' heavy and light' and then ' cubical and spherical,' for we now have the following classes—

heavy, cubical.   light, cubical.
heavy, spherical.   light, spherical.

But suppose that my two classes are in both cases discriminated by the same difference of light and heavy, then we have

$$\text{heavy heavy} = \text{heavy},$$
$$\text{heavy light} = 0,$$
$$\text{light heavy} = 0,$$
$$\text{light light} = \text{light}.$$

In short, *twice two is two* unless we take care that the second two has a different meaning from the first. But under similar circumstances logical terms would give exactly the like result, and it is not true that $A'A'' = A'$, identically where A'' is different in meaning from A'.

In an exactly similar manner it may be shown that the Law of Unity

$$A + A = A$$

holds true alike of logical and mathematical terms. It is absurd indeed to say that

$$x + x = x$$

except in the one case when $x =$ absolute zero. But this contradiction $x + x = x$ arises from the fact that we have already defined the unit in one x as differing from those in the other. Under such circumstances the Law of Unity does not apply. For if in

$$A' + A'' = A'$$

we mean that A'' is in any way different from A' the assertion of identity is evidently false.

The contrast then which seems to exist between logical and mathematical symbols is only apparent. It is because the Law of Simplicity and Unity must always be observed in the operation of counting that those laws can no longer be operative. This is the understood condition under which we use all numerical symbols. Whenever I use the symbol 5 I really mean

$$1 + 1 + 1 + 1 + 1,$$

and it is perfectly understood that each of these units is distinct from each other. If requisite I might mark them thus

$$1' + 1'' + 1''' + 1'''' + 1'''''.$$

Were this not the case and were the units really

$$1' + 1'' + 1'' + 1''' + 1''',$$

the Law of Unity would, as before remarked, apply, and

$$1'' + 1'' = 1''.$$

Mathematical symbols then obey all the laws of logical

symbols, but two of these laws seem to be inapplicable simply because they are presupposed in the definition of the mathematical unit. Logic thus lays down the conditions of number, and the science of arithmetic developed as it is into all the wondrous branches of mathematical calculus is but an outgrowth of logical discrimination.

Principle of Mathematical Inference.

As I have asserted, the universal principle of all reasoning is that which allows us to substitute like for like. I have now to point out that in the mathematical sciences this principle is involved in each step of reasoning. It is in these sciences indeed that we meet with the clearest cases of substitution, and it is the simplicity with which the principle can be applied which probably led to the comparatively early perfection of the sciences of geometry and arithmetic. Euclid, and the Greek mathematicians from the first, recognised *equality* as the fundamental relation of quantitative thought, but Aristotle rejected the exactly analogous, but far more general relation of identity, and thus crippled the formal science of logic as it has descended to the present day.

Geometrical reasoning starts from the Axiom that 'things equal to the same thing are equal to each other.' Two equalities enable us to infer a third equality ; and this is true not only of lines and angles, but of areas, volumes, numbers, intervals of time, forces, velocities, degrees of intensity, or, in short, anything which is capable of being equal or unequal. Two stars equally bright with the same star must be equally bright with each other, and two forces equally intense with a third force are equally intense with each other. It is remarkable that Euclid has not expressly stated two other axioms, the truth of which is necessarily implied. The second axiom should be that 'Two things of which one is equal and the other unequal to a third common thing, are unequal to each other.' An equality and inequality, in short, may give an inequality, and this is equally true with the first axiom of all kinds of quantity. If Venus, for instance, agrees with Mars in density, but Mars differs from Jupiter, then Venus differs from Jupiter. A third axiom must exist to the effect that 'Things unequal to the same thing may or may not be equal to each other.' *Two inequalities give no ground of inference whatever.* If we only know, for instance, that Mercury and Jupiter differ in density from Mars, we cannot say whether or not they agree between themselves. As a fact they do not agree; but Venus and Mars on the other hand both differ from Jupiter and yet closely agree with each other. The force of the axioms can be most clearly illustrated by drawing lines[d].

The general conclusion must be then that where there is equality there may be inference, but where there is not equality there cannot be inference   A plain induction will lead us to believe that *equality is the condition of inference concerning quantity.* All the three axioms may in fact be summed up in one, to the effect, that '*in whatever relation one quantity stands to another, it stands in the same relation to the equal of that other.*'

The active power is always the substitution of equals, and it is an accident that in a pair of equalities we can make the substitution in two ways. From $a = b = c$ we can infer $a = c$, either by substituting in $a = b$ the value of b as given in $b = c$, or else by substituting in $b = c$ the value of b as given in $a = b$. In $a = b \backsim d$ we can make but the one substitution of a for b. In $e \backsim f \backsim g$ we can make no substitution and get no inference.

In mathematics the relations in which terms may stand to each other are far more varied than in pure logic, yet

'Elementary Lessons in Logic' (Macmillan), p. 123.

our principle of substitution always holds true. We may say in the most general manner that *In whatever relation one quantity stands to another, it stands in the same relation to the equal of that other.* In this axiom we sum up a number of axioms which have been stated in more or less detail by algebraists[e]. Thus, 'If equal quantities be added to equal quantities, the sums will be equal.' To explain this, let

$$a = b, \qquad c = d.$$

Now $a + c$, whatever it means, must be identical with itself, so that

$$a + c = a + c.$$

In one side of this equation substitute for the quantities their equivalents, and we have the axiom proved

$$a + c = b + d.$$

The similar axiom concerning subtraction is equally evident, for whatever $a - c$ may mean it is equal to $a - c$, and therefore by substitution to $b - d$. Again, 'if equal quantities be multiplied by the same or equal quantities, the products will be equal.' For evidently

$$ac = ac,$$

and if for c in one side we substitute its equal d, we have

$$ac = ad,$$

and a second similar substitution gives us

$$ac = bd.$$

We might prove a like axiom concerning division in an exactly similar manner. I might even extend the list of axioms and say that 'Equal powers of equal number are equal.' For certainly, whatever $a \times a \times a$ may mean, it is equal to $a \times a \times a$; hence by our usual substitution

$$a \times a \times a = b \times b \times b,$$

or $\qquad a^3 = b^3.$

The truth will hold of roots, that is to say,

$$\sqrt[c]{a} = \sqrt[d]{b},$$

e Todhunter's 'Algebra,' 3rd ed. p. 40.

provided that the same roots are taken, that is that the root of a shall really be related to a as the root of b is to b. The ambiguity of meaning of an operation thus fails in any way to shake the universality of the principle.

We may go further and assert that, not only the above common relations, but all other known or conceivable mathematical relations obey the same principle. Let Pa denote in the most general manner that we do something with the quantity a; then if $a = b$ it follows that

$$Pa = Pb.$$

Let us make Pa, for instance, mean

$$a^3 - 3a^2 + 2a + 5;$$

then it necessarily follows that this quantity is exactly equal to $b^3 - 3b^2 + 2b + 5.$

The reader will also remember that one of the most frequent operations in mathematical reasoning is to substitute for a quantity its equal, as known either by assumed, natural, or self-evident condition. Whenever a quantity appears twice over in a problem, we may apply what we learn of its relations in one place to its relations in the other. All reasoning in mathematics, as in other branches of science, thus involves the principle of treating equals equally, or similars similarly. In whatever way we employ quantitative reasoning in the remaining parts of this work, we never can desert the simple principle on which we first set out.

Reasoning by Inequalities.

I have stated that all the processes of mathematical reasoning may be deduced from the principle of substitution. Exceptions to this assertion may seem to exist in the use of inequalities. The greater of a greater is undoubtedly a greater, and what is less than a less is certainly less. Snowdon is higher than the Wrekin, and Ben Nevis than

Snowdon ; therefore Ben Nevis is higher than the Wrekin. But a little consideration discloses much reason for believing that even in such cases, where equality does not apparently enter, the force of the reasoning entirely depends upon underlying and implied equalities.

In the first place, two statements of mere difference do not give any ground of inference. We learn nothing concerning the comparative heights of St. Paul's and Westminster Abbey from the assertions that they both differ in height from St. Peter's at Rome. Thus we need something more than mere inequality ; we require one identity in addition, namely the identity in direction of the two differences. Thus we cannot employ inequalities in the simple way in which we do equalities, and, when we try to express exactly what other conditions are requisite, we shall find ourselves lapsing into the use of equalities or identities.

In the second place, every argument by inequalities may be represented with at least equal clearness and force in the form of equalities. Thus we clearly express that a is greater than b by the equation

$$a = b + p, \qquad (1)$$

where p is an intrinsically positive quantity, denoting the difference of a and b. Similarly we express that b is greater than c by the equation

$$b = c + q, \qquad (2)$$

and substituting for b in (1) its value in (2) we have

$$a = c + q + p. \qquad (3)$$

Now as p and q are both positive, it follows that a is greater than c, and we have the exact amount of excess specified. It will be easily seen that the reasoning concerning that which is less than a less will result in an equation of the form

$$c = a - q - p.$$

Every argument by inequalities may then be thrown

into the form of an equality; but the converse is not true. We cannot possibly prove that two quantities are equal by merely asserting that they are both greater or both less than another quantity. From $e > f$ and $g > f$, or $e < f$ and $g < f$, we can infer no relation between e and g. And if the reader take the equations $x = y = 3$ and attempt to prove that therefore $x = 3$, by throwing them into inequalities, he will find it impossible to do so.

From these considerations I gather that reasoning in arithmetic or algebra by so-called inequalities is only an imperfectly expressed reasoning by equalities, and when we want to exhibit exactly and clearly the conditions of reasoning, we are obliged to use equalities explicitly. Just as in pure logic a negative proposition, as expressing mere difference, cannot be the means of inference, so inequality can never really be the true ground of inference. I do not deny that affirmation and negation, agreement and difference, equality and inequality, are pairs of equally fundamental relations, but I assert that inference is possible only where affirmation, agreement, or equality, some species of identity in fact, is present, explicitly or implicitly.

Arithmetical Reasoning.

It might seem somewhat inconsistent that I assert number to arise out of difference or discrimination, and yet hold that no reasoning can be grounded on difference. Number, of course, opens a most wide sphere for inference, and a little consideration shows that this is due to the unlimited series of identities which spring up out of numerical abstraction. If six people are sitting on six chairs, there is no resemblance between the chairs and the people in logical character. But if we overlook all the qualities both of a chair and a person, and merely remember that there are marks by which each of six chairs may be discriminated from each other, and similarly with the people, then there arises a resemblance between the chairs and people, and this resemblance in number may be the ground of inference. If on another occasion the chairs are filled by people again, we may infer that these people must resemble the others in number, though they need not resemble them in any other points.

Groups of units are what we really treat in arithmetic. The number *five* is really $1 + 1 + 1 + 1 + 1$, but for the sake of conciseness we substitute the more compact sign 5, or the name *five*. These names being arbitrarily imposed in any one manner, an indefinite variety of relations spring up between them which are not in the least arbitrary. If we define *four* as $1 + 1 + 1 + 1$, and *five* as $1 + 1 + 1 + 1 + 1$, then of course it follows that *five* = *four* + 1 ; but it would be equally possible to take this latter equality as a definition, in which case one of the former equalities would become an inference. It is hardly requisite to decide how we define the names of numbers, provided we remember that out of the infinitely numerous relations of one number to others, some one relation expressed in an equality must be a definition of the number in question and the other relations immediately become necessary inferences.

In the science of number the variety of classes which can be formed is altogether infinite, and statements of perfect generality may be made subject only to difficulty or exception at the lower end of the scale. Every existing number for instance belongs to the class

$$m + 7;$$

that is, every number must be the sum of another number and seven, except of course the first six or seven numbers, negative quantities not being here taken into account. Every number is the half of some other, and so on. The subject of generalization, as exhibited in arithmetical or mathematical truths, is an indefinitely wide one. In number we are only at the first step of an extensive series of generalizations. A number is general as compared with the particular things numbered, so we may have general symbols for numbers, or general symbols not for numbers, but for the relations between undetermined numbers. There is, in fact, an unlimited hierarchy of successive generalizations.

Numerically Definite Reasoning.

It was first discovered by Prof. de Morgan that many arguments are valid which combine logical and numerical reasoning, although they could in no way be included in the ancient logical formulas. He developed the doctrine of the 'Numerically Definite Syllogism,' fully explained in his 'Formal Logic' (pp. 141–170). Dr. Boole also devoted considerable attention to the determination of what he called 'Statistical Conditions,' meaning the numerical conditions of logical classes. In a paper published among the Memoirs of the Manchester Literary and Philosophical Society, Third Series, vol. IV. p. 330 (Session 1869–70), I have pointed out that we can apply arithmetical calculation to the Logical Abecedarium. Having given certain logical conditions and the numbers of objects in certain classes, we can either determine the number of objects in other classes governed by those conditions, or can show what further data are required to determine them. As an example of the kind of questions treated in numerical logic, and the mode of treatment, I give the following problem suggested by De Morgan, with my mode of representing its solution [f].

---

[f] It has been pointed out to me by Mr. A. J. Ellis, F.R.S., that my solution, as given in the Memoirs of the Manchester Philosophical Society, does not exactly answer to the conditions of the problem, and I therefore substitute above a more satisfactory solution.

'For every man in the house there is a person who is aged; some of the men are not aged. It follows that some persons in the house are not men[g].'

Now let   A = person in house,

                 B = male,

                 C = aged.

By enclosing logical symbols in brackets, let us denote the number of objects belonging to the class indicated by the symbol. Thus let

       (A) = number of persons in house,

       (AB) = number of male persons in house,

       (ABC) = number of aged male persons in house,

and so on. Now if we use w and w' to denote unknown and indefinite numbers, the conditions of the problem may be thus stated according to my interpretation of the words—

$$(AB) = (AC) - w, \qquad (1)$$

that is to say, the number of persons in the house who are aged is at least equal to, and may exceed, the number of male persons in the house ;

$$(ABc) = w', \qquad (2)$$

that is to say, the number of male persons in the house who are not aged is some unknown positive quantity.

If we develop the terms in (1) by the Law of Duality (pp. 87, 95, 97), we obtain

$$(ABC) + (ABc) = (ABC) + (AbC) - w.$$

Subtracting the common term (ABC) from each side and substituting for (ABc) its value as given in (2), we get at once

$$(AbC) = w + w',$$

and adding (Abc) to each side, we have

$$(Ab) = Abc + w + w'.$$

The meaning of this result is that the number of persons in the house who are not men is at least equal to $w + w'$,

---

[g] 'Syllabus of a proposed System of Logic,' p. 29.

and exceeds it by the number of persons in the house who are neither men nor aged (Abc).

It should be understood that this solution applies only to the terms of the example quoted above, and not to the general problem for which De Morgan intended it to serve as an illustration.

As a second instance, let us take the following question :—The whole number of voters in a borough is a; the number against whom objections have been lodged by liberals is b; and the number against whom objections have been lodged by conservatives is c; required the number, if any, who have been objected to on both sides. Taking

$$A = \text{voter},$$
$$B = \text{objected to by liberals},$$
$$C = \text{objected to by conservatives},$$

then we require the value of (ABC). Now the following equation in identically true—

$$(ABC) = (AB) + (AC) + (Abc) - (A). \qquad (1)$$

For if we develop all the terms on the second side we obtain

$$(ABC) = (ABC) + (ABc) + (ABC) + (AbC) + (Abc)$$
$$- (ABC) - (ABc) - (AbC) - (Abc);$$

and striking out the corresponding positive and negative terms, we have only left (ABC) = (ABC). Since then (1) is necessarily true, we have only to insert the known values, and we have

$$(ABC) = b + c - a + (Abc).$$

Hence the number who have received objections from both sides is equal to the excess, if any, of the whole number of objections over the number of voters together with the numbers of voters who have received no objections (Abc).

In many cases classes of objects may exist under special logical conditions, and we must consider how these conditions must be interpreted numerically.

Every logical proposition or equation now gives rise to a corresponding numerical equation. Sameness of qualities occasions sameness of numbers. Hence if

$$A = B$$

denotes the identity of the qualities of A and B, we may conclude that

$$(A) = (B).$$

It is evident that exactly those objects, and those objects only, which are comprehended under A must be comprehended under B. It follows that wherever we can draw an equation of qualities, we can draw a similar equation of numbers. Thus, from

$$A = B = C$$

we infer

$$A = C;$$

and similarly from

$$(A) = (B) = (C),$$

meaning the numbers of A's and C's are equal to the number of B's, we can infer

$$(A) = (C).$$

But, curiously enough, this does not apply to negative propositions and inequalities. For if

$$A = B \smile D$$

means that A is identical with B, which differs from D, it does not follow that

$$(A) = (B) \smile (D).$$

Two classes of objects may differ in qualities, and yet they may agree in number. This is a point which strongly confirms me in the opinion I have already expressed, that all inference really depends upon equations, not differences (p. 186).

The Logical Abecedarium thus enables us to make a complete analysis of any numerical problem, and though the symbolical statement may sometimes seem prolix, I conceive that it really represents the course which the mind must follow in solving the question. Although thought may seem to outstrip the rapidity with which the symbols can be written down, yet the mind does not really follow a different course from that indicated by the symbols. For a fuller explanation of this natural system of Numerically Definite Reasoning, with more abundant illustrations and an analysis of De Morgan's Numerically Definite Syllogism, I must refer the reader to the paper in the Memoirs of the Manchester Literary and Philosophical Society, as already referred to, portions of which, however, have been embodied in the present section.

The reader may be referred, also, to Boole's writings upon the subject in the 'Laws of Thought,' chap. xix. p. 295, and in a paper on 'Propositions Numerically Definite,' communicated by De Morgan, in 1868, to the Cambridge Philosophical Society, and printed in their 'Transactions,' vol. xi. part ii. Mr. Alexander J. Ellis treats the same subject in his 'Contributions to Formal Logic,' read to the Royal Society, in March, 1872, but as yet published only in the form of a brief abstract, in the Proceedings of the Society, vol. xx. p. 307.

CHAPTER IX.

THE VARIETY OF NATURE, OR THE DOCTRINE OF
COMBINATIONS AND PERMUTATIONS.

NATURE may be said to be evolved from the monotony of non-existence by the creation of diversity. It is plausibly asserted that we are conscious only so far as we experience difference. Life is change, and perfectly uniform existence would be no better than non-existence. Certain it is that life demands incessant novelty, and that nature though it probably never fails to obey the same fixed laws, yet presents to us an apparently unlimited series of varied combinations of events. It is the work of science to observe and record the kinds and comparative numbers of such combinations of phenomena, occurring spontaneously or produced by our interference. Patient and skilful examination of the records may then disclose the laws imposed on matter at its creation, and enable us more or less successfully to predict, or even to regulate, the future occurrence of any particular combination.

The Laws of Thought are the first and most important of all the laws which govern the combinations of phenomena; and, even though they be binding on the mind, they may also be regarded as verified in the external world. The Logical Abecedarium develops the utmost variety of things and events which may occur, and it is evident that as each new quality is introduced, the number of combinations is doubled. Thus four qualities may occur in 16 combinations; five qualities in 32; six qualities in 64; and so on. In general language, if n be-the number of qualities, 2^n is the number of varieties of things which may be formed from them, if there be no conditions but those of logic. This number, it need hardly be said, increases after the first few terms, in an extraordinary manner, so that it would require 302 figures, even to express the number of combinations in which 1000 qualities might conceivably present themselves.

If all the combinations allowed by the Laws of Thought occurred in nature, then science would begin and end with those laws. To observe nature would give us no additional knowledge, because no two qualities would in the long run be oftener associated than any other two. We could never predict events with more certainty than we now predict the throws of dice, and experience would be without use. But the universe, as actually created, presents a far different and much more interesting problem. The most superficial observation shows that some things are habitually associated with other things. The more mature our examination, the more we become convinced that each event depends upon the prior occurrence of some other series of events. Action and reaction are gradually discovered to underlie the whole scene, and an independent or casual occurrence does not exist except in appearance. Even dice as they fall are surely determined in their course by prior conditions and fixed laws. Thus the combinations of events which can really occur are found to be very restricted, and it is the work of science to detect these restricting conditions.

In the English alphabet, for instance, we have twenty-six letters. Were the combinations of such letters perfectly free, so that any letter could be indifferently sounded with any other, the number of words which could be formed without any repetition would be $2^{26} - 1$, or 67,108,863, equal in number to the combinations of the twenty-seventh column of the Abecedarium, excluding one for the case in which all the letters would be absent. But the formation of our vocal organs prevents our using the far greater part of these conjunctions of letters. At least one vowel must be present in each word; more than two consonants cannot usually be brought together; and to produce words capable of smooth utterance a number of other rules must be observed. To determine exactly how many words might exist in the English language under these circumstances, would be an exceedingly complex problem, the solution of which has never been attempted. The number of existing English words may perhaps be said not to exceed one hundred thousand, and it is only by investigating the combinations presented in the dictionary, that we can learn the Laws of Euphony or calculate the possible number. In this example we have an epitome of the work and method of science. The combinations of natural phenomena are limited by a great number of conditions which are in no way brought to our knowledge except so far as they are disclosed in the examination of nature.

It is often a very difficult matter to determine the numbers of permutations or combinations which may exist under various restrictions. Many learned men puzzled themselves in former centuries over what were called Protean verses, or Latin verses admitting many variations in accordance with the Laws of Metre. The most celebrated of these verses was that invented by Bernard Bauhusius, as follows [a]:—

    | 'Tot tibi sunt dotes, Virgo, quot sidera cœlo.'

One author, Ericius Puteanus, filled forty-eight pages of a work in reckoning up its possible transpositions, making them only 1022. Other calculators gave 2196, 3276, 2580 as their results. Dr. Wallis assigned 3096, but without

<hr>

[a] Montucla, 'Histoire,' &c., vol. iii. p. 388.

much confidence in the accuracy of his result.[b] It required the skill of James Bernouilli to decide the number of transpositions to be 3312, under the condition that the sense and metre of the verse shall be perfectly preserved.

In approaching the consideration of the great Inductive problem, it is very necessary that we should acquire correct notions as to the comparative number of combinations which may exist under different circumstances. The doctrine of combinations is that part of mathematical science which applies numerical calculation to determine the number of combinations under various conditions. It is a part of the science which really lies at the base not only of other sciences, but of other branches of mathematical science. The forms of algebraical expressions are determined by the principles of combination, and Hindenburg recognised this fact in his Combinatorial Analysis. The greatest mathematicians have, during the last three centuries, given their best powers to the treatment of this subject; it was the favourite study of Pascal; it early attracted the attention of Leibnitz, who wrote his curious essay, *De Arte Combinatoria,* at twenty years of age; James Bernouilli, one of the very profoundest mathematicians, devoted no small part of his life to the investigation of the subject as connected with that of Probability; and in his celebrated work, *De Arte Conjectandi,* he has so finely described the importance of the doctrine of combinations, that I need offer no excuses for quoting his remarks at full length. ' It is easy to perceive that the prodigious variety which appears both in the works of nature and in the actions of men, and which constitutes the greatest part of the beauty of the universe, is owing to the multitude of different ways in which its several parts are mixed with, or placed near, each other. But, because the number of causes

b Wallis, 'Of Combinations,' &c., p. 119.

that concur in producing a given event, or effect, is
oftentimes so immensely great, and the causes themselves are so different one from another, that it is extremely difficult to reckon up all the different ways in
which they may be arranged or combined together, it
often happens that men, even of the best understandings
and greatest circumspection, are guilty of that fault in
reasoning which the writers on logic call *the insufficient
or imperfect enumeration of parts or cases :* insomuch
that I will venture to assert, that this is the chief, and
almost the only, source of the vast number of erroneous
opinions, and those too very often in matters of great
importance, which we are apt to form on all the subjects
we reflect upon, whether they relate to the knowledge of
nature or the merits and motives of human actions. It
must therefore be acknowledged, that that art which
affords a cure to this weakness, or defect, of our understandings, and teaches us so to enumerate all the possible
ways in which a given number of things may be mixed
and combined together, that we may be certain that we
have not omitted any one arrangement of them that can
lead to the object of our inquiry, deserves to be considered as most eminently useful and worthy of our
highest esteem and attention. And this is the business
of *the art or doctrine of combinations.* Nor is this art
or doctrine to be considered merely as a branch of the
mathematical sciences. For it has a relation to almost
every species of useful knowledge that the mind of man
can be employed upon. It proceeds indeed upon mathematical principles, in calculating the number of the combinations of the things proposed : but by the conclusions
that are obtained by it, the sagacity of the natural
philosopher, the exactness of the historian, the skill and
judgment of the physician, and the prudence and foresight of the politician may be assisted ; because the business

of all these important profes io:s is but *to form reasonable conjectures* concerning the several objects which engage their attention, and all wise conjectures are the results of a just and careful examination of the several different effects that may possibly arise from the causes that are capable of producing them.' [c]

Distinction of Combinations and Permutations.

We must at once consider the deep difference which exists between Combinations and Permutations; a difference involving important logical principles, and influencing the form of all our mathematical expressions. In *permutation* we recognise varieties of order or arrangement, treating AB as a different group from BA. In *combination* we take notice only of the presence or absence of a certain thing, and pay no regard to its place in order of time or space. Thus the four letters *a, e, m, n* can form but one combination, but they occur in language in several permutations, as *name, amen, mean, mane.*

We have hitherto been dealing with purely logical questions, involving only combination of qualities. I have fully pointed out in more than one place that, though our symbols could not but be written in order of place and read in order of time, the relations expressed had no regard to place or time (pp. 40, 131). The Law of Commutativeness, in fact, expresses the condition that in logic we deal with combinations, and the same law is true of all the processes of algebra. In nature and art, order may be a matter of indifference; it makes no difference, for instance, whether gunpowder is a mixture of sulphur, carbon and nitre, or carbon, nitre and sulphur, or nitre, sulphur and carbon, provided that the substances are present in

[c] James Bernouilli, 'De Arte Conjectandi,' translated by Baron Maseres. London, 1795, pp. 35-36.

proper proportions and well mixed. But this indifference of order does not usually extend to the events of physical science or the operations of art. The change of mechanical energy into heat is not exactly the same as the change from heat into mechanical energy; thunder does not indifferently precede and follow lightning; it is a matter of some importance that we load, cap, present, and fire a rifle in this precise order. Time is the condition of all our thoughts, space of all our actions, and therefore both in art and science we are to a great extent concerned with permutations. All language, for instance, treats different permutations of letters as having different meanings.

Permutations of certain things are far more numerous than combinations of those things, for the obvious reason that each distinct thing is regarded differently according to its place. Thus the letters A, B, C, will make different permutations according as A stands first, second, or third; having decided the place of A, there are two places between which we may choose for B; and then there remains but one place for C. Accordingly the permutations of these letters will be altogether $3 \times 2 \times 1$ or 6 in number. With four things or letters, A, B, C, D, we shall have four choices of place for the first letter, three for the second, two for the third, and one for the fourth, so that there will be altogether $4 \times 3 \times 2 \times 1$, or 24 permutations. The same simple rule applies in all cases; beginning with the whole number of things we multiply at each step by a number decreased by a unit. In general language, if n be the number of things in a combination, the number of permutations is $n(n-1)(n-2)\ldots\ldots 4 \cdot 3 \cdot 2 \cdot 1$. Thus, if we were to re-arrange the names of the days of the week, the possible arrangements out of which we should have to choose the new order, would be no less than $7 \cdot 6 \cdot 5 \cdot 4 \cdot 3 \cdot 2 \cdot 1$, or 5040, or, excluding the existing order, 5039.

The reader will see that the numbers which we reach in questions of permutation, increase in a more extraordinary manner even than in combination. Each new object or term doubles the number of combinations (p. 195), but increases the permutations by a factor continually growing. Instead of $2 \times 2 \times 2 \times 2 \times \ldots \ldots$ we have $2 \times 3 \times 4 \times 5 \times \ldots \ldots$ and the products of the latter expression indefinitely exceed those of the former. These products of continually increasing factors are constantly employed, as we shall see, in questions both of permutation and combination. They are technically called *factorials*, that is to say, the product of all integer numbers, from unity up to any number n, is the *factorial* of n, and is often indicated symbolically by $\lfloor n$. I give below the factorials up to that of fifteen :—

$$6 = 1 \cdot 2 \cdot 3$$
$$24 = 1 \cdot 2 \cdot 3 \cdot 4$$
$$120 = 1 \cdot 2 \ldots 5$$
$$720 = 1 \cdot 2 \ldots 6$$
$$5,040 = \lfloor 7$$
$$40,320 = \lfloor 8$$
$$362,880 = \lfloor 9$$
$$3,628,800 = \lfloor 10$$
$$39,916,800 = \lfloor 11$$
$$479,001,600 = \lfloor 12$$
$$6,227,020,800 = \lfloor 13$$
$$87,178,291,200 = \lfloor 14$$
$$1,307,674,368,000 = \lfloor 15$$

The factorials up to $\lfloor 36$ are given in Rees' 'Cyclopædia,' art. *Cipher*, and the logarithms of products up to $\lfloor 265$ are given at the end of the table of logarithms published under the superintendence of the Society for the Diffusion of Useful Knowledge (p. 215). To express the factorial $\lfloor 265$ would require 529 places of figures.

Many writers have from time to time remarked upon

the extraordinary magnitude of the numbers with which we deal in this subject. Tacquet calculated[d] that the twenty-four letters of the alphabet may be arranged in more than 620 thousand trillions of orders; and Schottus estimated[e] that if a thousand millions of men were employed for the same number of years in writing out these arrangements, and each man filled each day forty pages with forty arrangements in each, they could not have accomplished the task, as they would have written only 584 thousand trillions instead of 620 thousand trillions.

In some questions the number of permutations may be restricted and reduced by various conditions. Some things in a group may be undistinguishable from others, so that change of order will produce no difference. Thus if we were to permutate the letters of the name *Ann*, according to our previous rule, we should obtain $3 \times 2 \times 1$, or 6 orders; but half of these arrangements would be identical with the other half, because the interchange of the two *n*'s has no effect. The really different orders will therefore be $\frac{3 \cdot 2 \cdot 1}{1 \cdot 2}$ or 3, namely *Ann, Nan, Nna.* In the word *utility* there are two *i*'s and two *t*'s, in respect of both of which pairs the number of permutations must be halved. Thus we obtain $\frac{7 \cdot 6 \cdot 5 \cdot 4 \cdot 3 \cdot 2 \cdot 1}{1 \cdot 2 \cdot 1 \cdot 2}$ or 1260, as the number of permutations. The simple rule evidently is that when some things or letters are undistinguished, proceed in the first place to calculate all the possible permutations as if all were different, and then divide by the number of possible permutations of those series of things which are not distinguished, and of which the permutations have therefore been counted in excess. Thus since the word *Utilitarianism* contains fourteen

d ' Arithmeticæ Theoria.' Ed. Amsterd. 1704, p. 517.
e Rees' ' Cyclopædia,' art. *Cipher*.

letters, of which four are *i*'s, two *a*'s, and two *t*'s, the
number of distinct arrangements will be found by
dividing the factorial of 14, by the factorials of 4, 2,
and 2, the result being 908,107,200. From the letters
of the word *Mississippi* we can get in like manner
$$\frac{\lfloor 11}{\lfloor 4 \times \lfloor 4 \times \lfloor 2}$$ or 34,650 permutations, or not one-thousandth
part of what we should obtain were all the letters
different.

Calculation of Number of Combinations.

Although in many questions both of art and science
we need to calculate the number of permutations on
account of their own interest, it far more frequently
happens in scientific subjects that they possess but an
indirect interest. As I have already pointed out, we
almost always deal in the logical and mathematical
sciences with *combinations,* and variety of order enters
only through the inherent imperfections of our symbols
and modes of calculation. Signs must be used in some
order, and we must withdraw our attention from this order
before the signs correctly represent the relations of things
which exist neither before nor after each other. Now, it
often happens that we cannot choose all the combinations
of things, without first choosing them subject to the
accidental variety of order, and we must then divide by
the number of possible variations of order, that we may
get to the true number of pure combinations.

Suppose that we wish to determine the number of
ways in which we can select three letters out of the
alphabet, without allowing the same letter to be repeated.
At the first choice we can take any one of 26 letters; at
the next step there remain 25 letters, any one of which
may be joined with that already taken ; at the third step
there will be 24 choices, so that apparently the whole

number of ways of choosing is $26 \times 25 \times 24$. But the fact that one choice succeeded another has caused us to obtain the same combinations of letters in different orders ; we should get, for instance, a, p, r at one time, and p, r, a at another, and every three distinct letters will appear six times over, because three things can be arranged in six permutations. Thus the true number of combinations will be $\frac{24 \times 23 \times 22}{1 \times 2 \times 3}$, or 2024.

It is apparent that we need the doctrine of permutations in order that we may in many questions counteract the exaggerating effect of successive selection. If out of a senate of 30 persons we have to choose a committee of 5, we may choose any of 30 first, any of 29 next, and so on, in fact there will be $30 \times 29 \times 28 \times 27 \times 26$ selections ; but as the actual character of the members of the committee will not be affected by the accidental order of their selection, we divide by $1 \times 2 \times 3 \times 4 \times 5$, and the possible number of different committees will be 142,506. Similarly if we want to calculate the number of ways in which the eight major planets may come into conjunction, it is evident that they may meet either two at a time or three at a time, or four or more at a time, and as nothing is said as to the relative order or place in the conjunction, we require the number of combinations. Now a selection of 2 out of 8 is possible in $\frac{8.7}{1.2}$ or 28 ways; of 3 out of 8 in $\frac{8.7.6}{1.2.3}$ or 56 ways ; of 4 out of 8 in $\frac{8.7.6.5}{1.2.3.4}$ or 70 ways; and it may be similarly shown that for 5, 6, 7, and 8 planets, meeting at one time, the number of ways is 56, 28, 8 and 1. Thus we have solved the whole question of the variety of conjunctions of eight planets; and adding all the numbers together, we find that 247 is the utmost possible number of modes of meeting.

In general algebraic language, we may say that a group

of m things may be chosen out of a total number of n things, in a number of combinations denoted by the formula

$$\frac{n \cdot (n-1)\ (n-2)\ (n-3) \ldots (n-m+1)}{1 \cdot 2 \cdot 3 \cdot 4 \ldots m}$$

The extreme importance and significance of this formula seems to have been first adequately recognised by Pascal, although its discovery is attributed by him to a friend, M. de Ganières.[f] We shall find it perpetually recurring in questions both of combinations and probability, and throughout the formulæ of mathematical analysis traces of its influence will be noticed.

The Arithmetical Triangle.

The Arithmetical Triangle is a name long since given to a series of remarkable numbers connected with the subject we are treating. According to Montucla[g] 'this triangle is in the theory of combinations and changes of order, almost what the table of Pythagoras is in ordinary arithmetic, that is to say, it places at once under the eyes, the numbers required in a multitude of cases of this theory.' As early as 1544 Stifels had noticed the remarkable properties of these numbers and the mode of their evolution.[h] Briggs, the inventor of the common system of logarithms, was so struck with their importance that he called them the Abacus Panchrestus. Pascal, however, was the first who wrote a distinct treatise on these numbers, and gave them the name by which they are still known. But Pascal did not by any means exhaust the subject, and it remained for James Bernouilli to demonstrate fully the importance of the *figurate numbers,* as they are also called. In his treatise *De Arte Conjectandi,* he points out their appli-

---

f 'Œuvres Complètes de Pascal' (1865), vol. iii. p. 302. Montucla states the name as De Gruières, 'Histoire des Mathématiques,' vol. iii. p. 389.

g 'Histoire des Mathématiques,' vol. iii. p. 387.

h Leslie, 'Dissertation on the Progress of Mathematical and Physical Science,' Encyclopædia Britannica.

cation in the theory of combinations and probabilities, and remarks of the Arithmetical Triangle, 'It not only contains the clue to the mysterious doctrine of combinations, but it is also the ground or foundation of most of the important and abstruse discoveries that have been made in the other branches of the mathematics.'[i]

The numbers of the triangle can be calculated in a very easy manner by successive additions. We commence with unity at the apex; in the next line we place a second unit to the right of this; to obtain the third line of figures we move the previous line one place to the right, and add them to the same figures as they were before removal, and we can then repeat the same process *ad infinitum*. The fourth line of figures, for instance, contains 1, 3, 3, 1; moving them one place and adding as directed we obtain :—

Fourth line . . . 1	3	3	1			
1	3	3	1			
Fifth line 1	4	6	4	1		
1	4	6	4	1		
Sixth line 1	5	10	10	5	1	
1	5	10	10	5	1	
Seventh line . . . 1	6	15	20	15	6	1

Carrying out this simple process through ten more steps we obtain the first seventeen lines of the Arithmetical Triangle as printed on the next page. Theoretically speaking the Triangle must be regarded as infinite in extent but the numbers increase so rapidly that it soon becomes almost impracticable to continue the table. The longest table of the numbers which I have found is given in Fortia's 'Traité des Progressions' (p. 80), where they are given up to the fortieth line and the ninth column.

i Bernouilli, 'De Arte Conjectandi,' translated by Francis Maseres, London, 1795, p. 75.

THE ARITHMETICAL TRIANGLE.

Line.	First Column.	Second Column.	Third Column.	Fourth Column.	Fifth Column.	Sixth Column.	Seventh Column.	Eighth Column.	Ninth Column.	Tenth Column.	Eleventh Column.	Twelfth Column.	Thirteenth Column.	Fourteenth Column.	Fifteenth Column.	Sixteenth Column.	Seventeenth Col.
1	1																
2	1	1															
3	1	2	1														
4	1	3	3	1													
5	1	4	6	4	1												
6	1	5	10	10	5	1											
7	1	6	15	20	15	6	1										
8	1	7	21	35	35	21	7	1									
9	1	8	28	56	70	56	28	8	1								
10	1	9	36	84	126	126	84	36	9	1							
11	1	10	45	120	210	252	210	120	45	10	1						
12	1	11	55	165	330	462	462	330	165	55	11	1					
13	1	12	66	220	495	792	924	792	495	220	66	12	1				
14	1	13	78	286	715	1287	1716	1716	1287	715	286	78	13	1			
15	1	14	91	364	1001	2002	3003	3432	3003	2002	1001	364	91	14	1		
16	1	15	105	455	1365	3003	5005	6435	6435	5005	3003	1365	455	105	15	1	
17	1	16	120	560	1820	4368	8008	11440	12870	11440	8008	4368	1820	560	120	16	1

On carefully examining these numbers, we shall find that they are connected with each other by an almost unlimited series of relations, a few of the more simple of which may be noticed.

1. Each vertical column of numbers exactly corresponds with an oblique series descending from left to right, so that the triangle is perfectly symmetrical in its contents.

2. The first column contains only *units;* the second column contains the natural numbers, 1, 2, 3, &c.; the third column contains a remarkable series of numbers, 1, 3, 6, 10, 15, &c., which have long been called *the triangular numbers,* because they correspond with the numbers of balls which may be arranged in a triangular form, thus—

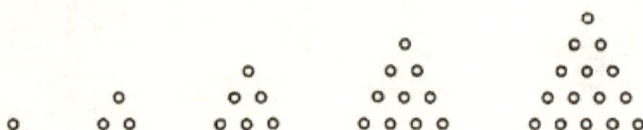

These numbers evidently differ each from the previous one by the series of natural numbers. Their employment has been explained, and the first 20,000 of the numbers calculated and printed by E. de Joncourt in a small quarto volume, which was published at the Hague, in 1762.

The fourth column contains the *pyramidal numbers,* so called because they correspond to the number of equal balls which can be piled in regular triangular pyramids. Their differences are the triangular numbers.

The numbers of the fifth column have the pyramidal numbers for their differences, but as there is no regular figure of which they express the contents, they have been arbitrarily called the *trianguli-triangular numbers.* The succeeding columns have, in a similar manner, been said to contain the *trianguli-pyramidal*, the *pyramidi-pyramidal* numbers, and so on.[k]

3. From the mode of formation of the table, it follows that the differences of the numbers in each column will be found in the preceding column to the left. Hence the *second differences,* or the *differences of differences* will be in the second column to the left of any given column, the third differences in the third column, and so on. Thus we may say that unity which appears in the first column is the *first difference* of the numbers in the second column; the *second difference* of those in the third column; the *third difference* of those in the fourth, and so on. The triangle is thus seen to be a complete classification of all numbers according as they have unity for any of their differences.

4. Every number in the table is equal to the sum of the numbers which stand higher in the next column to the left, beginning with the next line above; thus 84 is equal to the sum of 28, 21, 15, 10, 6, 3, 1.

5. Since each line is formed by adding the previous line to itself, it is evident that the sum of the numbers in each horizontal line must be double that of the line next above. Hence we know, without making any additions, that the successive sums must be 1, 2, 4, 8, 16, 32, 64, &c., the same as the numbers of combinations in the Logical Abecedarium. Speaking generally, the sum of the numbers in the nth line will be 2^{n-1}.

6. If the whole of the numbers down to any line be added together, we shall obtain a number less by unity than some power of 2; thus, the first line gives 1 or $2^1 - 1$; the first two lines give 3 or $2^2 - 1$; the first three lines 7 or $2^3 - 1$; the first six lines give 63 or $2^6 - 1$; or, speaking in general language, the sum of the first n lines is $2^n - 1$.

---

[k] Wallis's 'Algebra,' Discourse of Combinations, &c. p. 109.

7. It follows that the sum of the numbers in any one line is equal to the sum of those in all the preceding lines diminished by a unit. For the sum of the nth line is, as already shewn, 2^{n-1}, and the sum of the first $n-1$ lines is $2^{n-1}-1$, or less by a unit.

This enumeration of the properties of the figurate numbers does not approach completeness; a considerable, perhaps an unlimited, number of less simple and obvious relations might be traced out. Pascal, after giving many of the properties, exclaims[1] : ' Mais j'en laisse bien plus que je n'en donne ; c'est une chose étrange combien il est fertile en propriétés! Chacun peut s'y exercer.' The arithmetical triangle may be considered a natural classification of numbers, exhibiting, in the most complete manner, their evolution and relations in a certain point of view. It is obvious that in an unlimited extension of the triangle, each number will have at least two places.

Though the properties above explained are highly curious, the greatest value of the triangle arises from the fact that it contains a complete statement of the values of the formula (p. 206), for the number of combinations of m things out of n, for all possible values of m and n. Out of seven things one may be chosen in seven ways, and seven occurs in the eighth line of the second column. The combinations of two things chosen out of seven are $\frac{7\times6}{1\times2}$ or 21, which is the third number in the eighth line. The combinations of three things out of seven are $\frac{7\times6\times5}{1\times2\times3}$ or 35, which appears fourth in the eighth line. In a similar manner, in the fifth, sixth, seventh, and eighth columns of the eighth line I find it stated in how many ways I can select combinations of 4, 5, 6, and 7 things out of 7. Proceeding to the ninth line, I find in succession

the number of ways in which I can select 1, 2, 3, 4, 5, 6,
7, and 8 things, out of 8 things. In general language, if
I wish to know in how many ways m things can be
selected in combinations out of n things, I must look
in the $n + 1^{th}$ line, and take the $m + 1^{th}$ number, counting
from the left, as the answer. In how many ways, for
instance, can a sub-committee of five be chosen out of a
committee of nine. The answer is 126, and is the sixth
number in the tenth line; it will be found equal to
$\frac{9 \cdot 8 \cdot 7 \cdot 6 \cdot 5}{1 \cdot 2 \cdot 3 \cdot 4 \cdot 5}$, which our previous formula (p. 206) would
give.

The full utility of the figurate numbers will be more
apparent when we reach the subject of probabilities, but I
may give an illustration or two in this place. In how
many ways can we arrange four pennies as regards head
and tail? The question amounts to asking in how
many ways we can select 0, 1, 2, 3, or 4 heads out of 4
heads, and the *fifth* line of the triangle gives us the
complete answer, thus—

We can select No head and 4 tails in 1 way.
     ,,   1·head and 3 tails in 4 ways.
     ,,   2 heads and 2 tails in 6 ways.
     ,,   3 heads and 1 tail in 4 ways.
     ,,   4 heads and 0 tail in 1 way.

The total number of different cases is 16, or 2^4, and
when we come to the next chapter, it will be found that
these numbers give us the respective probabilities of all
throws with four pennies.

I gave in p. 205 a calculation of the number of ways in
which eight planets can meet in conjunction ; the reader
will find all the numbers detailed in the ninth line of the
arithmetical triangle. The sum of the whole line is 2^8 or
256 ; but we must subtract a unit for the case where no
planet appears, and 8 for the 8 cases in which only one

planet appears ; so that the total variety of conjunctions is $2^x - 1 - 8$ or 247.

If an organ has twelve stops, we find in the thirteenth line the numbers of combinations which we can draw, 0, 1, 2, 3, &c., at a time ; the total number of modes of varying the sound is no less than $2^{12} - 1$ or 4095[m]. If a number be the product of n prime factors, we find in the $n + 1^{th}$ line the numbers of divisors, being the product of 1, 2, 3, or more of the prime factors ; and the whole number of divisors of the number is the sum of the numbers in the line, subtracting unity, or $2^n - 1$.

One of the most important scientific uses of the arithmetical triangle, consists in the information which it gives concerning the comparative frequency of divergencies from an average. Suppose, for the mere sake of argument, that all persons were naturally of equal stature of five feet, but enjoyed during youth seven independent chances of growing one inch in addition. Of these seven chances, one, two, three, or more, may happen favourably to any individual, but as it does not matter what the chances are, so that the inch is gained, the question really turns upon the number of combinations of 0, 1, 2, 3, &c., things out of seven. Hence the eighth line of the triangle give us a complete answer to the question, as follows :—

Out of every 128 people—

	Feet.	Inches.
One person would have the stature of	5	0
7 persons ,, ,,	5	1
21 persons ,, ,,	5	2
35 persons ,, ,,	5	3
35 persons ,, ,,	5	4
21 persons ,, ,,	5	5
7 persons ,, ,,	5	6
1 person ,, ,,	5	7

[m] Bernouilli, 'De Arte Conjectandi,' trans. by Masères, p. 64.

By taking a proper line of the triangle, an answer may be had under any more natural supposition. This theory of comparative frequency of divergence from an average, was first adequately noticed by M. Quetelet, and has lately been employed in a very interesting and bold manner by Mr. Galton, in his work on 'Hereditary Genius.' We shall afterwards find that the theory of error, to which is made the ultimate appeal in cases of quantitative investigation, is founded upon the comparative numbers of combinations as displayed in the triangle.

Connection between the Arithmetical Triangle and the Logical Abecedarium.

There exists a close connection between the arithmetical triangle described in the last section, and the series of combinations of letters called the Logical Abecedarium. The one is to mathematical science what the other is to logical science. In fact the figurate numbers, or those exhibited in the triangle, are obtained by summing up the logical combinations. Accordingly, just as the total of the numbers in each line of the triangle was twice as great as that for the preceding line (p. 210), so each column of the Abecedarium (p. 109) contained twice as many combinations as the preceding one. The like correspondence would also exist between the sums of all the lines of figures down to any particular line, and of the combinations down to any particular column.

By examining any one column of the Abecedarium, we shall also find that the combinations naturally group themselves according to the figurate numbers. Take the combinations of the letters A, B, C, D; they consist of all the ways in which I can choose four, three, two, one, or none of the four letters, filling up the vacant spaces with negative terms. I may arrange the combinations as follows :—

ABCD . Four out of four . . 1 combination.

ABC*d*
AB*c*D
A*b*CD
*a*BCD } Three out of four . . 4 combinations.

AB*cd*
A*bc*D
A*b*C*d*
*a*BC*d*
*a*B*c*D
*ab*CD } Two out of four . . 6 combinations.

A*bcd*
*a*B*cd*
*ab*C*d*
*abc*D } One out of four . . . 4 combinations.

abcd . . None out of four . . 1 combination.

The numbers, it will be noticed, are exactly the same as those in the fifth line of the arithmetical triangle, and an exactly similar correspondence would be found to exist in the case of each other column of the Abece-darium.

Numerical abstraction, it has been asserted, consists in overlooking the kind of difference, and retaining only a consciousness of its existence (p. 177). While in logic, then, we have to deal with each combination as a separate kind of thing, in arithmetic we can distinguish only the classes which depend upon more or less positive terms being present, and the numbers of these classes imme-diately produce the numbers of the arithmetical triangle.

It may here be pointed out that there are two modes

in which we can calculate the whole number of combinations of certain things. Either we may take the whole number at once as shown in the Abecedarium, in which case the number will be some power of two, or else we may calculate successively, by aid of permutations, the number of combinations of none, one, two, three, and so on. Hence we arrive at a necessary identity between two series of numbers. In the case of four things we shall have

$$2^4 = 1 + \frac{4}{1} + \frac{4 \cdot 3}{1 \cdot 2} + \frac{4 \cdot 3 \cdot 2}{1 \cdot 2 \cdot 3} + \frac{4 \cdot 3 \cdot 2 \cdot 1}{1 \cdot 2 \cdot 3 \cdot 4}.$$

In a general form of expression we shall have

$$2^n = 1 + \frac{n}{1} + \frac{n \cdot (n-1)}{1 \cdot 2} + \frac{n(n-1)(n-2)}{1 \cdot 2 \cdot 3} + \&c.,$$

the terms being continued until they cease to have any value. Thus we have arrived at a proof of simple cases of the Binomial Theorem, of which each column of the Abecedarium is an exemplification. It may be shown that all other mathematical expansions likewise arise out of simple processes of combination, but the more complete consideration of this subject must be deferred.

Possible Variety of Nature and Art.

We cannot adequately understand the difficulties which beset us in certain branches of science, unless we gain a clear idea of the vast number of combinations or permutations which may be possible under certain conditions. Thus only can we learn how hopeless it would be to attempt to treat nature in detail, and exhaust the whole number of events which might arise. It is instructive to consider, in the first place, how immensely great are the numbers of combinations with which we deal in many arts and amusements.

In dealing a pack of cards, the number of hands, of thirteen cards each, which can be produced is

$$\frac{52 \cdot 51 \cdot 50 \cdot \cdot \cdot \cdot \cdot 40}{1 \cdot 2 \cdot 3 \cdot \cdot \cdot \cdot \cdot 13}$$

or 635,013,559,600.   But in whist four hands are simultaneously held, and the number of distinct deals becomes so vast that it would require twenty-eight figures to express it.   If the whole population of the world, say one hundred thousand millions of persons, were to deal cards day and night, for a hundred million of years, they would not in that time have exhausted one hundred-thousandth part of the possible deals.[o]   Now, even with the same hands the play may be almost infinitely varied, so that the complete variety of games which may exist is almost incalculably great.   It is in the highest degree improbable that any one game of whist was ever exactly like another, except by intention.

The end of novelty in art might well be dreaded, did we not find that nature at least has placed no attainable limit, and that the deficiency will lie in our inventive faculties.   It would be a cheerless time indeed when all possible varieties of melody were exhausted, but it is readily shown that if a peal of twenty-four bells had been rung continuously from the so-called beginning of the world to the present day, no approach could have been made to the completion of the possible changes.   Nay, had every single minute been prolonged to 10,000 years, still the task would have been unaccomplished.[p]   As regards ordinary melodies, the eight notes of a single octave give more than 40,000 permutations, and two octaves more than a million millions.   If we were to take

[o] 'Essay on Probability,' by Lubbock and Drinkwater, Useful Knowledge Society, 1833, p. 6.

[p] Wallis 'Of Combinations,' p. 116, quoting Vossius.

into account the semitones, it would become apparent that it is practically impossible to exhaust the variety of music.

Similar considerations apply to the possible number of natural substances, though we cannot always give precisely numerical results. It was recommended by Hatchett [q] that a systematic examination of all alloys of metals should be carried out, proceeding from the most simple binary ones to more complicated ternary or quaternary ones. He can hardly have been aware of the extent of his proposed inquiry. If we operated only upon thirty of the known metals, the number of possible selections of binary alloys would be 435, of ternary alloys 4060, of quaternary 27,405, without paying any regard to the varying proportions of the metals, and only regarding the kind of metal. If we varied all the ternary alloys by quantities not less than one per cent., the number of these alloys only would be 11,445,060. An exhaustive investigation of the subject is therefore out of the question, and unless some laws connecting the properties of the alloy and its components can be discovered, it is not apparent how our knowledge of them can be ever more than most incomplete.

The possible variety of definite chemical compounds, again, is enormously great. Chemists have already examined many thousands of inorganic substances, and a still greater number of organic·compounds; [r] they have nevertheless made no appreciable impression on the number which may exist. Taking the number of elements at sixty-one, the number of compounds containing different selections of four elements each would be more than half a million (521,855). As the same

q 'Philosophical Transactions' (1803), vol. xciii. p. 193.
r Hofmann's 'Introduction to Chemistry,' p. 36.

elements often combine in many different proportions, and some of them, especially carbon, have the power of forming an almost endless number of compounds, it would hardly be possible to assign any limit to the number of chemical compounds which may be formed. There are branches of physical science, therefore, of which it is unlikely that scientific men, with all their industry, can ever obtain a knowledge in any appreciable degree approaching to completeness.

Higher Orders of Variety.

The consideration of the facts already given in this chapter will not produce an adequate notion of the possible variety of existence, unless we consider the comparative numbers of combinations of different orders. By a combination of a higher order, I mean a combination of groups, which are themselves combinations of simpler groups. The almost unlimited number of compounds of carbon, hydrogen, and oxygen, described in organic chemistry, are combinations of a second order, for the atoms are groups of groups. The wave of sound produced by a musical instrument may be regarded as a combination of motions; the body of sound proceeding from a large orchestra is therefore a complex aggregate of sounds each in itself a complex combination of movements. All literature may be said to be developed out of the difference of white paper and black ink. From the almost unlimited number of marks which might be chosen we select twenty-six customary letters. The pronounceable combinations of letters are probably some trillions in number. Now, as a sentence is a certain selection of words, the possible sentences must be indefinitely more numerous than the words of which it may be composed. A book is a combination of sentences, and a library is a combination of books. A library, therefore, is a combination of the fifth order, and the powers of numerical expression would be almost exhausted in attempting to express the number of distinct libraries which might be constructed. The calculation would not be possible, because the union of letters in words, of words in sentences, and of sentences in books, are governed by conditions so complex as to defy calculation. I wish only to point out that there is no limit to the multitude of different sentences which may be developed out of the one difference of ink and paper. Galileo is said to have remarked that all truth is contained in the compass of the alphabet. We might add that it is all contained in the difference of ink and paper.

One consequence of this power of successive combination is that the simplest signals or marks will suffice to express any information. Francis Bacon proposed for secret writing a biliteral cipher, which resolves all letters of the alphabet into permutations of the two letters *a* and *b*. Thus A was *aaaaa*, B *aaaab*, X *babab*, and so on.[s] And in a similar way, as Bacon clearly saw, any one difference can be made the ground of a code of signals; we can express, as he says, *omnia per omnia.* The Morse alphabet uses only a succession of long and short marks, and other systems of telegraphic language employ right and left strokes. A single lamp obscured at various intervals, long or short, may be made to spell out any words, and with two lamps, distinguished by colour or position, we could at once represent Bacon's biliteral alphabet. Mr Babbage ingeniously suggested that every lighthouse in the world should be made to spell out its own name or number perpetually, by flashes or obscurations of

[s] 'Works,' edited by Shaw, vol. i. pp. 141–145, quoted in Rees' 'Encyclopædia,' art. *Cipher.*

various duration and succession, and the scheme would be easy of execution if needed.

Let us calculate the number of combinations of different orders which may arise out of the presence or absence of a single mark, say A. Thus in

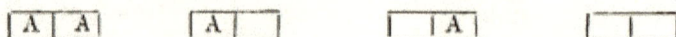

| A | A | | A | | | | A | | | |

we have four distinct varieties. Form them into a group of a higher order, and consider in how many ways we may vary that group by omitting one or more of the component parts. Now, as there are four parts, and any one may be present or absent, the possible varieties will be $2 \times 2 \times 2 \times 2$, or 16 in number. Form these into a new whole, and proceed again to create variety by omitting any one or more of the sixteen. The number of possible changes will now be 2.2.2.2.2.2.2.2.2.2.2.2.2.2.2.2, or 2^{16}, and we can repeat the process again and again if we wish. It will be easily seen that we are imagining the creation of objects, whose numbers are represented in the series of expressions—

$$
\begin{array}{cccc}
 & & & 2 \\
 & & 2 & 2 \\
 & 2 & 2 & 2 \\
2 & 2 & 2 & 2, \text{ and so on.}
\end{array}
$$

At the first step we have 2; at the next 2^2, or 4; at the third 2^{2^2}, or 16, numbers of very moderate amount. Let the reader calculate the next term, and he will be surprised to find it leap up to 65,536. But at the next step he has to calculate the value of 65,536 *two*'s multiplied together, and it is so great that we could not possibly compute it, the mere expression of the result requiring 19,729 places of figures. But go one step more and we pass the bounds of all reason. The sixth order of the powers of *two* becomes so great, that we could not even express the number of figures required in writing it down, without using about 19,729 figures for the purpose.

The successive orders of the powers of two have then the following values :—

First order   2
Second order   4
Third order   16
Fourth order   65,536
Fifth order, number expressed by 19,729 figures.
Sixth order, number expressed by
    figures, to express the number
    of which figures would require
    about   .   .   .   .   19,729 figures.

It may give us a powerful notion of infinity to remember that at this sixth step, having long surpassed all bounds of conception, we have made no approach to the goal. Nay, were we to make a hundred such steps, we should be as far away as ever from actual infinity.

It is well worth observing that our powers of expression rapidly overcome the possible multitude of finite objects which may exist in any assignable space. Archimedes showed long ago, in one of the most wonderful writings of antiquity,[t] that the grains of sand in the world could be numbered, or rather, that if numbered, the result could readily be expressed in arithmetical notation. Let us extend his problem, and ascertain whether we could express the number of atoms which could exist in the visible universe. The most distant stars which can now be seen by telescopes —those of the sixteenth magnitude—are supposed to have a distance of about 33,900,000,000,000,000 miles.[u] Sir W. Thomson, again, has shown reasons for supposing

[t] 'Liber de Arenæ Numero.'
[u] Chambers's 'Astronomy' (1861), p. 272.

that there do not exist more than from 3×10^{24} to 10^{26} molecules in a cubic centimetre of a solid or liquid substance.[x] Assuming these data to be true, for the sake of argument, a simple calculation enables us to show that the almost inconceivably vast sphere of our stellar system if entirely filled with solid matter, would not contain more than about 68×10^{90} atoms, that is to say, a number requiring for its expression 92 places of figures. Now, this number would be immensely less than the fifth order of the powers of two.

In the variety of logical relations, which may exist between a certain number of logical terms, we also meet a case of higher variations. Two terms, as it has been shewn (p. 154), may form four distinct combinations, but the possible selections from these series of combinations will be sixteen in number, or, excluding cases of contradiction, seven. Three terms may form eight combinations, allowing 256 selections, or with exclusion of contradictory cases, 193. Four terms give sixteen combinations, and no less than 65,536 possible selections from those combinations, the nature of which I naturally abstained from exhaustively examining. Five terms give thirty-two combinations, and 4,294,967,296 possible selections; and for six terms the corresponding numbers are sixty-four and 18,446,744,073,709,551,616. Considering that it is the most common thing in the world to use an argument involving six objects or terms, it may excite some surprise that the complete investigation of the relations in which six such terms may stand to each other, should involve an almost inconceivable number of cases. Yet these numbers of possible logical relations belong only to the second order of combinations.

CHAPTER X.

THE subject upon which we now enter must not be regarded as an isolated and curious branch of speculation. It is the necessary basis of nearly all the judgments and decisions we make in the prosecution of science, or the conduct of ordinary affairs. As Butler truly said, 'Probability is the very guide of life.' Had the science of numbers been developed for no other purposes, it must have been developed for the calculation of probabilities. All our inferences concerning the future are merely probable, and a due appreciation of the degree of probability depends entirely upon a due comprehension of the principles of the subject. I conceive that it is impossible even to expound the principles and methods of induction as applied to natural phenomena, in a sound manner, without resting them upon the theory of probability. Perfect knowledge alone can give certainty, and in nature perfect knowledge would be infinite knowledge, which is clearly beyond our capacities. We have, therefore, to content ourselves with partial knowledge—knowledge mingled with ignorance, producing doubt.

Almost the greatest difficulty in this subject consists in acquiring a precise notion of the matter treated. What is it that we number, and measure, and calculate in the theory of probabilities? Is it belief, or opinion, or doubt, or knowledge, or chance, or necessity, or want of art?

Does probability exist in the things which are probable, or in the mind which regards them as such ? The etymology of the name lends us no assistance : for, curiously enough, *probable* is ultimately the same word as *provable*, a good instance of one word becoming differentiated to two opposite meanings.

Chance cannot be the subject of the theory, because there is really no such thing as chance,[a] regarded as producing and governing events. This name signifies *falling*, and the notion is continually used as a simile to express uncertainty, because we can seldom predict how a die, or a coin, or a leaf will fall, or when a bullet will hit the mark. But every one knows, on a little reflection, that it is in our knowledge the deficiency lies, not in the certainty of nature's laws. There is no doubt in lightning as to the point it shall strike; in the greatest storm there is nothing capricious; not a grain of sand lies upon the beach, but infinite knowledge would account for its lying there ; and the course of every falling leaf is guided by the same principles of mechanics as rule the motions of the heavenly bodies.

Chance then exists not in nature, and cannot co-exist with knowledge; it is merely an expression for our ignorance of the causes in action, and our consequent inability to predict the result, or to bring it about infallibly. In nature the happening of a physical event has been pre-determined from the first fashioning of the universe. Probability belongs wholly to the mind ; this indeed is proved by the fact that different minds may regard the very same event at the same time with totally different degrees of probability. A steam-vessel, for instance, is missing and some persons believe that she has sunk in mid-ocean ; others think differently. In the

[a] Dufau, 'De la Méthode d'Observation,' chap. iii.

event itself there can be no such uncertainty ; the steam-vessel either has sunk or has not sunk, and no subsequent discussion of the probable nature of the event can alter the fact. Yet the probability of the event will really vary from day to day, and from mind to mind, according as the slightest information is gained regarding the vessels met at sea, the weather prevailing there, the signs of wreck picked up, or the previous condition of the vessel. Probability thus belongs to our mental condition, to the light in which we regard events, the occurrence or non-occurrence of which is certain in themselves. Many writers accordingly have asserted that probability is concerned with degree or quantity of belief. De Morgan says,[b] 'By degree of probability we really mean or ought to mean degree of belief.' The late Professor Donkin expressed the meaning of probability as 'quantity of belief;' but I have never felt satisfied with such a definition of probability. The nature of *belief* is not more clear to my mind than the notion it is used to define. But an all-sufficient objection is, that *the theory does not measure what the belief is, but what it ought to be.* Few minds think in close accordance with the theory, and there are many cases of evidence in which the belief existing is habitually different from what it ought to be. Even if the state of belief in any mind could be measured and expressed in figures, the results would be worthless. The very value of the theory consists in correcting and guiding our belief, and rendering our states of mind and consequent actions harmonious with our knowledge of exterior conditions.

This objection has been clearly perceived by some of those who still used quantity of belief as a definition of probability. Thus De Morgan adds—'Belief is but another name for imperfect knowledge.' Professor Donkin has well said that the quantity of belief is 'always relative to a particular state of knowledge or ignorance; but it must be observed that it is absolute in the sense of not being relative to any individual mind; since, the same information being presupposed, all minds *ought* to distribute their belief in the same way.'[c] Dr. Boole, too, seemed to entertain a like view, when he described the theory as engaged with 'the equal distribution of ignorance,'[d] but we may just as well say that it is engaged with the equal distribution of knowledge.

I prefer to dispense altogether with this obscure word belief, and to say that the theory of probability deals with *quantity of knowledge*, an expression of which a precise explanation and measure can presently be given. An event is only probable when our knowledge of it is diluted with ignorance, and exact calculation is needed to discriminate how much we do and do not know. The theory has been described by some as professing *to evolve knowledge out of ignorance*; but as Professor Donkin has admirably remarked, it is really 'a method of avoiding the erection of belief upon ignorance.'[e] It defines rational expectation by measuring the comparative amounts of knowledge and ignorance, and teaches us to regulate our action with regard to future events in a way which will, in the long run, lead to the least amount of disappointment and injury. It is, as Laplace as happily expressed it, *good sense reduced to calculation.*

This theory appears to me the noblest creation of human intellect, and it passes my conception how two men possessing such high intelligence as Auguste Comte and J. S. Mill, could have been found depreciating it, or even vainly attempting to question its validity. To

---

[c] 'Philosophical Magazine,' 4th Series, vol. i. p. 355.

[d] 'Transactions of the Royal Society of Edinburgh,' vol. xxi. part iv.

[e] 'Philosophical Magazine,' 4th Series, vol i. p. 355.

eulogise the theory is as needless as to eulogise reason itself.

Fundamental Principles of the Theory.

The calculation of probabilities is really founded, as I conceive, upon the principle of reasoning set forth in preceding chapters. We must treat equals equally, and what we know of one case may be affirmed of every other case resembling it in the necessary circumstances. The theory consists in putting similar cases upon a par, and distributing equally among them whatever knowledge we may possess. Throw a penny into the air, and consider what we know with regard to its mode of falling. We know that it will certainly fall upon a flat side, so that either the head or tail will be uppermost, but as to whether it will be head or tail, our knowledge is equally divided. Whatever we know concerning head, we know as much concerning tail, so that we have no reason for expecting one more than the other. The least predominance of belief to either side would be irrational, as it would consist in treating unequally things of which our knowledge is equal.

The theory does not in the least require, as some writers have erroneously supposed, that we should first ascertain by experiment the equal facility of the events we are considering. So far as we can examine and measure the causes in operation, events are removed out of the sphere of probability. The theory comes into play where ignorance begins, and the knowledge we possess requires to be distributed over many cases. Nor does the theory show that the coin will fall as often on one side as the other. It is almost impossible that this should happen, because some inequality in the form of the coin, or some uniform manner in throwing it up, is almost sure to occasion a slight preponderance in one direction. But as we do not previously know in which way a preponderance will exist, we have no more reason for expecting head than tail. Our state of knowledge will be changed, indeed, should we throw up the coin many times in succession and register the result. Every throw gives us some slight information as to the probable tendency of the coin, and in subsequent calculations we must take this into account. In other cases experience might show that we had been entirely mistaken; we might expect that a die would fall as often on each of the six sides as on each other one in the long run; trial might show that the die was a loaded one, and fell much the most often on a particular face. The theory would not have misled us: it treated correctly the information we had, which is all that any theory can do.

It may be asked, Why spend so much trouble in calculating from imperfect data, when a very little trouble would enable us to render a conclusion certain by actual trial? Why calculate the probability of a measurement being correct, when we can try whether it is correct? But I shall fully point out in later parts of this work that in measurement we never can attain perfect coincidence. Two measurements of the same base line in a survey may show a difference of some inches, and there may be no means of knowing which is the better result. A third measurement would probably agree with neither. To select any one of the measurements, would imply that we knew it to be the most nearly correct one, which we do not. In this state of ignorance, the only guide is the theory of probability, which proves that in the long run the mean of different quantities will come most nearly to the truth. In all other scientific operations whatsoever, perfect knowledge is impossible, and when we have exhausted all our instrumental means in the attainment of truth, there is a margin of error which can only be safely treated by the principles of probability.

The method which we employ in the theory consists in calculating the number of all the cases or events concerning which our knowledge is equal. If we have even the slightest reason for suspecting that one event is more likely to occur than another, we should take this knowledge into account. This being done, we must determine the whole number of events which are, so far as we know, equally likely. Thus, if we have no reason for supposing that a penny will fall more often one way than another, there are two cases, head and tail, equally likely. But if from trial or otherwise we know, or think we know, that of 100 throws 55 will give tail, then the probability is measured by the ratio of 55 to 100.

The mathematical formulæ of the theory are exactly the same as those of the theory of combinations. In this latter theory, we determine in how many ways events may be joined together, and we now proceed to use this knowledge in calculating the number of ways in which a certain event may come about, and thus defining its probability. If we throw three pennies into the air, what is the probability that two of them will fall tail uppermost? This amounts to asking in how many possible ways can we select two tails out of three, compared with the whole number of ways in which the coins can be placed. Now, the fourth line of the Arithmetical Triangle (p. 208) gives us the answer. The whole number of ways in which we can select or leave three things is eight, and the possible combinations of two things at a time is three; hence the probability of two tails is the ratio of three to eight. From the numbers in the triangle we may draw all the following probabilities :—

One combination gives 0 tail.   Probability $\frac{1}{8}$.

Three combinations give 1 tail.   Probability $\frac{3}{8}$.

Three combinations give 2 tails.　Probability $\frac{3}{8}$.

One combination gives 3 tails.　Probability $\frac{1}{8}$.

We could apply the same considerations to the imaginary causes of the difference of stature, the combinations of which were shown in p. 213. There are altogether 128 ways in which seven causes can be combined together. Now, twenty-one of these combinations give an addition of two inches, so that the probability of a person under the circumstances being five feet two inches is $\frac{21}{128}$. The probability of five feet three inches is $\frac{35}{128}$; of five feet one inch is $\frac{7}{128}$; of five feet $\frac{1}{128}$, and so on. Thus the eighth line of the Arithmetical Triangle gives all the probabilities arising out of the combinations of seven causes or things.

Rules for the Calculation of Probabilities.

I will now explain as simply as possible the rules for calculating probabilities. The principal rule is as follows :—

Calculate the number of events which may happen independently of each other, and which are as far as is known equally probable. Make this number the denominator of a fraction, and take for the numerator the number of such events as imply or constitute the happening of the event, whose probability is required.

Thus, if the letters of the word *Roma* be thrown down casually in a row, what is the probability that they will form a significant Latin word? The possible arrangements of four letters are 4 × 3 × 2 × 1, or 24 in number (p. 201), and if all the arrangements be examined, seven of these will be found to have meaning, namely *Roma, ramo, oram, mora, maro, armo,* and *amor.* Hence the probability of a significant result is $\frac{7}{24}$.[f]

---

[f] Wallis 'Of Combinations,' p. 117.

We must distinguish comparative from absolute probabilities. In drawing a card casually from a pack, there is no reason to expect any one card more than any other. Now, there are four kings and four queens in a pack, so that there are just as many ways of drawing one as the other, and the probabilities are equal. But there are thirteen diamonds, so that the probability of a king is to that of a diamond as four to thirteen. Thus the probabilities of each are proportional to their respective numbers of ways of happening. Now, I can draw a king in four ways, and not draw one in forty-eight, so that the probabilities are in this proportion, or, as is commonly said, the odds against drawing a king are forty-eight to four. The odds are seven to seventeen in favour, or seventeen to seven against the letters R,o,m,a, accidentally forming a significant word. The odds are five to three against two tails appearing in three throws of a penny. Conversely, when the odds of an event are given, and the probability is required, *take the number in favour of the event for numerator, and the sum of the numbers for denominator.*

It is obvious that an event is certain when all the combinations of causes which can take place produce that event. Now, if we were to represent the probability of any such event according to our rule, it would give the ratio of some number to itself, or unity. An event is certain not to happen when no possible combination of causes gives the event, and the ratio by the same rule becomes that of o to some number. Hence it follows that in the theory of probability certainty is expressed by 1, and impossibility by o; but no mystical meaning should be attached to these symbols, as they merely express the fact that *all* or *no* possible combinations give the event.

By a compound event, we mean an event which may be

distinguished into two or more simpler events. Thus
the firing of a gun may be distinguished into pulling the
trigger, the fall of the hammer, the explosion of the cap,
&c. In this example the simple events are not *independent*, because if the trigger is pulled, the other events
will under proper conditions necessarily follow, and their
probabilities are therefore the same as that of the first
event. Events are *independent* when the happening of
one does not render the other either more or less probable
than before. Thus the death of a person is neither more
nor less probable because the planet Mars happens to be
visible. When the component events are independent,
a simple rule can be given for calculating the probability
of the compound event, thus—*Multiply together the fractions expressing the probabilities of the independent
component events.*

The probability of throwing tail twice with a penny
is $\frac{1}{2} \times \frac{1}{2}$, or $\frac{1}{4}$; the probability of throwing it three times
running is $\frac{1}{2} \times \frac{1}{2} \times \frac{1}{2}$, or $\frac{1}{8}$; a result agreeing with that
obtained in an apparently different manner (p. 230). In
fact when we multiply together the denominators, we get
the whole number of ways of happening of the compound
event, and when we multiply the numerators, we get the
number of ways favourable to the required event.

Probabilities may be added to or subtracted from each
other under the important condition that the events in
question are exclusive of each other, so that not more than
one of them can happen. It might be argued that as
the probability of throwing head at the first trial is $\frac{1}{2}$, and
at the second trial also $\frac{1}{2}$, the probability of throwing
it in the first two throws is $\frac{1}{2} + \frac{1}{2}$, or certainty. Not only
is this result evidently absurd, but a repetition of the
process would lead us to a probability of $1\frac{1}{2}$ or of any
greater number, results which could have no meaning
whatever. The probability we wish to calculate is that of

one head in two throws, but in our addition we have involved the case in which two heads also appear. The true result is $\frac{1}{2} + \frac{1}{2} \times \frac{1}{2}$ or $\frac{3}{4}$, or the probability of head at the first throw, added to the exclusive probability that if it does not come at the first, it will come at the second. Some of the greatest difficulties of the theory and the subtlest errors arise from the confusion of exclusive and unexclusive alternatives. I may remind the reader that the possibility of unexclusive alternatives was a point previously discussed (p. 81), and to the reasons then given for considering alternation as logically unexclusive, may be added the existence of these difficulties in the theory of probability. The expression

Head first throw or *head second throw*

ought to be interpreted in our logical system as including both cases at once, and so it is in practice.

Employment of the Logical Abecedarium in questions of Probability.

When the probabilities of certain events are given, and it is required to deduce the probabilities of compound events, the Logical Abecedarium may give assistance, provided that there are no special logical conditions and all the combinations are possible. Thus, if there be three events A, B, C, of which the probabilities are a, β, γ, then the negatives of those events, expressing the absence of the events, will have the probabilities $1 - a$, $1 - \beta$, $1 - \gamma$. We have only to insert these values for the letters of the combinations and multiply, and we obtain the probability of each combination. Thus the probability of ABC is $a\beta\gamma$; of Abc, $a(1 - \beta)(1 - \gamma)$.

We can now clearly distinguish between the probabilities of exclusive and unexclusive events. Thus if A and B are events which may happen together like rain and high

tide, or an earthquake and a storm, the probability of
A or B happening is not the sum of their separate probabilities. For by the Laws of Thought we develop A + B
into AB + Ab + aB, and substituting a and β, the proba-
bilities of A and B respectively we obtain $a.\beta + a.(1 - \beta) + (1 - a).\beta$ or $a + \beta - a.\beta$. But if events are incompossible or
incapable of happening together, like a clear sky and rain,
or a new moon and a full moon, then the events are not
really A or B but A not–B, or B not–A or in symbols
Ab + aB. Now if we take

$$\mu = \text{probability of } Ab$$
$$\nu = \text{probability of } aB,$$

then we may add simply, and probability of $Ab + aB = \mu + \nu$.

Let the reader observe that since the combination AB
cannot exist, the probability of Ab is not the product of
the probabilities of A and b.

But when certain combinations are logically impossible,
it is no longer allowable to substitute the probability of
each term for the term, because the multiplication of
probabilities presupposes the independence of the events.
A large part of the late Dr. Boole's Laws of Thought is
devoted to an attempt to overcome this difficulty and
produce a General Method in Probabilities, by which from
certain logical conditions and certain given probabilities it
would be possible to deduce the probability of any other
combinations of events under those conditions. Boole
pursued his task with wonderful ingenuity and power, but
after spending much study on his work, I am compelled to
adopt the conclusion that his method is fundamentally
erroneous. As pointed out by Mr. Wilbraham [g], Boole
obtains his results by an arbitrary assumption, which is
only the most probable, and not the only possible assump-

[g] 'Philosophical Magazine,' 4th Series, vol. vii. p. 465; vol. viii.
p. 91.

tion. The answer obtained is therefore not the real probability, which is usually indeterminate, but only, as it were, the most probable probability. Certain problems solved by Boole are free from logical conditions and therefore may admit of valid answers. These as I have shown [h] may also be solved by the simple combinations of the Abecedarium, but the remaind er of the problems do not admit of a determinate answer, at least by Boole's method.

Comparison of the Theory with Experience.

The Laws of Probability rest upon the simplest principles of reasoning, and cannot be really negatived by any possible experience. It might happen that a person should always throw a coin head uppermost, and appear incapable of getting tail by chance. The theory would not be falsified, because it contemplates the possibility of the most extreme runs of luck. Our actual experience might be counter to all that is probable ; the whole course of events might seem to be in complete contradiction to what we should expect, and yet a casual conjunction of events might be the real explanation. It is just possible that some regular coincidences which we attribute to fixed laws of nature, are due to the accidental conjunction of phenomena in the cases to which our attention is directed. All that we can learn from finite experience is capable, according to the theory of probabilities, of misleading us, and it is only infinite experience that could assure us of any inductive truths.

At the same time, the probability that any extreme runs of luck will occur is so excessively slight, that it would be absurd seriously to expect their occurrence. It

[h] 'Memoirs of the Manchester Literary and Philosophical Society,' 3rd Series, vol. iv. p. 347.

is almost impossible, for instance, that any whist player should have played in any two games where the distribution of the cards was exactly the same, by pure accident (p. 217). Such a thing as a person always losing at a game of pure chance, is wholly unknown. Coincidences of this kind are not impossible, as I have said, but they are so unlikely that the lifetime of any person, or indeed the whole duration of history does not give any appreciable probability of their being encountered. Whenever we make any extensive series of trials of chance results, as in throwing a die or coin, the probability is great that the results will agree nearly with the predictions yielded by theory. Precise agreement must not be expected, for that, as the theory could show, is highly improbable. Several attempts have been made to test, in this way, the accordance of theory and experience. The celebrated naturalist, Buffon, caused the first trial to be made by a young child who threw a coin many times in succession, and he obtained 1992 tails to 2048 heads. A pupil of Professor De Morgan repeated the trial for his own satisfaction, and obtained 2044 tails to 2048 heads. In both cases the coincidence with theory is as close as could be expected, and the details may be found in De Morgan's 'Formal Logic,' p. 185.

Quetelet also tested the theory in a rather more complete manner, by placing 20 black and 20 white balls in an urn and drawing a ball out time after time in an indifferent manner, each ball being replaced before a new drawing was made. He found, as might be expected, that the greater the number of drawings made the more nearly were the white and black balls equal in number. At the termination of the experiment he had registered 2066 white and 2030 black balls, the ratio being 1·02[i].

i 'Letters on the Theory of Probabilities,' translated by Downes, 1849, pp. 36, 37.

I have made a series of experiments in a third manner, which seemed to me even more interesting, and capable of more extensive trial. Taking a handful of ten coins, usually shillings, I threw them up time after time, and registered the numbers of heads which appeared each time. Now the probability of obtaining 10, 9, 8, 7, &c., heads is proportional to the number of combinations of 10, 9, 8, 7, &c., things out of 10 things. Consequently the results ought to approximate to the numbers in the eleventh line of the Arithmetical Triangle. I made altogether 2048 throws, in two sets of 1024 throws each, and the numbers obtained are given in the following table :—

Character of Throw.	Theoretical Numbers.	First Series.	Second Series.	Average.	Divergence.
10 Heads 0 Tail	1	3	1	2	+ 1
9 " 1 "	10	12	23	17½	+ 7½
8 " 2 "	45	57	73	65	+ 20
7 " 3 "	120	129	123	126	+ 6
6 " 4 "	210	181	190	185½	− 25½
5 " 5 "	252	257	232	244½	− 7½
4 " 6 "	210	201	197	199	−11
3 " 7 "	120	111	119	115	− 5
2 " 8 "	45	52	50	51	+ 6
1 " 9 "	10	21	15	18	+ 8
0 " 10 "	1	0	1	½	− ½
Totals.	1024	1024	1024	1024	− 1

The whole number of single throws of coins amounted to 10×2048 or 20,480 in all, one half of which or 10,240 should theoretically give head. The total number of heads obtained was actually 10,353, or 5222 in the first series, and 5131 in the second. The coincidence with theory is pretty close, but considering the large number of throws there is some reason to suspect a tendency in favour of heads.

The special interest of this trial consists in the exhibition, in a practical form, of the results of Bernouilli's theorem, and the law of error or divergence from the mean to be afterwards more fully considered. It illustrates the connection between combinations and permutations, which is exhibited in the Arithmetical Triangle, and which underlies many of the most important theorems of science.

Probable Deductive Arguments.

With the aid of the theory of probabilities, we may extend the sphere of deductive argument. Hitherto we have treated propositions as certain, and on the hypothesis of certainty have deduced conclusions equally certain. But the information on which we reason in ordinary life is seldom or never certain, and almost all reasoning is really a question of probability. We ought therefore to be fully aware of the mode and degree in which the forms of deductive reasoning are affected by the theory of probability, and many persons might be surprised at the results which must be admitted. Many controversial writers appear to consider, as De Morgan remarked[k], that an inference from several equally probable premises is itself as probable as any of them, but the true result is very different. If a fact or argument involves many propositions, and each of them is uncertain, the conclusion will be of very little force.

The truth of a conclusion may be regarded as a compound event, depending upon the premises happening to be true; thus, to obtain the probability of the conclusion, we must multiply together the fractions expressing the probabilities of the premises. Thus, if the probability is $\frac{1}{2}$ that A is B, and also $\frac{1}{2}$ that B is C, the conclusion that A is C, on the ground of these premises, is $\frac{1}{2} \times \frac{1}{2}$ or $\frac{1}{4}$. Similarly if there be any number of premises requisite to

[k] 'Encyclopædia Metrop.' art. *Probabilities*, p. 396.

the establishment of a conclusion and their probabilities be m, n, p, q, r, &c., the probability of the conclusion on the ground of these premises is $m \times n \times p \times q \times r \times \ldots\ldots$ This product has but a small value, unless each of the quanties m, n, &c., be nearly unity.

But it is particularly to be noticed that the probability thus calculated is not the whole probability of the conclusion, but that only which it derives from the premises in question. Whately's[1] remarks on this subject might mislead the reader into supposing that the calculation is completed by multiplying together the probabilities of the premises. But it has been fully explained by De Morgan[m] that we must take into account the antecedent probability of the conclusion; A may be C for other reasons besides its being B, and as he remarks, 'It is difficult, if not impossible, to produce a chain of argument of which the reasoner can rest the result on those arguments only.' We must also bear in mind that the failure of one argument does not, except under special circumstances, disprove the truth of the conclusion it is intended to uphold, otherwise there are few truths which could survive the ill considered arguments adduced in their favour. But as a rope does not necessarily break because one strand in it is weak, so a conclusion may depend upon an endless number of considerations besides those immediately in view. Even when we have no other information we must not consider a statement as devoid of all probability. The true expression of complete doubt is a ratio of equality between the chances in favour of and against it, and this ratio is expressed in the probability $\frac{1}{2}$.

Now if A and C are wholly unknown things, we have no reason to believe that A is C rather than A is not C. The antecedent probability is then $\frac{1}{2}$. If we also have the

[1] 'Elements of Logic,' Book III, sections, 11 and 18.

[m] 'Encyclopædia Metrop.' art. *Probabilities*, p. 400.

probabilities that A is B, $\frac{1}{2}$ and that B is C, $\frac{1}{2}$, we have no right to suppose that the probability of A being C is reduced by the argument in its favour. If the conclusion is true on its own grounds, the failure of the argument does not affect it; thus its total probability is its antecedent probability, added to the probability that this failing, the new argument in question establishes it. There is a probability $\frac{1}{2}$ that we shall not require the special argument; a probability $\frac{1}{2}$ that we shall, and a probability $\frac{1}{4}$ that the argument does in that case establish it. Thus the complete result is $\frac{1}{2} + \frac{1}{2} \times \frac{1}{4}$, or $\frac{5}{8}$. In general language, if a be the probability formed on a particular argument, and c the antecedent probability, then the general result is

$$1 - (1-a)(1-c), \text{ or } a + c - ac.$$

We may put it still more generally in this way :—Let a, b, c, d, &c., be the probabilities of a conclusion grounded on various arguments or considerations of any kind. It is only when all the arguments fail that our conclusion proves finally untrue; the probabilities of each failing are respectively $1-a$, $1-b$, $1-c$, &c.; the probability that they will all fail $(1-a)(1-b)(1-c)...$; therefore the probability that the conclusion will not fail is $1 - (1-a)(1-b)(1-c)...$&c. On this principle it follows that every argument in favour of a fact, however flimsy and slight, adds probability to it. When it is unknown whether an overdue vessel has foundered or not, every slight indication of a lost vessel will add some probability to the belief of its loss, and the disproof of any particular evidence will not disprove the event.

We must apply these principles of evidence with great care, and observe that in a great proportion of cases the adducing of a weak argument does tend to the disproof of its conclusion. The assertion may have in itself great inherent improbability as being opposed to other evidence

or to the supposed laws of nature, and every reasoner may
be assumed to be dealing plainly, and putting forward the
whole force of evidence which he possesses in its favour.
If he brings but one argument, and its probability a is
small, then in the formula $1 - (1 - a)(1 - c)$ both a and c
are small, and the whole expression has but little value.
The whole effect of an argument thus turns upon the
question whether other arguments remain so that we can
introduce other factors $(1 - b)$, $(1 - c)$, &c., into the above
expression. In a court of justice, in a publication having
an express purpose, and in many other cases, it is doubtless
right to assume that the whole evidence considered to
have any value as regards the conclusion asserted, is
put forward.

To assign the antecedent probability of any proposition, may be a matter of great difficulty or impossibility, and one with which logic and the theory of probability has little concern. From the general body of
science or evidence in our possession, we must in each
case make the best judgment we can. But in the absence
of all knowledge the probability should be considered $= \frac{1}{2}$,
for if we make it less than this we incline to believe it
false rather than true. Thus before we possessed any
means of estimating the magnitudes of the fixed stars, the
statement that Sirius was greater than the sun had
a probability of exactly $\frac{1}{2}$; for it was as likely that it
would be greater as that it would be smaller; and so of
any other star. This indeed was the assumption which
Michell made in his admirable speculations.[o] It might
seem indeed that as every proposition expresses an agreement, and the agreements or resemblances between phenomena are infinitely fewer than the differences (p. 52),
every proposition should in the absence of other information be infinitely improbable, or $c = 0$. But in our logical

---

[o] 'Philosophical Transactions' (1767). Abridg. vol. xii. p. 435.

system every term may be indifferently positive or negative, so that we express under the form A is B or A = AB as many differences as agreements. It is impossible therefore that we should have any reason to disbelieve rather than to believe it. We can hardly indeed invent a proposition concerning the truth of which we are absolutely ignorant, except when we are absolutely ignorant of the terms used. If I ask the reader to assign the odds that a 'Platythliptic Coefficient is positive'[p] he will hardly see his way to doing so, unless he regard them as even.

The assumption that complete doubt is properly expressed by $\frac{1}{2}$ has been called in question by Bishop Terrot,[q] who proposes instead the indefinite symbol $\frac{0}{0}$; and he considers that 'the *à priori* probability derived from absolute ignorance has no effect upon the force of a subsequently admitted probability.' But a writer of far greater power, the late Professor Donkin, has strongly defended the commonly adopted expression of complete doubt. If we grant that the probability may have any value between o and 1, and that every separate value is equally likely, then n and $1 - n$ are equally likely, and the average is always $\frac{1}{2}$. Or we may take $p \,.\, dp$ to express the probability that our estimate concerning any proposition should lie between p and $p + dp$. The complete probability of the proposition is then the integral taken between the limits 1 and o, or again $\frac{1}{2}$[r].

Difficulties of the Theory.

The doctrine of probability, though undoubtedly true, requires very careful application. Not only is it a branch

---

[p] 'Philosophical Transactions,' vol. 146. part i. p. 273.

[q] 'Transactions of the Edinburgh Philosophical Society,' vol. xxi. p. 375.

[r] 'Philosophical Magazine,' 4th Series, vol. i. p. 361.

of mathematics in which positive blunders are frequently committed, but it is a matter of great difficulty in many cases, to be sure that the formulæ correctly represent the data of the problem. These difficulties often arise from the logical complexity of the conditions, which might be, perhaps to some extent cleared up by constantly bearing in mind the system of combinations as developed in the Indirect Logical Method. In the study of probabilities, mathematicians had unconsciously employed logical processes far in advance of those in possession of logicians, and the Indirect Method is but the full statement of these processes.

It is very curious how often the most acute and powerful intellects have gone astray in the calculation of probabilities. Seldom was Pascal mistaken, yet he inaugurated the science with a mistaken solution.[s] Leibnitz fell into the extraordinary blunder of thinking that the number twelve was as probable a result in the throwing of two dice as the number eleven.[t] In not a few cases the false solution first obtained seems more plausible to the present day than the correct one since demonstrated. James Bernouilli candidly records two false solutions of a problem which he at first thought self-evident ;[u] and he adds an express warning against the risk of error, especially when we attempt to reason on this subject without a rigid adherence to the methodical rules and symbols.[x] Montmort was not free from similar mistakes,[y] and as to D'Alembert, great though his reputation was, and perhaps is, he constantly fell into blunders which must diminish the weight of his opinions.[z] He could not perceive, for

[s] Montucla, 'Histoire des Mathématiques,' vol. iii. p. 386.

[t] Leibnitz 'Opera,' Dutens' Edition, vol. vi. part i. p. 217. Todhunter's 'History of the Theory of Probability,' p. 48.

[u] Todhunter, pp. 67–69.   [x] Ibid. p. 63.   [y] Ibid. p. 100.

[z] Ibid. pp. 258–59, 286.

instance, that the probabilities would be the same when coins are thrown successively as when thrown simultaneously.[a] Some men of high ability, such as Ancillon, Moses Mendelssohn, Garve,[b] Auguste Comte[c] and J. S. Mill,[d] have so far misapprehended the theory, as to question its value or even to dispute altogether its validity.

Many persons have a fallacious tendency to believe that when a chance event has happened several times together in an unusual conjunction, it is less likely to happen again. D'Alembert seriously held that if head was thrown three times running with a coin, tail would more probably appear at the next trial.[e] Bequelin adopted the same opinion, and yet there is no reason for it whatever. If the event be really casual, what has gone before cannot in the slightest degree influence it.

As a matter of fact, the more often the most casual event takes place the more likely it is to happen again; because there is some slight empirical evidence of a tendency, as will afterwards be pointed out. The source of the fallacy is to be found entirely in the feelings of surprise with which we witness an event happening by apparent chance, in a manner which seems to proceed from design.

Misapprehension may also arise from overlooking the difference between permutations and combinations. To throw ten heads in succession with a coin is no more unlikely than to throw any other particular succession of heads and tails, but it is much less likely than five heads and five tails without regard to their order, be-

[a] Todhunter, p. 279.   [b] Ibid. p. 453.

[c] 'Positive Philosophy,' translated by Martineau, vol. ii. p. 120.

[d] 'System of Logic,' bk. iii. chap. 18. 5th Ed. vol. ii. p. 61.

[e] Montucla, 'Histoire,' vol. iii. p. 405. Todhunter, p. 263.

cause there are no less than 252 different particular throws which will give this result, when we abstract the difference of order.

Difficulties arise in the application of the theory from our habitual disregard of slight probabilities. We are obliged practically to accept truths as certain which are nearly so, because it ceases to be worth while to calculate the difference. No punishment could be inflicted if absolutely certain evidence of guilt were required, and as Locke remarks, 'He that will not stir till he infallibly knows the business he goes about will succeed, will have but little else to do but to sit still and perish.'[f] There is not a moment of our lives when we do not lie under a slight danger of death, or some most terrible fate. There is not a single action of eating, drinking, sitting down, or standing up which has not proved fatal to some person. Several philosophers have tried to assign the limit of the probabilities which we regard as zero; Buffon named $\frac{1}{10,000}$, because it is the probability that a man of 56 years of age would die the next day, and is practically disregarded. Pascal had remarked that a man would be esteemed a fool for hesitating to accept death when three dice gave sixes twenty times running, if his reward in case of a different result was to be a crown; but as the chance of death in question is only $1 \div 6^{60}$, or unity divided by a number of 47 places of figures, we may be said every day to incur greater risks for less motives. There is far greater risk of death, for instance, in a game of cricket.

Nothing is more requisite than to distinguish carefully between the truth of a theory and the truthful application of the theory to actual circumstances. As a general rule, events in nature or art will present a complexity of

[f] 'Essay on the Human Understanding,' bk. IV. ch. 14. § 1.

relations exceeding our powers of treatment. The infinitely intricate action of the mind often intervenes and renders complete analysis hopeless. If, for instance, the probability that a marksman shall hit the target in a single shot be 1 in 10, we might seem to have no difficulty in calculating the probability of any succession of hits ; thus the probability of three successive hits would be one in a thousand. But, in reality, the confidence and experience derived from the first successful shot would render a second success more probable. The events are not really independent, and there would generally be a far greater preponderance of runs of apparent luck, than a simple calculation of probabilities could account for. In many persons, however, a remarkable series of successes will produce a degree of excitement rendering continued success almost impossible.

Attempts to apply the theory of probabilities to the results of judicial proceedings have proved of little value, simply because the conditions are far too intricate. As Laplace said,[g] 'Tant de passions, d'intérêts divers et de circonstances compliquent les questions relatives à ces objets, qu'elles sont presque toujours insolubles.' Men acting on a jury, or giving evidence before a court, are subject to so many complex influences that no mathematical formulæ can be framed to express the real conditions. Jurymen or even judges on the bench cannot be regarded as acting independently, with a definite probability in favour of each delivering a correct judgment. Each man of the jury is more or less influenced by the opinion of the others, and there are subtle effects of character and manner and strength of mind which defy human analysis. Even in physical science we shall in comparatively few cases be able to apply the theory in a definite manner, because the

[g] Quoted by Todhunter, 'History of the Theory of Probability,' p. 410.

data required for the estimation of probabilities are too complicated and difficult to obtain. But such failures in no way diminish the truth and beauty of the theory itself; for in reality there is no branch of science in which, as we shall afterwards fully consider, our symbols can cope with the complexity of Nature. As the late Professor Donkin excellently said,—

'I do not see on what ground it can be doubted that every definite state of belief concerning a proposed hypothesis, is in itself capable of being represented by a numerical expression, however difficult or impracticable it may be to ascertain its actual value. It would be very difficult to estimate in numbers the *vis viva* of all the particles of a human body at any instant; but no one doubts that it is capable of numerical expression.'[h]

The difficulty, in short, is merely relative to our knowledge and skill, and is not absolute or inherent in the subject. We must distinguish between what is theoretically conceivable and what is practicable with our present mental resources. Provided that our aspirations are pointed in a right direction, we must not allow them to be damped by the consideration that they pass beyond what can now be turned to immediate use. In spite of its immense difficulties of application, and the aspersions which have been mistakenly cast upon it, the theory of probabilities, I repeat, is the noblest, as it will in course of time prove, perhaps the most fruitful branch of mathematical science. It is the very guide of life, and hardly can we take a step or make a decision of any kind without correctly or incorrectly making an estimation of probabilities. In the next chapter we proceed to consider how the whole cogency of inductive reasoning, as applied to physical science rests upon probability. The truth or untruth of a natural law, when carefully investigated, resolves itself into a high or low degree of probability, and this is the case whether or not we are capable of producing precise numerical data.

CHAPTER XI.

WE have inquired into the nature of the process of perfect induction, whereby we pass backwards from certain observed combinations of qualities or events, to the logical conditions governing such combinations. We have also investigated the grounds of that theory of probability, which must be our guide when we leave certainty behind us, and dilute knowledge with ignorance. There is now before us the difficult task of endeavouring to decide how, by the aid of that theory, we can ascend from the facts to the laws of nature; and may then with more or less success anticipate the future course of events. All our knowledge of natural objects must be ultimately derived from observation, and the difficult question arises—How can we ever know anything which we have not directly observed through one of our senses, the apertures of the mind? The practical utility of reasoning is to assure ourselves that, at a determinate time or place, or under specified conditions, a certain phenomenon may be observed. When we can use our senses and perceive that the phenomenon does occur, reasoning is superfluous. If the senses cannot be used, because the event is in the future, or out of reach, how can reasoning take their place? Apparently, at least, we must infer the unknown from the known, and the mind must itself create an addition to the sum of knowledge. But I hold that it is quite impossible to make any real additions to the contents of our knowledge, except through new impressions upon the senses, or upon some seat of feeling. I shall attempt to show that inference, whether inductive or deductive, is never more than an unfolding of the contents of our experience, and that it always proceeds upon the assumption that the future and the unperceived will be governed by the same conditions as the past and the perceived, an assumption which will often prove to be mistaken.

In inductive just as in deductive reasoning, the conclusion never passes beyond the premises. Reasoning adds no more to the implicit contents of our knowledge, than the arrangement of the specimens in a museum adds to the number of those specimens. This arrangement adds to our knowledge in a certain sense : it allows us to perceive the similarities and peculiarities of the individual specimens, and on the assumption that the museum is an adequate representation of nature, it enables us to judge of the prevailing forms of natural objects. Bacon's first aphorism holds perfectly true, that man knows nothing but what he has observed, provided that we include his whole sources of experience, and the whole implicit contents of his knowledge. Inference but unfolds the hidden meaning of our observations, and the theory of probability shows how far we go beyond our data in assuming that new specimens will resemble the old ones, or that the future may be regarded as proceeding uniformly with the past.

Various Classes of Inductive Truths.

It will be desirable, in the first place, to distinguish between the several kinds of truths which we endeavour to establish by induction. Although there is a certain common and universal element in all our processes of reasoning, yet a diversity arises in their application. Similarity of conditions between the events from which we argue, and those to which we argue, must always be the ground of inference; but this similarity may have regard either to time or place, or the simple logical combination of events, or to any conceivable junction of circumstances involving quality, time, and place. Having met with many pieces of substance possessing ductility, and a bright yellow colour, and having discovered, by perfect induction, that they all possess in addition a high specific gravity, and a freedom from the corrosive action of acids, we are led to expect that every piece of substance, possessing like ductility, and a similar yellow colour, will have an equally high specific gravity, and a like freedom from corrosion by acids. This is a case of the co-existence of qualities; for the character of the specimens examined alters not with time or place.

In a second class of cases, time will enter as a principal ground of similarity. When we hear a clock pendulum beat moment after moment, at equal intervals, and with a uniform sound, we confidently expect that the stroke will continue to be repeated uniformly. A comet having appeared several times at nearly equal intervals, we infer that it will probably appear again at the end of another like interval. A man who has returned home evening after evening for many years, and found his house standing, may, on like grounds, expect that it will be standing the next evening, and on many succeeding evenings. Even the continuous existence of an object in an unaltered state, or the finding again of that which we have hidden, is but a matter of inference to be decided by experience.

A still larger and more complex class of cases involves the relations of space, in addition to those of time and quality. Having observed that every triangle drawn upon the diameter of a circle, with its apex upon the circumference, apparently contains a right angle, we may ascertain that all triangles in similar circumstances will contain right angles. This is a case of pure space reasoning, apart from circumstances of time or quality, and it seems to be governed by different principles of reasoning. I shall endeavour to show, however, that geometrical reasoning differs but in degree from that which applies to other natural relations. If we observe that the components of a binary star have moved for a length of time in elliptic curves, we have reason to believe that they will continue so to move. Time and space relations are here complicated together.

The Relation of Cause and Effect.

In a very large part of the scientific investigations which must be considered, we deal with events which follow from previous events, or with existences which succeed existences. Science, indeed, might arise even were material nature a fixed and changeless whole. Endow mind with the power to travel about, and compare part with part, and it could certainly draw inferences concerning the similarity of forms, the co-existence of qualities, or the preponderance of a particular kind of matter in a changeless world. A solid universe, in at least approximate equilibrium, is not inconceivable, and then the relation of cause and effect would evidently be no more than the relation of before and after. As nature exists, however, it is a progressive existence, ever moving and changing as time, the great independent variable, proceeds. Hence it arises that we must continually compare what is happening now with what happened a moment before, and a moment before that moment, and so on, until we reach indefinite periods of past time. A comet

is seen moving in the sky, or its constituent particles
illumine the heavens with their tails of fire.   We cannot
explain the present movements of such a body without
supposing its prior existence, with a determinate amount
of energy and direction of motion ; nor can we validly
suppose that our task is concluded when we find that it
came wandering to our solar system through the un-
measured vastness of surrounding space.   Every event
must have a cause, and that cause again a cause, until
we are lost in the obscurity of the past, and are driven
to the belief in one First Cause, by whom the whole
course of nature was determined.

Fallacious Use of the Term Cause.

The words Cause and Causation have given rise to infinite trouble and obscurity, and have in no slight degree
retarded the progress of science.   From the time of
Aristotle, the work of philosophy has been often described as the discovery of the causes of things, and
Francis Bacon adopted the notion when he said [a] '*vere
scire esse per causas scire.*'   Even now it is not uncommonly supposed that the knowledge of causes is something different from other knowledge, and consists, as it
were, in getting possession of the keys of nature.   A
single word may thus act as a spell, and throw the
clearest intellect into confusion, as I have often thought
that Locke was thrown into confusion when endeavouring
to find a meaning for the word *power.*[b]   In Mr. Mill's
' System of Logic' the term *cause* seems to have re-
asserted its old noxious power.   Not only does Mr. Mill
treat the Laws of Causation as almost co-extensive with

---

[a] 'Novum Organum,' bk. ii. Aphorism 2.
[b] 'Essay on the Human Understanding,' bk. ii. chap. xxi.

science, but he so uses the expression as to imply that when once we pass within the circle of causation we deal with certainties.

The philosophical danger which attaches to the use of this word may be thus described. A cause is defined as the necessary or invariable antecedent of an event, so that when the cause exists the effect will also exist or soon follow. If then we know the cause of an event, we know when it will certainly happen ; and as it is implied that science, by a proper experimental method, may attain to a knowledge of causes, it follows that experience may give us a certain knowledge of future events. Now, nothing is more unquestionable than that finite experience can never give us certain knowledge of the future, so that either a cause is not an invariable antecedent, or else we can never gain certain knowledge as to causes. The first horn of this dilemma is hardly to be accepted. Doubtless there is in nature some invariably acting mechanism, such that from certain fixed conditions an invariable result always emerges. But we, with our finite minds and short experience, can never penetrate the mystery of those existences which embody the Will of the Creator, and evolve it throughout time. We are in the position of spectators who witness the productions of a complicated machine, but are not allowed to examine its intimate structure. We learn what does happen and what does appear, but if we ask for the reason, the answer would involve an infinite depth of mystery. The simplest bit of matter, or the most trivial incident, such as the stroke of two billiard balls, offers infinitely more to learn than ever the human intellect can fathom. The word cause covers just as much untold meaning as any of the words *substance, matter, thought, existence.*

Confusion of Two Questions.

The subject is much complicated, too, by the confusion of two distinct questions. An event having happened, we may ask—

(1) Is there any cause for the event?

(2) Of what kind is that cause?

No one would assert that the mind possesses any faculty capable of inferring, prior to experience, that the occurrence of a sudden noise with flame and smoke indicates the combustion of a black powder, formed by the mixture of black, white, and yellow powders. The greatest upholder of *à priori* doctrines will allow that the particular aspect, shape, size, colour, texture, and other qualities of a cause must be gathered from experience and through the senses.

The question whether there is any cause at all for an event, is of a totally different kind. If an explosion could happen without any prior existing conditions, it must be a new creation—a distinct addition to the universe. It may be plausibly held that we can imagine neither the creation nor annihilation of anything. As regards matter, this has long been held true ; as regards force, it is now almost universally assumed as an axiom that energy can neither come into nor go out of existence without distinct acts of Creative Will. That there exists any instinctive belief to this effect, indeed, seems doubtful. We find Lucretius, a philosopher of the utmost intellectual power and cultivation, gravely assuming that his raining atoms could turn aside from their straight paths in a self-determining manner, and by this spontaneous origination of energy determine the form of the universe.[c] Sir George Airy, too, seriously discussed the mathematical conditions

[c] 'De Rerum Natura,' bk. ii. ll. 216–293.

under which a perpetual motion, that is, a perpetual source of self-created energy might exist.[d] The larger part of the philosophic world has long held that in mental acts there is free will—in short, self-causation. It is in vain to attempt to reconcile this doctrine with that of an intuitive belief in causation, as Sir W. Hamilton candidly allowed.

It is quite obvious, moreover, that to assert the existence of a cause for every event, cannot do more than remove into the indefinite past the inconceivable fact and mystery of creation. At any given moment matter and energy were equal to what they are at present, or they were not ; if equal, we may make the same inquiry concerning any other moment, however long prior, and we are thus obliged to accept one horn of the dilemma—existence from infinity, or creation at some moment. This is but one of the many cases in which we are compelled to believe in one or other of two alternatives, both inconceivable. My present purpose, however, is to point out that we must not confuse this supremely difficult question with that into which inductive science inquires on the foundation of facts. By induction we gain no certain knowledge ; but by observation, and the inverse use of deductive reasoning, we estimate the probability that an event which has occurred was preceded by conditions of specified character, or that such conditions will be followed by the event.

Definition of the Term Cause.

Clear definitions of the word cause have been given by several philosophers. Hobbes has said, ‘A cause is the sum or aggregate of all such accidents both in the agents

---

d ‘Cambridge Philosophical Transactions,’ [1830] vol. iii. pp. 369–372.

and the patients, as concur in the producing of the effect propounded; all which existing together, it cannot be understood but that the effect existeth with them; or that it can possibly exist if any of them be absent.' Dr. Brown, in his 'Essay on Causation,' gave a nearly corresponding statement. 'A cause,' he says [e], 'may be defined to be the object or event which immediately precedes any change, and which existing again in similar circumstances will be always immediately followed by a similar change.' Of the kindred word *power*, he likewise says : [f] ' Power is nothing more than that invariableness of antecedence which is implied in the belief of causation.'

These definitions may be accepted with the qualification that our knowledge of causes in such a sense can be probable only. The work of science consists in ascertaining the combinations in which phenomena present themselves. Concerning every event we shall have to determine its probable conditions, or group of antecedents from which it probably follows. An antecedent is anything which exists prior to an event; a consequent is anything which exists subsequently to an antecedent. It will not usually happen that there is any probable connection between an antecedent and consequent. Thus nitrogen is an antecedent to the lighting of a common fire; but it is so far from being a cause of the lighting, that it renders the combustion less active. Daylight is an antecedent to all fires lighted during the day, but it probably has no appreciable effect one way or the other. But in the case of any given event it is usually possible to discover a certain number of antecedents which

---

[e] 'Observations on the Nature and Tendency of the Doctrine of Mr. Hume, concerning the Relation of Cause and Effect.'   Second ed. p. 44.   [f] Ibid. p. 97.

seem to be always present, and with more or less probability we conclude that when they exist the event will follow.

Let it be observed that the utmost latitude is at present enjoyed in the use of the term *cause*. Not only may a cause be an existent thing endowed with powers, as oxygen is the cause of combustion, gunpowder the cause of explosion, but the very absence or removal of a thing may also be a cause. It is quite correct to speak of the dryness of the Egyptian atmosphere, or the absence of moisture, as being the cause of the preservation of mummies, and other remains of antiquity. The cause of a mountain elevation, Ingleborough for instance, is the excavation of the surrounding valleys by denudation. It is not so usual to speak of the existence of a thing at one moment as the cause of its existence at the next, but to me it seems the commonest case of causation which can occur. The cause of motion of a billiard ball may be the stroke of another ball; and recent philosophy leads us to look upon all motions and changes, as but so many manifestations of prior existing energy. In all probability there is no creation of energy and no destruction, so that as regards both mechanical and molecular changes, the cause is really the manifestation of existing energy. In the same way I see not why the prior existence of matter is not also a cause as regards its subsequent existence. All science tends to show us that the existence of the universe in a particular state at one moment, is the condition of its existence at the next moment, in an apparently different state. When we analyse the meaning which we can attribute to the word *cause*, it amounts to the existence of suitable portions of matter endowed with suitable quantities of energy. If we may accept Horne Tooke's assertion, *cause* has etymologically the meaning of *thing before*. Though, indeed, the origin of the word is very obscure, its derivatives the Italian *cosa*, and the French *chose*, mean simply *thing*. In the German equivalent *ursache*, we have plainly the original meaning of *thing before*, the *sache* denoting 'interesting or important object,' the English *sake*, and *ur* being the equivalent of the English *ere, before* [h]. We abandon, then, both etymology and philosophy, when we attribute to the *laws of causation* any meaning beyond that of the *conditions* in which an event may be expected to happen, according to our observation of the previous course of nature.

I have no objection to use the words cause and causation, provided they are never allowed to lead us to imagine that our knowledge of nature can attain to certainty. I repeat that if a cause is an invariable and necessary condition of an event, we can never know certainly whether the cause exists or not. To us, then, a cause is not to be distinguished from the group of positive or negative conditions which, with more or less probability, precede an event. In this sense, there is no particular difference between knowledge of causes and our general knowledge of the combinations, or succession of combinations, in which the phenomena of nature are presented to us, or found to occur in experimental inquiry.

Distinction of Inductive and Deductive Results.

We must carefully avoid confusing together inductive investigations which terminate in the establishment of general laws, and those which seem to lead directly to the knowledge of future particular events. That process only can be called induction which gives general laws, and it is by the subsequent employment of deduction that we can alone anticipate particular events. If the observation of a number of cases shews that alloys of metals

[h] Leslie, 'Inquiry into the Nature of Heat,' Note xvi. p. 521.

fuse at lower temperatures than their constituent metals,
I may with more or less probability draw a general inference to that effect, and may thence deductively ascertain the probability that the next alloy examined will fuse
at a lower temperature than its constituents. It has been
asserted, indeed, by Mr. J. S. Mill[i], and partially admitted
by Mr. Fowler[k], that we can argue directly from case to
case, so that what is true of some alloys will be true of
the next. Doubtless, this is the usual result of our
reasoning, regard being had to degrees of probability ; but
these logicians fail entirely to give any explanation of the
process by which we get from case to case. To point, as
Mr. Mill has done, to the reasoning, if such it can be
called, of brute animals, is little better than to parody
philosophy[l]. It may well be allowed, indeed, that the
knowledge of future particular events is one main purpose
of our investigations, and if there were any process of
thought by which we could pass directly from event to
event without ascending into general truths, this method
would be sufficient, and certainly the most brief and
simple. It is true, also, that the laws, of mental association lead the mind always to expect the like again in
apparently like circumstances, and even animals of very
low intelligence must have some trace of such powers of
association, serving to guide them more or less correctly,
in the absence of true reasoning faculties. But it is the
very purpose of logic, according to Mr. Mill, to ascertain
whether inferences have been correctly drawn, rather than
to discover them[m]. Even if we can, then, by habit,

i 'System of Logic,' bk. II. chap. iii. Mr. Bain has not adopted the
views of Mr. Mill, on this particular point, so far as I can ascertain. See
his 'Inductive Logic,' p. 1.

k 'Inductive Logic,' pp. 13–14.

l 'System of Logic,' bk. II. chap. 3, § 3. Fifth ed. pp. 212–213.

m Ibid., Introduction, § 4. Fifth ed. pp. 8–9.

association, or any rude process of inference, infer the future directly from the past, it is the work of logic to analyse the conditions on which the correctness of this inference depends. Even Mr. Mill would admit that such analysis involves the consideration of general truths[n], and in this, as in several other important points, we might controvert Mr. Mill's own views by his own statements.

On the Grounds of Inductive Inference.

I hold that, in all cases of inductive inference, we must invent hypotheses, until we fall upon some hypothesis which yields deductive results in accordance with experience. Such accordance renders the chosen hypothesis more or less probable, and we may then deduce, with some degree of likelihood, the nature of our future experience, on the assumption that no arbitrary change takes place in the conditions of nature. We can only argue from the past to the future, on the general principle set forth in the commencement of this work, that what is true of a thing will be true of the like. So far then as one object or event differs from another, all inference is impossible ; particulars as particulars can no more make an inference than grains of sand can make a rope. We must always rise to something which is general or same in the cases, and assuming that sameness to be extended to new cases we learn their nature. Hearing a clock tick five thousand times without exception or variation, we adopt the very probable hypothesis that there is some invariably acting machine which produces those uniform sounds, and which will, in the absence of change, go on producing them. Meeting twenty times with a bright yellow ductile substance, and finding it to be always very heavy and incorrodible, I infer that there was some natural condition,

***

[n] 'System of Logic,' bk. II. chap. iii. § 5. pp. 225, &c.

which tended, in the creation of things, to associate these properties together, and I expect to find them associated in the next instance. But there always is the possibility that some unknown change may take place between past and future cases. The clock may run down, or be subject to any one of a hundred accidents altering its condition. There is no reason in the nature of things, so far as known to us, why yellow colour, ductility, high specific gravity, and incorrodibility, should always be associated together ; and in other like cases, if not in this, men's expectations have been deceived. Our inferences, therefore, always retain more or less of a hypothetical character, and are so far open to doubt. Only in proportion as our induction approximates to the character of perfect induction, does it approximate to certainty. The amount of uncertainty corresponds to the probability that other objects than those examined, may exist and falsify our inferences ; the amount of probability corresponds to the amount of information yielded by our examination ; and the theory of probability will be needed to prevent our over-estimating or under-estimating the knowledge we possess.

Illustrations of the Inductive Process.

To illustrate the passage from the known to the apparently unknown, let us suppose that the phenomena under investigation consist of numbers, and that the following six numbers being exhibited to us, we are required to infer the character of the next in the series :—

5, 15, 35, 45, 65, 95.

The question first of all arises, How may we describe this series of numbers ? What is uniformly true of them ? The reader cannot fail to perceive at the first glance that they all end in five, and the problem is, from the properties of these six numbers, to infer the properties of the next number ending in five. If we proceed to test their properties by the process of perfect induction, we soon perceive that they have another common property, namely that of being *divisible by five without remainder*. May we then assert that the next number ending in five is also divisible by five, and, if so, upon what grounds? Or extending the question, Is every number ending in five divisible by five? Does it follow that because six numbers obey a supposed law, therefore 376,685,975 or any other number, however large, obeys the law? I answer *certainly not*. The law in question is undoubtedly true; but its truth is not proved by any finite number of examples. All that these six numbers can do, is to suggest to my mind the possible existence of such a law; and I then ascertain its truth, by proving deductively from the rules of decimal numeration, that any number ending in five must be made up of multiples of five, and must therefore be itself a multiple.

To make this more plain, let the reader now examine the numbers—

7, 17, 37, 47, 67, 97.

They all obviously end in 7 instead of 5, and though not at equal intervals, the intervals are exactly the same as in the previous case. After a little consideration, the reader will perceive that these numbers all agree in being *prime numbers*, or multiples of unity only. May we then infer that the next, or any other number ending in 7, is a prime number? Clearly not, for on trial we find that 27, 57, 117 are not primes. Six instances, then, treated empirically, lead us to a true and universal law in one case, and mislead us in another case. We ought, in fact to have no confidence in any law until we have treated it deductively, and have shown that from the conditions supposed the results expected must ensue. From the principles of number, no one can show that numbers ending in 7 should be primes.

From the history of the theory of numbers some good examples of false induction can be adduced. Taking the following series of prime numbers

$$41, 43, 47, 53, 61, 71, 83, 97, 113, 131, 151, \&c.,$$

it will be found that they all agree in being values of the general expression $x^2 + x + 41$, putting for x in succession the values, 0, 1, 2, 3, 4, &c. We thus seem always to obtain a prime number, and the induction is apparently very strong, to the effect that this expression always will give primes. Yet a few more trials will disprove this false conclusion. Put $x = 40$, and we obtain $40 \times 40 + 40 + 41$, or 41×41. Now such a failure could never have happened, had we shown any deductive reason why $x^2 + x + 41$ should give primes.

There can be no doubt that what here happens with forty instances, might happen with forty thousand or forty million instances. An apparent law never once failing up to a certain point may then suddenly break down, so that inductive reasoning, as it has been described by some writers, can give no sure knowledge of what is to come. Mr. Babbage admirably pointed out, in his Ninth Bridgewater Treatise, that a machine could be constructed to give a perfectly regular series of numbers, through a vast series of steps, and yet to break the law of progression suddenly at any required point. No number of particular cases as particulars enables us to pass by inference to any new case. It is hardly needful to inquire here what can be inferred from an infinite series of facts, because they are never practically within our power ; but we may unhesitatingly accept the conclusion, that no finite number of instances can ever prove a general law, or can give us sure knowledge of even one other instance.

General mathematical theorems have indeed been dis-

covered by the observation of particular cases, and may
again be so discovered. We have Newton's own statement, to the effect that he was thus led to the all-important Binomial Theorem, the basis of the whole structure
of mathematical analysis. Speaking of a certain series of
terms, expressing the area of a circle or hyperbola, he says,
' I reflected that the denominators were in arithmetical
progression ; so that only the numerical co-efficients of
the numerators remained to be investigated. But these,
in the alternate areas, were the figures of the powers of
the number eleven, namely 11^0, 11^1, 11^2, 11^3, 11^4; that is,
in the first 1 ; in the second $1, 1$; in the third $1, 2, 1$; in
the fourth $1, 3, 3, 1$; in the fifth $1, 4, 6, 4, 1.$[o] I inquired,
therefore, in what manner all the remaining figures could
be found from the first two ; and I found that if the first
figure be called m, all the rest could be found by the
continual multiplication of the terms of the formula

$$\frac{m-0}{1} \times \frac{m-1}{2} \times \frac{m-2}{3} \times \frac{m-3}{4} \times \&c.'[p]$$

It is pretty evident, from this most interesting statement,
that Newton having simply observed the succession of the
numbers, tried various formulæ until he found one which
agreed with them all. He was so little satisfied with this
process, however, that he verified particular results of his
new theorem by comparison with the results of common
multiplication, and the rule for the extraction of the
square root. Newton, in fact, gave no demonstration of
his theorem ; and a number of the first mathematicians
of the last century, James Bernouilli, Maclaurin, Landen,
Euler, Lagrange, &c., occupied themselves with discovering
a conclusive method of deductive proof.

[o] These are the figurate numbers considered in pages 206–216.

[p] 'Commercium Epistolicum. Epistola ad Oldenburgum,' Oct. 24,
1676. Horsley's 'Works of Newton', vol. iv. p. 541. See De Morgan
in 'Penny Cyclopædia', art. Binomial Theorem, p. 412.

Sir George Airy has also recorded a curious case, in which he accidentally fell by trial on a new geometrical property of the sphere.[q] Many of the most important and now trivial propositions in geometry, were probably thus discovered by the ancient Greek geometers; and we have pretty clear evidence of this in the Commentaries of Proclus.[r] But discovery in such cases means nothing more than suggestion, and it is always by pure deduction that the general law is really established. As Proclus puts it, *we must pass from sense to consideration.*

Given, for instance, the series of figures in the accompanying diagram, a little examination and measurement will show that the curved lines approximate to semicircles, and the rectilineal figures to right-angled triangles. These figures may seem to suggest to the mind the general law that angles inscribed in semicircles are right angles; but no number of instances, and no possible accuracy of measurement would really establish the truth of that general law. Availing ourselves of the suggestion furnished by the figures, we can only investigate deductively the consequences which flow from the definition of a circle, until we discover among them the property of containing right angles. Many persons, after much labour, have thought that they had discovered a method of trisecting angles by plane geometrical construction, because a certain complex arrangement of lines and circles had appeared to trisect an angle in every case tried by them, and they inferred, by a

supposed act of induction, that it would succeed in all other cases. Professor de Morgan has recorded a proposed mode of trisecting the angle which could not be discriminated by the senses from a true general solution, except when it was applied to very obtuse angles.[s] In all such cases, it has always turned out either that the angle was not trisected at all, or that only certain particular angles could be thus trisected. They were misled by some apparent or special coincidence, and only deductive proof could establish the truth and generality of the result. In this case, deductive proof shows that the problem, as attempted, is impossible, and that angles generally cannot be trisected by common geometrical methods.

Geometrical Reasoning.

This view of the matter is strongly supported by the further consideration of geometrical reasoning. No skill and care could ever enable us to verify absolutely any one geometrical proposition. Rousseau, in his *Emile*,[t] tells us that we should teach a child geometry by causing him to measure and compare figures by superposition. While a child was yet incapable of general reasoning, this would doubtless be an instructive exercise ; but it never could teach geometry, nor prove the truth of any one proposition. All our figures are rude approximations, and they may happen to seem unequal when they should be equal, and equal when they should be unequal. Moreover, figures may from chance be equal in case after case, and yet there may be no general reason why they should be so. The results of deductive geometrical reasoning are

[s] 'Budget of Paradoxes,' p. 257.
[t] 12mo. Amsterdam, 1762, vol. i. p. 401.

absolutely certain, and are either exactly true or capable of being carried to any required degree of approximation. In a perfect triangle, the angles must be equal to one half-revolution precisely ; even an infinitesimal divergence would be impossible ; and I believe with equal confidence, that however many are the angles of a figure, provided there are no re-entrant angles, the sum of the angles will be precisely and absolutely equal to twice as many right-angles as the figure has sides, less by four right-angles. In such cases, the deductive proof is absolute and complete ; empirical verification can at the most guard against accidental oversights.

There is a second class of geometrical truths which can only be proved by approximation ; but, as the mind sees no reason why that approximation should not always go on, we arrive at complete conviction. We thus learn that the surface of a sphere is equal exactly to two-thirds of the whole surface of the circumscribing cylinder, or to four times the area of the generating circle. The area of a parabola is exactly two-thirds of that of the circumscribing parallelogram. The area of the cycloid is exactly three times that of the generating circle. These are truths that we could never ascertain, nor even verify by observation ; for any finite amount of difference, vastly less than what the senses can discern, would falsify them. There are again geometrical relations which we cannot assign exactly, but can carry to any desirable degree of approximation. Thus, the ratio of the circumference to the diameter of a circle is that of 3·14159265358979323846…. to 1, and the approximation may be carried to any extent by the expenditure of sufficient labour, as many as 607 places of figures having been calculated.[u] Some years since, I amused myself by trying how near I could get to this ratio, by the careful use of compasses, and I did not come

---

[u] 'English Cyclopædia,' art. *Tables.*

nearer than 1 part in 540. We might imagine measurements so accurately executed as to give us eight or ten places correctly. But the power of the hands and senses must soon stop, whereas the mental powers of deductive reasoning can proceed to an unlimited degree of approximation. Geometrical truths, then, are incapable of verification; and, if so, they cannot even be learnt by observation. How can I have learnt by observation a proposition of which I cannot even prove the truth by observation, when I am in possession of it? All that observation or empirical trial can do is to suggest propositions, of which the truth may afterwards be proved deductively. By drawing a number of right-angled triangles on paper, with squares upon their sides, and cutting out and weighing these squares very accurately, I might have reason to suspect the existence of the relation of equality proved in Euclid's 47th Proposition; but no process of weighing or measuring could ever prove it, nor could it ever assure me that the like degree of approximation would exist in untried cases.

Much has been said about the peculiar certainty of mathematical reasoning, but it is only certainty of deductive reasoning, and equal certainty attaches to all correct logical deduction. If a triangle be right-angled, the square on the hypothenuse will undoubtedly equal the sum of the two squares on the other sides; but I can never be sure that a triangle is right-angled: so I can be certain that nitric acid will not dissolve gold, provided I know that the substances employed really correspond to those on which I tried the experiment previously. Here is like certainty of inference, and like doubt as to the facts.

Discrimination of Certainty and Probability in the Inductive Process.

We can never recur too often to the truth that our knowledge of the laws and future events of the external world is only probable. The mind itself is quite capable of possessing certain knowledge, and it is well to discriminate carefully between what we can and cannot know with certainty. In the first place, whatever feeling is actually present to the mind is certainly known to that mind. If I see blue sky, I may be quite sure that I do experience the sensation of blueness. Whatever I do feel, I do feel beyond all doubt. We are indeed very likely to confuse what we really feel with what we are inclined to associate with it, or infer inductively from it; but the whole of our consciousness, as far as it is the result of pure intuition and free from inference, is certain knowledge beyond all doubt.

In the second place, we may have certainty of inference; the first axiom of Euclid, the fundamental laws of thought, and the rule of substitution (p. 11), are certainly true; and if my senses could inform me that A was indistinguishable in colour from B, and B from C, then I should be equally certain that A was indistinguishable from C. In short, whatever truth there is in the premises, I can certainly embody in their correct logical result. But practically the certainty generally assumes a hypothetical character. I never can be quite sure that two colours are exactly alike, that two magnitudes are exactly equal, or that two bodies whatsoever are identical even in their apparent qualities. Almost all our judgments involve quantitative relations, and, as will be shown in succeeding chapters, we can never attain exactness and certainty where continuous quantity enters. Judgments concerning discontinuous quantity or numbers, however, allow of certainty; for I may establish beyond doubt, for instance, that the difference of the squares of 17 and 13 is the product of 17 + 13 and 17−13, and is therefore 30 × 4, or 120.

Inferences which we draw concerning natural objects are never certain except in a hypothetical point of view. It might seem, indeed, to be certain that iron is magnetic, or that gold is incapable of solution in nitric acid; but, if we carefully investigate the meanings of these statements, they will be found to involve no certainty but that of consciousness and that of hypothetical inference. For what do I mean by iron or gold? If I choose a remarkable piece of yellow substance, call it gold, and then immerse it in a liquid which I call nitric acid, and find that there is no change called solution, then consciousness has certainly informed me that with my meaning of the terms, 'Gold is insoluble in nitric acid.' I may further be certain of something else; for if this gold and nitric acid remain what they were, I may be sure there will be no solution on again trying the experiment. If I take other portions of gold and nitric acid, and am sure that they really are identical in properties with the former portions, I can be certain that there will be no solution. But at this point my knowledge becomes purely hypothetical; for how can I be sure without trial that the gold and acid are really identical in nature with what I formerly called gold and nitric acid. How do I know gold when I see it? If I judge by the apparent qualities—colour, ductility, specific gravity, &c., I may be misled, because there may always exist a substance which to the colour, ductility, specific gravity, and other specified qualities, joins others which we do not expect. Similarly, if iron is magnetic, as shown by an experiment with objects answering to those names, then all iron is magnetic, meaning all pieces of matter identical with my assumed piece. But in trying to identify iron, I am always open to mistake. Nor is this liability to mistake a matter of speculation only [v].

The history of chemistry shows that the most confident inferences may have been falsified by the confusion of one substance with another. Thus strontia was never discriminated from baryta until Klaproth and Haüy detected differences between some of their properties [x]. Accordingly chemists must often have inferred concerning strontia what was only true of baryta, and *vice versâ*. There is now no doubt that the recently discovered substances, cæsium and rubidium were long mistaken for potassium [y]. Other elements have often been confused together, for instance, tantalum and niobium; sulphur and selenium; cerium, lanthanum, and didymium; yttrium and erbium.

Even the best-established laws of physical science do not exclude false inference. No law of nature has been better established than that of universal gravitation, and we believe with the utmost confidence that any body capable of affecting the senses will attract other bodies, and fall to the earth if not prevented. Euler remarks that, although he had never made trial of the stones which compose the church of Magdeburg, yet he had not the least doubt that all of them were heavy, and would fall if unsupported. But he adds, that it would be extremely difficult to give any satisfactory explanation of this confident belief [z]. The fact is, that the belief ought not to amount to certainty until the experiment has been tried, and in the meantime a slight amount of uncer-

[v] Professor Bowen has excellently stated this view. 'Treatise on Logic.' Cambridge, U.S.A., 1866. P. 354.

[x] Whewell's 'History of the Inductive Sciences,' vol. iii. p. 174.

[y] Roscoe's 'Spectrum Analysis,' 1st edit. p. 99.

[z] Euler's 'Letters to a German Princess,' translated by Hunter. 2nd ed. vol. ii. pp. 17–18.

T

tainty enters, because we cannot be sure that the stones of the Magdeburg church resemble other stones in all their properties.

In like manner, not one of the inductive truths which men have established, or think they have established, is really safe from exception or reversal. Lavoisier, when laying the foundations of chemistry, met with so many instances tending to show the existence of oxygen in all acids, that he adopted a general conclusion to that effect, and devised the name oxygen accordingly. He entertained no appreciable doubt that the acid existing in sea salt also contained oxygen[a]; yet subsequent experience falsified his expectations.

This instance refers to a science in its infancy, speaking relatively to the possible achievements of men. But all sciences are and will ever remain in their infancy, relatively to the extent and complexity of the universe which they undertake to investigate. Euler expresses no more than the truth when he says that it would be impossible to fix on any one thing really existing, of which we could have so perfect a knowledge as to put us beyond the reach of mistake[b].

Like remarks may be made concerning all other inductive inferences. We may be quite certain that a comet will go on moving in a similar path *if* all circumstances remain the same as before ; but if we leave out this extensive qualification, our predictions will always be subject to the chance of falsification by some wholly unexpected event, such as the division of Biela's comet, or the unforeseen interference of some planetary or other gravitating body.

Inductive inference might attain to certainty if our

[a] Lavoisier's 'Chemistry,' translated by Kerr.   3rd edit. pp. 114, 121, 123.   [b] Euler's 'Letters,' vol. ii. p. 21.

knowledge of the agents existing throughout the universe were complete, and if we were at the same time certain that the same Power which created the universe would allow it to proceed without arbitrary change. There is always a possibility of causes being in existence without our knowledge, and these may at any moment produce an unexpected effect. Even when by the theory of probabilities we succeed in forming some notion of the comparative confidence with which we should receive inductive results, it yet appears to me that we must make an assumption. Events come out like balls from the vast ballot-box of nature, and close observation will enable us to form some notion, as we shall see in the next chapter, of the contents of that ballot-box. But we must still assume that between the time of an observation and that to which our inferences relate, no change in the ballot-box shall have been made.

CHAPTER XII.

WE have hitherto considered the theory of probability only in its simple deductive employment, by which it enables us to determine from given conditions the probable character of events happening under those conditions. But as deductive reasoning when inversely applied constitutes the process of induction, so the calculation of probabilities may be inversely applied; from the known character of certain events we may argue backwards to the probability of a certain law or condition governing those events. Having satisfactorily accomplished this work, we may indeed calculate forwards to the probable character of future events happening under the same conditions; but this part of the process is a direct use of deductive reasoning (p. 260).

Now it is highly instructive to find that whether the theory of probabilities be deductively or inductively applied, the calculation is always performed according to the principles and rules of deduction. The probability that an event has a particular condition entirely depends upon the probability that if the condition existed the event would follow. If we take up a pack of common playing cards, and observe that they are arranged in perfect numerical order, we conclude beyond all reasonable doubt that they have been thus intentionally arranged by some person acquainted with the usual order of sequence. This conclusion is quite irresistible, and rightly so ; for there are but two suppositions which we can make as to the reason of the cards being in that particular order :—

1. They have been intentionally arranged by some one who would probably prefer the numerical order.

2. They have fallen into that order by chance, that is, by some series of conditions which, being wholly unknown in nature, cannot be known to lead by preference to the particular order in question.

The latter supposition is by no means absurd, for any one order is as likely as any other when there is no preponderating tendency. But we can readily calculate by the doctrines of permutation the probability that fifty-two objects would fall by chance into any one particular order. Fifty-two objects can be arranged in—

$$52 \times 51 \times 50 \times \ldots \times 4 \times 3 \times 2 \times 1 \text{ or } 8066 \times (10)^{64}$$

possible orders, the number obtained requiring 68 places of figures for its full expression. Hence it is excessively unlikely, and, in fact, practically impossible, that any one should ever meet with a pack of cards arranged in perfect order by pure accident. If we do meet with a pack so arranged, we inevitably adopt the other supposition, that some person having reasons for preferring that special order, has thus put them together.

We know that of the almost infinite number of possible orders the numerical order is the most remarkable ; it is useful as proving the perfect constitution of the pack, and it is the intentional result of certain games. At any rate, the probability that intention should produce that order is incomparably greater than the probability that chance should produce it ; and as a certain pack exists in that order, we rightly prefer the supposition which most probably leads to the observed result.

By a similar mode of reasoning we every day arrive,
and validly arrive, at conclusions approximating to certainty. Whenever we observe a perfect resemblance
between two objects, as, for instance, two printed pages,
two engravings, two coins, two foot-prints, we are warranted
in asserting that they proceed from the same type, the
same plate, the same pair of dies, or the same boot. And
why ? Because it is almost impossible that with different
types, plates, dies, or boots some minute distinction of
form should not be discovered. It is barely possible for
the hand of the most skilful artist to make two objects
alike, so that mechanical repetition is the only probable
explanation of exact similarity. We can often establish
with extreme probability that one document is copied
from another. Suppose that each document contains
10,000 words, and that the same word is incorrectly
spelt in each. There is then a probability of less than
1 in 10,000 that the same mistake should be made in
each.

If we meet with a second error occurring in each document, the probability is less than 1 in 10,000 × 9999, that
such two coincidences should occur by chance, and the
numbers grow with extreme rapidity for more numerous
coincidences. We cannot indeed make any precise calculations without taking into account the character of the
errors committed, concerning the conditions of which we
have no accurate means of estimating probabilities.
Nevertheless, abundant evidence may thus be obtained
as to the derivation of documents from each other. In
the examination of many sets of logarithmic tables, six
remarkable errors were found to be present in all but
two, and it was proved that tables printed at Paris, Berlin,
Florence, Avignon, and even in China, besides thirteen
sets printed in England, between the years 1633 and
1822, were derived directly or indirectly from some

common source[a]. With a certain amount of labour, it is possible to establish beyond reasonable doubt the relationship or genealogy of any number of copies of one document, proceeding possibly from parent copies now lost. Tischendorf has thus investigated the relations between the manuscripts of the New Testament now existing, and the same work has been performed by German scholars for several classical writings.

Principle of the Inverse Method.

The inverse application of the rules of probability entirely depends upon a proposition which may be thus stated, nearly in the words of Laplace[b]. *If an event can be produced by any one of a certain number of different causes, the probabilities of the existence of these causes as inferred from the event, are proportional to the probabilities of the event as derived from these causes.* In other words, the most probable cause of an event which has happened is that which would most probably lead to the event supposing the cause to exist; but all other possible causes are also to be taken into account with probabilities proportional to the probability that the event would have happened if the cause existed. Suppose, to fix our ideas clearly, that E is the event, and C_1 C_2 C_3 are the three only conceivable causes. If C_1 exist, the probability is p_1 that E would follow; if C_2 and C_3 exist, the like probabilities are respectively p_2 and p_3. Then as p_1 is to p_2, so is the probability of C_1 being the actual cause to the probability of C_2 being it; and, similarly, as p_2 is to p_3, so is the probability of C_2 being the actual cause to the probability of C_3 being it. By a very simple mathematical

---

[a] Lardner, 'Edinburgh Review,' July 1834, p. 277.

[b] 'Mémoires par divers Savans,' tom. vi.; quoted by Todhunter in his 'History of Theory of Probability,' p. 458.

process we arrive at the conclusion that the actual probability of C_1 being the cause is

$$\frac{p_1}{p_1 + p_2 + p_3};$$

and the similar probabilities of the existence of C_2 and
C_3 are,

$$\frac{p_2}{p_1 + p_2 + p_3} \quad \text{and} \quad \frac{p_3}{p_1 + p_2 + p_3}.$$

The sum of these three fractions amounts to unity, which
correctly expresses the certainty that one cause or other
must be in operation.

We may thus state the result in general language.
If it is certain that one or other of the supposed causes
exists, the probability that any one does exist is the
probability that if it exists the event happens, divided by
the sum of all the similar probabilities. There may seem
to be an intricacy in this subject which may prove dis-
tasteful to some readers; but this intricacy is essential
to the subject in hand. No one can possibly understand
the principles of inductive reasoning, unless he will take
the trouble to master the meaning of this rule, by which
we recede from an event to the probability of each of its
possible causes.

This rule or principle of the indirect method is that
which common sense leads us to adopt almost instinctively,
before we have any comprehension of the principle in its
general form. It is easy to see, too, that it is the rule
which will, out of a great multitude of cases, lead us most
often to the truth, since the most probable cause of an
event really means that cause which in the greatest
number of cases produces the event. But I have only
met with one attempt at a general demonstration of the
principle. Poisson imagines each possible cause of an
event to be represented by a distinct ballot-box, containing
black and white balls, in such ratio that the probability of
a white ball being drawn is equal to that of the event

happening. He further supposes that each box, as is possible, contains the same total number of balls, black and white; and then, mixing all the contents of the boxes together, he shows that if a white ball be drawn from the aggregate ballot-box thus formed, the probability that it proceeded from any particular ballot-box is represented by the number of white balls in that particular box, divided by that total number of white balls in all the boxes. This result corresponds to that given by the principle in question[c].

Thus, if there be three boxes, each containing ten balls in all, and respectively containing seven, four, and three white balls, then on mixing all the balls together we have fourteen white ones; and if we draw a white ball, that is if the event happens, the probability that it came out of the first box is $\frac{7}{14}$; which is exactly equal to $\dfrac{\frac{7}{10}}{\frac{7}{10} + \frac{4}{10} + \frac{3}{10}}$, the fraction given by the rule of the Inverse Method.

Simple Applications of the Inverse Method.

In many cases of scientific induction we may apply the principle of the inverse method in a simple manner. If only two, or at the most a few hypotheses, may be made as to the origin of certain phenomena, or the connection of one phenomenon with another, we may sometimes easily calculate the respective probabilities of these hypotheses. It was thus that Professors Bunsen and Kirchhoff established, with a probability little short of certainty, that iron exists in the sun. On comparing the spectra of sunlight and of the light proceeding from the incandescent vapour of iron, it became apparent that at least sixty bright lines in the spectrum of iron coincided with dark

---

[c] Poisson, 'Recherches sur la Probabilité des Jugements,' Paris, 1837, pp. 82, 83.

lines in the sun's spectrum. Such coincidences could never be observed with certainty, because, even if the lines only closely approached, the instrumental imperfections of the spectroscope would make them apparently coincident, and if one line came within half a millemetre of another, on the map of the spectra, they could not be pronounced distinct. Now the average distance of the solar lines on Kirchhoff's map is 2 millemetres, and if we throw down a line, as it were, by pure chance on such a map, the probability is about one-half that the new line will fall within $\frac{1}{2}$ millemetre on one side or the other of some one of the solar lines. To put it in another way, we may suppose that each solar line, either on account of its real breadth or the defects of the instrument, possesses a breadth of $\frac{1}{2}$ millemetre, and that each line in the iron spectrum has a like breadth. The probability then is just one-half that the centre of each iron line will come by chance within 1 millemetre of the centre of a solar line, so as to appear to coincide with it. The probability of casual coincidence of each iron line with a solar line is in like manner $\frac{1}{2}$. Coincidence in the case of each of the sixty iron lines is a very unlikely event if it arises casually, for it would have a probability of only $(\frac{1}{2})^{60}$ or less than 1 in a trillion. The odds, in short, are more than a million million millions to unity against such casual coincidence [d]. But on the other hypothesis, that iron exists in the sun, it is highly probable that such coincidences would be observed ; it is immensely more probable that sixty coincidences would be observed if iron existed in the sun, than that they should arise from chance. Hence by our principle it is immensely probable that iron does exist in the sun.

All the other interesting results given by the comparison of spectra, rest upon the same principle of proba-

[d] Kirchhoff's 'Researches on the Solar Spectrum.' First part, translated by Professor Roscoe, pp. 18, 19.

bility. The almost complete coincidence between the spectra of solar, lunar, and planetary light renders it practically certain that the light is all of solar origin, and is reflected from the surfaces of the moon and planets, suffering only slight alteration from the atmospheres of some of the planets. A fresh confirmation of the truth of the Copernican theory is thus furnished.

A vast probability may be shown to exist that the heat, light, and chemical effects of the sun are due to the same rays, and are so many different manifestations of the same undulations. For a photograph of the spectrum corresponds exactly with what the eye observes, allowance being made for the great differences of chemical activity in different parts of the spectrum ; and delicate experiments with the thermopile also show that, where there is a dark line, there also the heat of the rays is absent.

Sir J. Herschel proved the connexion between the direction of the oblique faces of symmetrical quartz crystals, and the direction in which the same crystals rotate the plane of the polarisation of light. For if it is found in a second crystal that the relation is the same as in the first, the probability of this happening by chance is $\frac{1}{2}$; the probability that in another crystal also the direction would be the same is $\frac{1}{4}$, and so on. The probability that in $n + 1$ crystals there would be casual agreement of direction is the nth power of $\frac{1}{2}$. Thus, if in examining fourteen crystals the same relation of the two phenomena is discovered in each, the probability that it proceeds from uniform conditions is more than 8000 to 1 [e]. Now, since the first observations on this subject were made in 1820, no exceptions have been observed, so that the probability of invariable connexion is incalculably great.

[e] 'Edinburgh Review,' No. 185, vol. xcii. July 1850, p. 32 ; Herschel's 'Essays,' p. 421; 'Transactions of the Cambridge Philosophical Society,' vol. i. p. 43.

A good instance of this method is furnished by the
agreement of numerical statements with the truth. Thus,
in a manuscript of Diodorus Siculus, as Dr. Young states[g],
the ceremony of an ancient Egyptian funeral is described
as requiring the presence of forty-two persons sitting in
judgment on the merits of the deceased, and in many
ancient papyrus rolls the same number of persons are
found delineated. The probability is but slight that Diodorus, if inventing his statements or writing without
proper information, would have chosen such a number as
forty-two, and though there are not the data for an exact
calculation, Dr. Young considers that the probability in
favour of the correctness of the manuscript and the
veracity of the writer on this ground alone, is at least
100 to 1.

It is exceedingly probable that the ancient Egyptians
had exactly recorded the eclipses occurring during long
periods of time, for Diogenes Laertius mentions that 373
solar and 832 lunar eclipses had been observed, and the
ratio between these numbers exactly expresses that which
would hold true of the eclipses of any long period, of
say 1200 or 1300 years, as estimated on astronomical
grounds[h].

It is evident that an agreement between small numbers,
or customary numbers, such as seven, one hundred, a
myriad, &c., is much more likely to happen from chance,
and therefore gives much less presumption of dependence.
If two ancient writers spoke of the sacrifice of oxen, they
would in all probability describe it as a hecatomb, and
there would be nothing remarkable in the coincidence.

On similar grounds, we must inevitably believe in the
human origin of the flint flakes so copiously discovered of
late years. For though the accidental stroke of one stone

[g] Young's 'Works,' vol. ii. pp. 18, 19.
[h] 'History of Astronomy,' Library of Useful Knowledge, p. 14.

against another may often produce flakes, such as are occasionally found on the sea-shore, yet when several flakes are found in close company, and each one bears evidence, not of a single blow only, but of several successive blows, all conducing to form a symmetrical knife-like form, the probability of a natural and accidental origin becomes incredibly small, and the contrary supposition, that they are the work of intelligent beings, approximately certain [i].

An interesting calculation concerning the probable connexion of languages, in which several or many words are similar in sound and meaning, was made by Dr. Young [k].

Application of the Theory of Probabilities in Astronomy.

The science of astronomy, occupied with the simple relations of distance, magnitude, and motion of the heavenly bodies, admits more easily than almost any other science of interesting conclusions founded on the theory of probability. More than a century ago, in 1767, Michell showed the extreme probability of bonds connecting together systems of stars. He was struck by the unexpected number of fixed stars which have companions close to them. Such a conjunction might happen casually by one star, although possibly at a great distance from the other, happening to lie on the same straight line passing near the earth. But the probabilities are so greatly against such an optical union happening often in the expanse of the heavens, that Michell asserted the existence of a bond between most of

[i] Evans' 'Ancient Stone Implements of Great Britain.' London, 1872 (Longmans).

[k] 'Philosophical Transactions,' 1819; Young's 'Works,' vol. ii. pp. 15–18.

the double stars. It has since been estimated by Struve, that the odds are 95,70 to 1 against any two stars of not less than the seventh magnitude falling within the apparent distance of four seconds of each other by chance, and yet ninety-one such cases were known when the estimation was made, and many more cases have since been discovered. There were also four known triple stars, and yet the odds against the appearance of any one such conjunction are 173,524 to 1[1]. The conclusions of Michell have been entirely verified by the discovery that many double stars are in connexion under the law of gravitation.

Michell also investigated the probability that the six brightest stars in the Pleiades should have come by accident into such striking proximity. Estimating the number of stars of equal or greater brightness at 1500, he found the odds to be nearly 500,000 to 1 against casual conjunction. Extending the same kind of argument to other clusters, such as that of Præsepe, the nebula in the hilt of Perseus' sword, he says[m] : ' We may with the highest probability conclude, the odds against the contrary opinion being many million millions to one, that the stars are really collected together in clusters in some places, where they form a kind of system, while in others there are either few or none of them, to whatever cause this may be owing, whether to their mutual gravitation, or to some other law or appointment of the Creator.'

The calculations of Michell have been called in question by the late James D. Forbes[n], and Mr. Todhunter vaguely

[1] Herschel, 'Outlines of Astronomy,' 1849, p. 565 ; but Todhunter, in his 'History of the Theory of Probability,' p. 335, states that the calculations do not agree with those published by Struve.

[m] 'Philosophical Transactions,' 1767, vol. lvii. p. 431.

[n] 'Philosophical Magazine,' 3rd Series, vol. xxxvii. p. 401, December, 1850 ; also August, 1849.

countenances his objections [o], otherwise I should not have thought them of much weight.   Certainly Laplace accepts Michell's views [p], and if Michell be in error, it is in the methods of calculation, not in the general validity of his conclusions.

Similar calculations might no doubt be applied to the peculiar drifting motions which have been detected by Mr. R. A. Proctor in some of the constellations [q].   Against a general tendency of stars to move in one direction by chance, the odds are very great.   It is on a similar ground that a considerable proper motion of the sun is found to exist with immense probability, because on the average the fixed stars show a tendency to move apparently from one point of the heavens towards that diametrically opposite.   The sun's motion in the contrary direction would explain this tendency, otherwise we must believe that myriads of stars accidentally agree in their direction of motion, or are urged by some common force from which the sun is exempt.   It may be said that the rotation of the earth is proved in like manner, because it is immensely more probable that one body would revolve than that the sun, moon, planets, comets, and the whole of the stars of the heavens should be whirled round the earth daily, with a uniform motion superadded to their own peculiar motions.   This appears to be nearly the reason which led Gilbert, one of the earliest English Copernicans, and in every way an admirable physicist, to admit the rotation of the earth, while Francis Bacon denied it [r].

In contemplating the planetary system, we are struck with the similarity in direction of nearly all its move-

[o] 'History,' &c., p. 334.

[p] 'Essai Philosophique,' p. 57.

[q] 'Proceedings of the Royal Society,' 20 January, 1870. 'Philosophical Magazine,' 4th Series, vol. xxxix. p. 381.

[r] Hallam's 'Literature of Europe,' 1st ed. vol. ii. p. 464.

ments.   Newton remarked upon the regularity and uniformity of these motions, and contrasted them with the
eccentricity and irregularity of the cometary orbits[s].
Could we, in fact, look down upon the system from the
northern side, we should see all the planets moving round
from west to east, the satellites moving round their
primaries and the sun, planets, and all the satellites
rotating in the same direction, with some exceptions on
the verge of the system.   Now in the time of Laplace
eleven planets were known, and the directions of rotation
were known for the sun, six planets, the satellites of Jupiter,
Saturn's ring, and one of his satellites.   Thus there were
altogether 43 motions all concurring, namely :—

Orbital motions of eleven planets   .   .   11
Orbital motions of eighteen satellites .   .   18
Axial rotations   14
                                         ——
                                           43

The probability that 43 motions independent of each
other would coincide by chance is the 42nd power of $\frac{1}{2}$, so
that the odds are about 4,400,000,000,000 to 1 in favour
of some common cause for the uniformity of direction. This
probability, as Laplace observes[t], is higher than that of
many historical events which we undoubtingly believe.
In the present day, the probability is much increased by
the discovery of additional planets, and the rotation of
other satellites, and it is only slightly weakened by the
fact that some of the outlying satellites are exceptional in
direction, there being considerable evidence of an accidental disturbance in the more distant parts of the
system.

Hardly less remarkable than the uniformity of motion

[s] 'Principia,' bk. ii.   General scholium.
[t] 'Essai Philosophique,' p. 55. Laplace appears to count the rings of
Saturn as giving two independent movements.

is the near approximation of all the orbits of the planets
to a common plane.   Daniel Bernouilli roughly estimated
the probability of such an agreement arising from accident
at $\frac{1}{(12)^6}$, the greatest inclination of any orbit to the sun's
equator being 1-12th part of a quadrant.   Laplace de-
voted to this subject some of his most ingenious investigations.   He found the probability that the sum of the
inclinations of the planetary orbits would not exceed by
accident the actual amount ('914187 of a right angle for
the ten planets known in 1801) to be $\frac{1}{10}$ ('914187)^{10}, or
about '0000001235.   This probability may be combined
with that derived from the direction of motion, and it
then becomes immensely probable that the constitution of
the planetary system arose out of uniform conditions, or,
as we say, from some common cause[u].

If the same kind of calculation be applied to the orbits
of comets the result is very different[y].   Of the orbits
which have been determined 48·9 per cent. only are direct
or in the same direction as the planetary motions[z].   Hence
it becomes apparent that comets do not properly belong
to the solar system, and it is probable that they are stray
portions of nebulous matter which have become accidently
attached to the system by the attractive powers of the
sun or Jupiter.

Statement of the General Inverse Problem.

In the instances described in the preceding sections,
we have been occupied in receding from the occurrence

[u] Lubbock, 'Essay on Probability,' p. 14.   De Morgan, 'Encyc.
Metrop.' art. *Probability,* p. 412.   Todhunter's 'History of the Theory of
Probability,' p. 543.   Concerning the objections raised to these conclusions by the late Dr. Boole, see the 'Philosophical Magazine,' 4th Series,
vol. ii. p. 98.   Boole's 'Laws of Thought,' pp. 364–375.

[y] Laplace, 'Essai Philosophique,' pp. 55, 56.

[z] Chambers's 'Astronomy,' 2nd ed. pp. 346-49.

of certain similar events to the probability that there must have been a condition or cause for such events. We have found that the theory of probability, although never yielding a certain result, often enables us to establish an hypothesis beyond the reach of reasonable doubt. There is, however, another method of applying the theory, which possesses for us even greater interest, because it illustrates, in the most complete manner, the theory of inference adopted in this work, which theory indeed it suggested. The problem to be solved is as follows :—

An event having happened a certain number of times, and failed a certain number of times, required the probability that it will happen any given number of times in the future under the same circumstances.

All the larger planets hitherto discovered move in one direction round the sun ; what is the probability that, if a new planet exterior to Neptune be discovered, it will move in the same direction ? All known permanent gases, except chlorine, are colourless ; what is the probability that, if some new permanent gas should be discovered, it will be colourless ? In the general solution of this problem, we wish to infer the future happening of any event from the number of times that it has already been observed to happen. Now, it is very instructive to find that there is no known process by which we can pass directly from the data to the conclusion. It is always requisite to recede from the data to the probability of some hypothesis, and to make that hypothesis the ground of our inference concerning future happenings. Mathematicians, in fact, make every hypothesis which is applicable to the question in hand ; they then calculate, by the inverse method, the probability of every such hypothesis according to the data, and the probability that if each hypothesis be true, the required future event will happen The total probability that the event will happen, is the sum of the separate probabilities contributed by each distinct hypothesis.

To illustrate more precisely the method of solving the problem, it is desirable to adopt some concrete mode of representation, and the ballot-box, so often employed by mathematicians, will best serve our purpose. Let the happening of any event be represented by the drawing of a white ball from a ballot-box, while the failure of an event is represented by the drawing of a black ball. Now, in the inductive problem we are supposed to be ignorant of the contents of the ballot-box, and are required to ground all our inferences on our experience of those contents as shown in successive drawings. Rude common sense would guide us nearly to a true conclusion. Thus if we had drawn twenty balls, one after another, replacing the ball after each drawing, and the ball had in each case proved to be white, we should believe that there was a considerable preponderance of white balls in the urn, and a probability in favour of drawing a white ball on the next occasion. Though we had drawn white balls for thousands of times without fail, it would still be possible that some black balls lurked in the urn and would at last appear, so that our inferences could never be certain. On the other hand, if black balls came at intervals, I should expect that after a certain number of trials the future results would agree more or less closely with the past ones.

The mathematical solution of the question consists in nothing more than a close analysis of the mode in which our common sense proceeds. If twenty white balls have been drawn and no black ball, my common sense tells me that any hypothesis which makes the black balls in the urn considerable compared with the white ones is improbable; a preponderance of white balls is a more probable hypothesis, and as a deduction from this more

probable hypothesis, I expect a recurrence of white balls.
The mathematician merely reduces this process of thought
to exact numbers. Taking, for instance, the hypothesis
that there are 99 white and one black ball in the urn,
he can calculate the probability that 20 white balls
should be drawn in succession in those circumstances; he
thus forms a definite estimate of the probability of this
hypothesis, and knowing at the same time the probability
of a white ball reappearing if such be the contents of the
urn, he combines these probabilities, and obtains an exact
estimate that a white ball will recur in consequence of
this hypothesis. But as this hypothesis is only one out
of many possible ones, since the ratio of white and black
balls may be 98 to 2, or 97 to 3, or 96 to 4, and so on,
he has to repeat the estimate for every such possible
hypothesis. To make the method of solving the problem
perfectly evident, I will describe in the next section a
very simple case of the problem, originally devised for the
purpose by Condorcet, which was also adopted by Lacroix[a],
and has passed into the works of De Morgan, Lubbock,
and others.

Simple Illustration of the Inverse Problem.

Suppose it to be known that a ballot-box contains only
four black or white balls, the ratio of black and white balls
being unknown. Four drawings having been made with
replacement, and a white ball having appeared on each
occasion but one, it is required to determine the probability that a white ball will appear next time. Now the
hypotheses which can be made as to the contents of the
urn are very limited in number, and are at most the
following five :—

[a] 'Traité élémentaire du Calcul des Probabilités,' 3rd ed. (1833),
p. 148.

4	white	and	o	black	balls
3	„	„	I	„	„
2	„	„	2	„	„
I	„	„	3	„	„
o	„	„	4	„	„

The actual occurrence of black and white balls in the drawings renders the first and last hypotheses out of the question, so that we have only three left to consider.

If the box contains three white and one black, the probability of drawing a white each time is $\frac{3}{4}$, and a black $\frac{1}{4}$; so that the compound event observed, namely, three white and one black, has the probability $\frac{3}{4} \times \frac{3}{4} \times \frac{3}{4} \times \frac{1}{4}$, by the rule already given (p. 233). But as it is indifferent to us in what order the balls are drawn, and the black ball might come first, second, third, or fourth, we must multiply by four, to obtain the probability of three white and one black in any order, thus getting $\frac{27}{64}$.

Taking the next hypothesis of two white and two black balls in the urn, we obtain for the same probability the quantity $\frac{1}{2} \times \frac{1}{2} \times \frac{1}{2} \times \frac{1}{2} \times 4$, or $\frac{16}{64}$, and from the third hypothesis of one white and three black we deduce likewise $\frac{1}{4} \times \frac{1}{4} \times \frac{1}{4} \times \frac{3}{4} \times 4$, or $\frac{3}{64}$. According, then, as we adopt the first, second, or third hypothesis, the probability that the result actually noticed would follow is $\frac{27}{64}$, $\frac{16}{64}$, and $\frac{3}{64}$. Now it is certain that one or other of these hypotheses must be the true one, and their absolute probabilities are proportional to the probabilities that the observed events would follow from them (see p. 279). All we have to do, then, in order to obtain the absolute probability of each hypothesis, is to alter these fractions in a uniform ratio, so that their sum shall be unity, the expression of certainty. Now since 27 + 16 + 3 = 46, this will be effected by dividing each fraction by 46 and multiplying by 64. Thus the probabilities of the first, second, and third hypotheses are respectively—

$$\frac{27}{46}, \quad \frac{16}{46}, \quad \frac{3}{46}.$$

The inductive part of the problem is now completed, since we have found that the urn most likely contains three white and one black ball, and have assigned the exact probability of each possible supposition. But we are now in a position to resume deductive reasoning, and infer the probability that the next drawing will yield, say a white ball. For if the box contains three white and one black ball, the probability of drawing a white one is certainly $\frac{3}{4}$; and as the probability of the box being so constituted is $\frac{27}{46}$, the compound probability that the box will be so filled and will give a white ball at the next trial, is

$$\frac{27}{46} \times \frac{3}{4} \text{ or } \frac{81}{184}.$$

Again, the probability is $\frac{16}{46}$ that the box contains two white and two black, and under those conditions the probability is $\frac{1}{2}$ that a white ball will appear; hence the probability that a white ball will appear in consequence of that condition, is

$$\frac{16}{46} \times \frac{1}{2} \text{ or } \frac{32}{184}.$$

From the third supposition we get in like manner the probability

$$\frac{3}{46} \times \frac{1}{4} \text{ or } \frac{3}{184}.$$

Now since one and not more than one hypothesis can be true, we may add together these separate probabilities, and we find that

$$\frac{81}{184} + \frac{32}{184} + \frac{3}{184} \text{ or } \frac{116}{184}$$

is the complete probability that a white ball will be next drawn under the conditions and data supposed.

General Solution of the Inverse Problem.

In the instance of the inverse method described in the last section, a very few balls were supposed to be in the ballot-box for the purpose of simplifying the calculation. In order that our solution may apply to natural phenomena, we must render our hypothesis as little arbitrary as possible. Having no *à priori* knowledge of the conditions of the phenomena in question, there is no limit to the variety of hypotheses which might be suggested. Mathematicians have therefore had recourse to the most extensive suppositions which can be made, namely, that the ballot-box contains an infinite number of balls; they have thus varied the proportion of white balls to black balls continuously, from the smallest to the greatest possible proportion, and estimated the aggregate probability which results from this comprehensive supposition.

To explain their procedure, let us imagine that, instead of an infinite number, the ballot-box contained a large finite number of balls, say 1000. Then the number of white balls might be 1 or 2 or 3 or 4, and so on, up to 999. Supposing that three white and one black ball have been drawn from the urn as before, there is a certain very small probability that this would have occurred in the case of a box containing one white and 999 black balls; there is also a small probability that from such a box the next ball would be white. Compound these probabilities, and we have the probability that the next ball really will be white, in consequence of the existence of that proportion of balls. If there be two white and 998 black balls in the box, the probability is greater, and will increase until the balls are supposed to be in the proportion of those drawn. Now 999 different hypotheses are possible, and the calculation is to be made for each of these, and their aggregate taken as the final result. It is apparent that as the number of balls in the box is increased, the absolute probability of any one hypothesis concerning the exact proportion of balls is decreased, but the aggregate results of all the hypotheses will assume the character of a wide average.

When we take the step of supposing the balls within the urn to be infinite in number, the possible proportions of white and black balls also become infinite, and the probability of any one proportion actually existing is infinitely small. Hence the final result that the next ball drawn will be white is really the sum of an infinite number of infinitely small quantities. It might seem, indeed, utterly impossible to calculate out a problem having an infinite number of hypotheses, but the wonderful resources of the integral calculus enable this to be done with far greater facility than if we supposed any large finite number of balls, and then actually computed the results. I will not attempt to describe the processes by which Laplace finally accomplished the complete solution of the problem. They are to be found described in several English works, especially De Morgan's 'Treatise on Probabilities,' in the 'Encyclopædia Metropolitana,' and Mr. Todhunter's 'History of the Theory of Probability.' The abbreviating power of mathematical analysis was never more strikingly shown. But I may add that though the integral calculus is employed as a means of summing infinitely numerous results, we in no way abandon the principles of combinations already treated. We calculate the values of infinitely numerous factorials, not, however, obtaining their actual products, which would lead to an infinite number of figures, but obtaining the final answer to the problem by devices which can only be comprehended after study of the integral calculus.

It must be allowed that the hypothesis adopted by Laplace is in some degree arbitrary, so that there was some opening for the doubt which Boole has cast upon it[b]. But it may be replied, (1) that the supposition of an infinite number of balls treated in the manner of Laplace is less arbitrary and more comprehensive than any other that could be suggested. (2) The result does not differ much from that which would be obtained on the hypothesis of any very large finite number of balls. (3) The supposition leads to a series of simple formulæ which can be applied with ease in many cases, and which bear all the appearance of truth so far as it can be independently judged by a sound and practiced understanding.

Rules of the Inverse Method.

By the solution of the problem, as described in the last section, we obtain the following series of simple rules.

1. *To find the probability that an event which has not hitherto been observed to fail will happen once more, divide the number of times the event has been observed increased by one, by the same number increased by two.*

If there have been m occasions on which a certain event might have been observed to happen, and it has happened on all those occasions, then the probability that it will happen on the next occasion of the same kind is $\frac{m+1}{m+2}$. For instance, we may say that there are nine places in the planetary system where planets might exist obeying Bode's law of distance, and in every place there is a planet obeying the law more or less exactly, although no reason is known for the coincidence. Hence the probability that the next planet beyond Neptune will conform to the law is $\frac{10}{11}$.

2. *To find the probability that an event which has not hitherto failed will not fail for a certain number of new occasions, divide the number of times the event has hap-*

pened increased by one, by the same number increased by one and the number of times it is to happen.

An event having happened m times without fail, the probability that it will happen n more times is $\dfrac{m+1}{m+n+1}$. Thus the probability that three new planets would obey Bode's law is $\frac{10}{13}$, but it must be allowed that this, as well as the previous result, would be much weakened by the fact that Neptune can barely be said to obey the law.

3. *An event having happened and failed a certain number of times, to find the probability that it will happen the next time, divide the number of times the event has happened increased by one, by the whole number of times the event has happened or failed increased by two.*

Thus, if an event has happened m times and failed n times, the probability that it will happen on the next occasion is $\dfrac{m+1}{m+n+2}$.

Thus, if we assume that of the elements yet discovered 50 are metallic and 14 non-metallic, then the probability that the next element discovered will be metallic is $\frac{51}{66}$.

Again since of 37 metals which have been sufficiently examined only four, namely, sodium, potassium, lanthanum and lithium, are of less density than water, the probability that the next metal examined or discovered will be less dense than water is $\dfrac{4+1}{37+2}$ or $\dfrac{5}{39}$.

We may state the results of the method in a more general manner thus,—If under given circumstances certain events A, B, C, &c., have happened respectively m, n, p, &c., times, and one or other of these events must happen, then the probabilities of these events are proportional to $m+1$, $n+1$, $p+1$, &c., so that the probability of A will be $\dfrac{m+1}{m+1+n+1+p+1+ \&c.}$.   But if new events may happen in addition to those which have been observed, we must assign unity for the probability of such new event. The proportional probabilities then become 1 for a new event, $m + 1$ for A, $n + 1$ for B, and so on, and the absolute probability of A is $\dfrac{m + 1}{1 + m + 1 + n + 1 + \&c.}$[c]

It is very interesting to trace out the variations of probability according to these rules under diverse circumstances. Thus the first time a casual event happens it is 1 to 1, or as likely as not that it will happen again; if it does happen it is 2 to 1 that it will happen a third time; and on successive occasions of the like kind the odds become 3, 4, 5, 6, &c., to 1. The odds of course will be discriminated from the probabilities which are successively $\frac{1}{2}$, $\frac{1}{3}$, $\frac{1}{4}$, &c. Thus on the first occasion on which a person sees a shark, and notices that it is accompanied by a little pilot fish, the odds are 1 to 1, or the probability $\frac{1}{2}$, that the next shark will be so accompanied.

When an event has happened a very great number of times, its happening once again approaches nearly to certainty. Thus if we suppose the sun to have risen demonstratively one thousand million times, the probability that it will rise again, on the ground of this knowledge merely, is $\dfrac{1,000,000,000 + 1}{1,000,000,000 + 1 + 1}$. But then the probability that it will continue to rise for as long a period as we know it to have risen is only $\dfrac{1,000,000,000 + 1}{2,000,000,000 + 1}$, or almost exactly $\frac{1}{2}$. The probability that it will continue so rising a thousand times as long is only about $\frac{1}{1001}$. The lesson which we may draw from these figures is quite that which we should adopt on other grounds, namely that experience never affords certain knowledge, and that it is exceedingly improbable that events will always happen as we observe

c De Morgan's 'Essay on Probabilities,' Cabinet Cyclopædia, p. 67.

them.   Inferences pushed far beyond their data soon lose
any considerable probability.   De Morgan has said[d], 'No
finite experience whatsoever can justify us in saying that
the future shall coincide with the past in all time to come,
or that there is any probability for such a conclusion.'   On
the other hand, we gain the assurance that experience
sufficiently extended and prolonged will give us the
knowledge of future events with an unlimited degree of
probability, provided indeed that those events are not
subject to arbitrary interference.

It must be clearly understood that these probabilities are
only such as arise from the mere happening of the events,
irrespective of any knowledge derived from other sources
concerning those events or the general laws of nature.
All our knowledge of nature is indeed founded in like
manner upon observation, and is therefore only probable.
The law of gravitation itself is only probably true.   But
when a number of different facts, observed under the most
diverse circumstances, are found to be harmonized under a
supposed law of nature, the probability of the law approximates closely to certainty.   Each science rests upon so
many observed facts, and derives so much support from
analogies or direct connections with other sciences, that
there are comparatively few cases where our judgment of
the probability of an event depends entirely upon a few
antecedent events, disconnected from the general body of
physical science.

Events may often again exhibit a regularity of succession or preponderance of character, which the simple
formula will not take into account.   For instance, the
majority of the elements recently discovered are metals,
so that the probability of the next discovery being that of
a metal, is doubtless greater than we calculated (p. 298).
At the more distant parts of the planetary system, there

---

[d] 'Treatise on Probability,' Cabinet Cyclopædia, p. 128.

are symptoms of disturbance which would prevent our placing much reliance on any inference from the prevailing order of the known planets to those undiscovered ones which may possibly exist at great distances. These and all like complications in no way invalidate the theoretic truth of the formulæ, but render their sound application much more difficult.

Erroneous objections have been raised to the theory of probability, on the ground that we ought not to trust to our *à priori* conceptions of what is likely to happen, but should always endeavour to obtain precise experimental data to guide us[e]. This course, however, is perfectly in accordance with the theory, which is our best and only guide, whatever data we possess. We ought to be always applying the inverse method of probabilities so as to take into account all additional information. When we throw up a coin for the first time, we are probably quite ignorant whether it tends more to fall head or tail upwards, and we must therefore assume the probability of each event as $\frac{1}{2}$. But if it shows head, for instance, in the first throw, we now have very slight experimental evidence in favour of a tendency to show head. The chance of two heads is now slightly greater than $\frac{1}{4}$, which it appeared to be at first[f], and as we go on throwing the coin time after time, the probability of head appearing next time constantly varies in a slight degree according to the character of our previous experience. As Laplace remarks, we ought always to have regard to such considerations in common life. Events when closely scrutinized will hardly ever prove to be quite independent, and the slightest preponderance one way or the other is some evidence of connexion, and in the absence of better evidence should be taken into account.

e J. S. Mill, 'System of Logic,' 5th Edition, bk. iii. chap. xviii. § 3.

f Todhunter's 'History,' pp. 472, 598.

The grand object of seeking to estimate the probability
of future events from past experience, seems to have been
entertained by James Bernouilli and De Moivre, at least
such was the opinion of Condorcet; and Bernouilli may be
said to have solved one case of the problem[g]. The English
writers Bayes and Price are, however, undoubtedly the
first who put forward any distinct rules on the subject[h].
Condorcet and several other eminent mathematicians advanced the mathematical theory of the subject; but it was
reserved to the immortal Laplace to bring to the subject
the full power of his genius, and carry the solution of the
problem almost to perfection. It is instructive to observe
that a theory which arose from the consideration of the
most petty games of chance, the rules and the very names
of which are in many cases forgotten, gradually advanced,
until it embraced the most sublime problems of science,
and finally undertook to measure the value and certainty
of all our inductions.

Fortuitous Coincidences.

We should have studied the theory of probability to
very little purpose, if we thought that it would furnish
us with an infallible guide. The theory itself points out
the possibility, or rather the approximate certainty, that
we shall sometimes be deceived by extraordinary, but
fortuitous coincidences. There is no run of luck so extreme that it may not happen, and it may happen to us,
or in our time, as well as to other persons or in other
times. We may be forced by all correct calculation to
refer such coincidences to some necessary cause, and yet
we may be deceived. All that the calculus of probability

g Todhunter's 'History,' pp 378, 79.
h 'Philosophical Transactions' [1763], vol. liii. p. 370, and [1764],
vol. liv. p. 296. Todhunter, pp. 294-300.

pretends to give, is *the result in the long run*, as it is called, and this really means in *an infinity of cases*. During any finite experience, however long, chances may be against us. Nevertheless the theory is the best guide we can have. If we always think and act according to its well interpreted indications, we shall have the best chance of escaping error; and if all persons, throughout all time to come, obey the theory in like manner, they will undoubtedly thereby reap the greatest advantage.

No rule can be given for descriminating between coincidences which are casual and those which are the effect of law or common conditions. By a fortuitous or casual coincidence, we mean an agreement between events, which nevertheless arise from wholly independent and different causes or conditions, and which will not always so agree. It is a fortuitous coincidence, if a penny thrown up repeatedly in various ways always falls on the same side; but it would not be fortuitous if there were any similarity in the motions of the hand, and the height of the throw, so as to cause or tend to cause a uniform result. Now among the infinitely numerous events, objects, or relations in the universe, it is quite likely that we shall occasionally notice casual coincidences. There are seven intervals in the octave, and there is nothing very improbable in the colours of the spectrum happening to be apparently divisible into the same or similar series of seven intervals. It is hardly yet decided whether this apparent coincidence, with which Newton was much struck, is well founded or not[i], but the question will probably be decided in the negative.

It is certainly a casual coincidence which the ancients noticed between the seven vowels, the seven strings of the lyre, the seven Pleiades, and the seven chiefs at Thebes[k].

[i] 'Nature,' vol. i. p. 286.
[k] Aristotle's 'Metaphysics,' xiii. 6. 3.

The accidents connected with the number seven have misled the human intellect throughout the historical period. Pythagoras imagined a connection between the seven planets, and the seven intervals of the monochord. The alchemists were never tired of drawing inferences from the coincidence in numbers of the seven planets and the seven metals, not to speak of the seven days of the week.

A singular circumstance was pointed out concerning the dimensions of the earth, sun, and moon; the sun's diameter was almost exactly 110 times as great as the earth's diameter, while in almost exactly the same ratio the mean distance of the earth was greater than the sun's diameter, and the mean distance of the moon from the earth was greater than the moon's diameter[1]. The agreement was so close that it might have proved more than casual, but its fortuitous character is sufficiently shown by the fact, that the coincidence ceases to be remarkable when we adopt the amended dimensions of the planetary system.

A considerable number of the elements have atomic weights, which are apparently exact multiples of that of hydrogen. If this be not a law to be ultimately extended to all the elements, as supposed by **Prout**, it is a most remarkable coincidence. But, as I have observed, we have no means of absolutely discriminating accidental coincidences from those which imply a deep producing cause. A coincidence must either be very strong in itself, or it must be corroborated by some explanation or connection with other laws of nature. Little attention was ever given to the coincidence concerning the dimensions of the sun, earth, and moon, because it was not very strong in itself, and had no apparent connexion with the

---

[1] Chambers's 'Astronomy,' 1st. ed. p. 23.

principles of physical astronomy. Prout's Law bears more probability because it would bring the constitution of the elements themselves in close connexion with the atomic theory, representing them as built up out of a simpler substance.

In historical and social matters, coincidences are frequently pointed out which are due to chance, although there is always a strong popular tendency to regard them as the work of design, or as having some hidden cause. It has been pointed out that if to 1794, the number of the year in which Robespierre fell, we add the sum of its digits, the result is 1815, the year in which Napoleon fell; the repetition of the process gives 1830, the year in which Charles the Tenth abdicated. Again, the French Chamber of Deputies, in 1830, consisted of 402 members, of whom 221 formed the party called, 'La queue de Robespierre,' while the remainder, 181 in number, were named 'Les honnêtes gens.' If we give to each letter a numerical value corresponding to its place in the alphabet, it will be found that the sum of the values of the letters in each name exactly indicates the number of the party[m].

A number of such coincidences, often of a very curious character, might be adduced, and the probability against the occurrence of each may be enormously great. They must be attributed to chance, because they cannot be shown to have the slightest connexion with the general laws of nature ; but persons are often found to be greatly influenced by such coincidences, regarding them as evidence of fatality, that is of a system of causation governing human affairs independently of the ordinary laws of nature. Let it be remembered that there are an infinite number of opportunities in life for some strange coincidence to present itself, so that it is quite to be expected that remarkable conjunctions will sometimes happen.

[m] S. B. Gould's 'Curious Myths,' p. 222.

X

In all matters of judicial evidence, we must bear in mind the necessary occurrence from time to time of unaccountable coincidences. The Roman jurists refused for this reason to invalidate a testamentary deed, the witnesses of which had sealed it with the same seal. For witnesses independently using their own seals might be found to possess identical ones by accident[n]. It is well known that circumstantial evidence of apparently overwhelming completeness will sometimes lead to a mistaken judgment, and as absolute certainty is never really attainable, every court must act upon probabilities of a very high amount, and in a certain small proportion of cases they must almost of necessity condemn the innocent victims of a remarkable conjuncture of circumstances[o]. Popular judgments usually turn upon probabilities of far less amount, as when the palace of Nicomedia, and even the bedchamber of Diocletian, having been on fire twice within fifteen days, the people entirely refused to believe that it could be the result of accident. The Romans believed that there was a fatality connected with the name of Sextus.

'Semper sub Sextis perdita Roma fuit.'

The utmost precautions will not provide against all contingencies. To avoid errors in important calculations, it is usual to have them repeated by different computers, but a case is on record in which three computers made exactly the same calculations of the place of a star, and yet all did it wrong in precisely the same manner, for no apparent reason[p].

---

[n] Possunt autem omnes testes et uno annulo signare testamentum. Quid enim si septem annuli una sculptura fuerint, secundum quod Pomponio visum est?—'Justinian,' ii. tit. x. 5.

[o] See Wills on 'Circumstantial Evidence,' p. 148.

[p] 'Memoirs of the Royal Astronomical Society,' vol. iv. p. 290, quoted by Lardner, 'Edinburgh Review,' July 1834, p. 278.

Summary of the Theory of Inductive Inference.

The theory of inductive inference adopted in this and the previous chapter, was chiefly suggested by the study of the Inverse Method of Probabilities, but it also bears much resemblance to the so-called Deductive Method described by Mr. J. S. Mill, in his well known 'System of Logic[q].' Mr. Mill's views concerning the Deductive Method, probably form the most original and valuable part of his treatise, and I should have ascribed the doctrine entirely to him, had I not found that the opinions put forward in other parts of his work are entirely inconsistent with the theory here upheld. As this subject is the most important and difficult one with which we have to deal, I will try to remedy the imperfect manner in which I have treated it, by giving a brief recapitulation of the views adopted.

All inductive reasoning is but an inverse application of deductive reasoning. Being in possession of certain particular facts or events expressed in propositions, we imagine some more general proposition expressing the existence of a law or cause; and, deducing the particular results of that supposed general proposition, we observe whether they agree with the facts in question. Hypothesis is thus always employed, consciously or unconsciously. The sole conditions to which we need conform in framing any hypothesis is, that we both have and exercise the power of inferring deductively from the hypothesis, to the particular logical combinations or results, which are to be compared with the known facts. Thus there are but three steps in the process of induction :—

(1) Framing of some hypothesis as to the character of the general law.

(2) Deducing consequences from that law.

(3) Observing whether the consequences agree with the particular facts under consideration.

In very simple cases of inverse reasoning, hypothesis may sometimes seem altogether needless. Thus, to take numbers again as a convenient illustration, I have only to look at the series,

$$1, \ 2, \ 4, \ 8, \ 16, \ 32, \ \&c.,$$

to know at once that the general law is that of geometrical progression ; I need no successive trial of various hypotheses, because I am familiar with the series, and have long since learnt from what general formula it proceeds. In the same way a mathematician becomes acquainted with the integrals of a number of common formulæ, so that we have no need to go through any process of discovery. But it is none the less true that whenever previous reasoning does not furnish the knowledge, hypotheses must be framed and tried. (See p. 142.)

There naturally arise two different cases, according as the nature of the subject admits of certain or only probable deductive reasoning. Certainty, indeed, is but a singular case of probability, and the general principles of procedure are always the same. Nevertheless, when certainty of inference is possible the process is simplified. Of several mutually inconsistent hypotheses, the results of which can be certainly compared with fact, but one hypothesis can ultimately be entertained. Thus in the inverse logical problem, two logically distinct conditions could not yield the same series of possible combinations. Accordingly in the case of two terms we had to choose one of seven different kinds of propositions, or in the case of three terms, our choice lay among 192 possible distinct hypotheses (pp. 154–164). Natural laws, however, are often quantitative in character, and the possible hypotheses are then infinite in variety.

When deduction is certain, comparison with fact is needed only to assure ourselves that we have rightly selected the hypothetical conditions. The law establishes itself, and no number of particular verifications can add to its probability. Having once deduced from the principles of algebra that the difference of the squares of two numbers is equal to the product of their sum and difference, no number of particular trials of its truth will render it more certain. On the other hand, no finite number of particular verifications of a supposed law will render that law certain. In short, certainty belongs only to the deductive process, and to the teachings of direct intuition; and as the conditions of nature are not given by intuition, we can only be certain that we have got a correct hypothesis when, out of a limited number conceivably possible, we select that one which alone agrees with the facts to be explained.

In geometry and kindred branches of mathematics, deductive reasoning is conspicuously certain, and it would often seem as if the consideration of a single diagram yields us certain knowledge of a general proposition. But in reality all this certainty is of a purely hypothetical character. Doubtless if we could ascertain that a supposed circle was a true and perfect circle, we could be certain concerning a multitude of its geometrical properties. But geometrical figures are physical objects, and the senses can never assure us as to their exact forms. The figures really treated in Euclid's 'Elements' are imaginary, and we never can verify in practice the conclusions which we draw with certainty in inference; questions of degree and probability enter.

Passing now to subjects in which deduction is only probable, it ceases to be possible to adopt one hypothesis to the exclusion of the others. We must entertain at the same time all conceivable hypotheses, and regard each

with the degree of esteem proportionate to its probability. We go through the same steps as before.

(1) We frame an hypothesis.

(2) We deduce the probability of various series of possible consequences.

(3) We compare the consequences with the particular
facts, and observe the probability that such facts would
happen under the hypothesis.

The above processes must be performed for every conceivable hypothesis, and then the absolute probability of
each will be yielded by the principle of the inverse
method (p. 279). As in the case of certainty we accept
that hypothesis which certainly gives the required results,
so now we accept as most probable that hypothesis which
most probably gives the results; but we are obliged to
entertain at the same time all other hypotheses with
degrees of probability proportionate to the probabilities
that they would give the results.

So far we have treated only of the process by which
we pass from special facts to general laws, that inverse
application of deduction which constitutes induction.
But the direct employment of deduction is often combined with the inverse. No sooner have we established
a general law, than the mind rapidly draws other particular
consequences from it. In geometry we may almost seem
to infer that *because* one equilateral triangle is equiangular, therefore another is so. In reality it is not
because one is that another is, but because all are. The
geometrical conditions are perfectly general, and by what is
sometimes called *parity of reasoning* whatever is true of
one equilateral triangle, so far as it is equilateral, is true
of all equilateral triangles.

Similarly, in all other cases of inductive inference,
where we seem to pass from some particular instances to
a new instance, we go through the same process. We

form an hypothesis as to the logical conditions under which the given instances might occur; we calculate inversely the probability of that hypothesis, and compounding this with the probability that a new instance would proceed from the same conditions, we gain the absolute probability of occurrence of the new instance in virtue of this hypothesis. But as several, or many, or even an infinite number of mutually inconsistent hypotheses may be possible, we must repeat the calculation for each such conceivable hypothesis, and then the complete probability of the future instance will be the sum of the separate probabilities. The complication of this process is often very much reduced in practice, owing to the fact that one hypothesis may be nearly certainly true, and other hypotheses, though conceivable, may be so improbable as to be neglected without appreciable error. But when we possess no knowledge whatever of the conditions from which the events proceed, we may be unable to form any probable hypotheses as to their mode of origin. We have now to fall back upon the general solution of the problem effected by Laplace, which consists in admitting on an equal footing every conceivable ratio of favourable and unfavourable chances for the production of the event, and then accepting the aggregate result as the best which can be obtained. This solution is only to be accepted in the absence of all better means, but like other results of the calculus of probabilities, it comes to our aid where knowledge is at an end and ignorance begins, and it prevents us from over-estimating the knowledge we possess. The general results of the solution are in accordance with common sense, namely, that the more often an event has happened the more probable, as a general rule, is its subsequent occurrence. With the extension of experience this probability indefinitely increases, but at the same time the probability is slight that events will long continue to happen as they have previously happened.

We have now pursued the theory of inductive inference, as far as can be done with regard to simple logical or numerical relations. The laws of nature deal with time and space, which are indefinitely, or rather infinitely, divisible. As we passed from pure logic to numerical logic, so we must now pass from questions of discontinuous, to questions of continuous quantity, encountering fresh considerations of much difficulty. Before, therefore, we consider how the great inductions and generalizations of physical science illustrate the views of inductive reasoning just explained, we must break off for a time, and review the means which we possess of measuring and comparing magnitudes of time, space, mass, force, momentum, energy, and the various manifestations of energy in motion, heat, electricity, chemical change, and the other phenomena of nature.

BOOK III.

CHAPTER XIII.

THE EXACT MEASUREMENT OF PHENOMENA.

As physical science advances, it becomes more and more accurately quantitative. Questions of simple logical fact after a time resolve themselves into questions of degree, time, distance, or weight. Forces hardly suspected to exist by one generation, are clearly recognised by the next, and precisely measured by the third generation. But one condition of this rapid advance is the invention of suitable instruments of measurement. We need what Francis Bacon called *Instantiæ citantes*, or *evocantes*, methods of rendering minute phenomena perceptible to the senses; and we also require *Instantiæ radii* or *curriculi*, that is measuring instruments[a]. Accordingly, the introduction of a new instrument often forms an epoch in the history of science. As Davy said, 'Nothing tends so much to the advancement of knowledge as the application of a new instrument. The native intellectual powers of men in different times, are not so much the causes of the different success of their labours, as the peculiar nature of the means and artificial resources in their possession[b]'.

In the absence indeed of advanced theory and analyti-

---

[a] 'Novum Organum,' bk. ii. Aphorisms 40, 45 and 46.
[b] 'Chemical Philosophy,' Works, vol. iv. p. 39. Quoted by Young, Works, vol. i. p. 576.

cal power, a very precise instrument would be useless. Measuring apparatus and mathematical theory should advance *pari passu*, and with just such precision as the theorist can anticipate results, the experimentalist should be able to compare them with experience. The laborious and scrupulously accurate observations of Flamsteed, were the proper complement to the intense mathemetical powers of Newton.

Every branch of knowledge commences with quantitative notions of a very rude character. After we have far progressed, it is often amusing to look back into the infancy of the science, and contrast present with past methods. At Greenwich Observatory in the present day, the hundredth part of a second is not thought an inconsiderable portion of time. The ancient Chaldæans recorded an eclipse to the nearest hour, and even the early Alexandrian astronomers thought it superfluous to distinguish between the edge and centre of the sun. By the introduction of the astrolabe, Ptolemy and the later Alexandrian astronomers could determine the places of the heavenly bodies within about ten minutes of arc. But little progress then ensued for thirteen centuries, until Tycho Brahe made the first great step towards accuracy, not only by employing better instruments, but even more by ceasing to regard an instrument as correct. Tycho, in fact, determined the errors of his instruments, and corrected his observations. He also took notice of the effects of atmospheric refraction, and succeeded in attaining an accuracy often sixty times as great as that of Ptolemy. Yet Tycho and Hevelius often erred several minutes in the determination of a star's place, and it was a great achievement of Rœmer and Flamsteed to reduce this error to seconds. Bradley, the modern Hipparchus, carried on the improvement, his errors in right ascension being under one second of time, and those of declination under four seconds of arc according to Bessel.

In the present day the average error of a single observation is probably reduced to the half or quarter of what it
was in Bradley's time; and further extreme accuracy is
attained by the multiplication of observations, and their
skilful combination according to the theory of error.

Some of the more important constants, for instance that
of nutation, have been determined within the tenth part
of a second of space[c].

It would be a matter of great interest to trace out the
dependence of this vast progress upon the introduction of
new instruments. The astrolabe of Plotemy, the telescope of Galileo, the pendulum of Galileo and Huygens,
the micrometer of Horrocks, and the telescopic sights and
micrometer of Gascoygne and Picard, Rœmer's transit instrument, Newton's and Hadley's quadrant, Dollond's
achromatic lenses, Harrison's chronometer, and Ramsden's
dividing engine—such were some of the principal additions to astronomical apparatus. The result is, that we
now take note of quantities, 300,000 or 400,000 times as
small as in the time of the Chaldæans.

It would be interesting again to compare the scrupulous
accuracy of a modern trigonometrical survey with Eratosthenes' rude but ingenious guess at the difference of latitude between Alexandria and Syene—or with Norwood's
measurement of a degree of latitude in 1635. ' Sometimes
I measured, sometimes I paced,' said Norwood; ' and I
believe I am within a scantling of the truth.' Such was
the germ of those elaborate geodesical measurements
which have made the dimensions of the globe known to
us within a few hundred yards.

In other branches of science, the invention of an instrument has usually marked, if it has not made, an epoch.
The science of heat might be said to commence with the

[c] Baily, 'British Association Catalogue of Stars,' pp. 7, 23.

construction of the thermometer, and it has recently been advanced by the introduction of the thermo-electric pile. Chemistry has been created chiefly by the careful use of the balance, which forms a unique instance of an instrument remaining substantially in the form in which it was first applied to scientific purposes by Archimedes. The balance never has been and probably never can be improved, except in details of construction. On the other hand, the torsion balance, introduced by Coulomb towards the end of last century, has rapidly become essential in many branches of investigation. In the hands of Cavendish and Baily, it gave a determination of the earth's density; applied in the galvanometer, it gave a delicate measure of electrical forces, and was essential to the introduction of the thermo-electric pile. This balance is made by simply suspending any light rod by a thin wire or thread attached to the middle point. And we owe to it almost all the more delicate investigations in the theories of heat, electricity, and magnetism.

Though we can now take note of the millionth of an inch in space, and the millionth of a second in time, we must not overlook the fact that in other operations of science we are yet in the position of the Chaldæans. Not many years have elapsed since the magnitudes of the stars, meaning the amount of light they send to the observer's eye, were guessed at in the rudest manner, and the astronomer adjudged a star to this or that order of magnitude by a rough comparison with other stars of the same order. To the late Sir John Herschel we owe an attempt to introduce an uniform method of measurement and expression, bearing some relation to the real photometric magnitudes of the stars[d]. Previous to the re-

---

[d] 'Outlines of Astronomy,' 4th ed. sect. 781, p. 522. 'Results of Observations at the Cape of Good Hope,' &c., p. 371.

searches of Bunsen and Roscoe on the chemical action of light, we were absolutely devoid of any mode of measuring the energy of light; even now the methods are tedious, and it is not clear that they give the energy of light so much as one of its special effects. Many natural phenomena have hardly yet been made the subject of measurement at all, such as the intensity of sound, the phenomena of taste and smell, the magnitude of atoms, the temperature of the electric spark or of the sun's photosphere.

To suppose, then, that quantitative science treats only of exactly measurable quantities, is a gross if it be a common mistake. Whenever we are treating of an event which either happens altogether or does not happen at all, we are engaged with a non-quantitative phenomenon, a matter of fact, not of degree; but whenever a thing may be greater or less, or twice or thrice as great as another, whenever, in short, ratio enters even in the rudest manner, there science will have a quantitative character. There can be little doubt, indeed, that every science as it progresses will become gradually more and more quantitative. Numerical precision is doubtless the very soul of science, as Herschel said[e], and as all natural objects exist in space, and involve molecular movements, measurable in velocity and extent, there is no apparent limit to the ultimate extension of quantitative science. But the reader must not for a moment suppose that, because we depend more and more upon mathematical methods, we leave logical methods behind us. Number, as I have endeavoured to show, is logical in its origin, and quantity is but a development of number, or is analogous thereto.

[e] 'Preliminary Discourse on the Study of Natural Philosophy,' p. 122.

Division of the Subject.

The general subject of quantitative investigation will have to be divided into several parts. We shall, firstly, consider the means at our disposal for measuring phenomena, and thus rendering them more or less amenable to mathematical treatment. This task will involve an analysis of the principles on which accurate methods of measurement are founded, forming the subject of the remainder of the present chapter. As measurement, however, only yields ratios, we have in the next chapter (XIV) to consider the establishment of unitary magnitudes, in terms of which our results may be expressed. As every phenomenon is usually the sum of several distinct quantities proceeding from different causes, we have next to investigate in Chapter XV the methods by which we may disentangle complicated effects, and refer each part of the joint effect to its separate cause.

It yet remains for us in subsequent chapters to treat of quantitative induction, properly so called. We must follow out the inverse logical method, as it presents itself in problems of a far higher degree of difficulty than those which treat of objects related in a simple logical manner, and incapable of merging into each other by addition and subtraction.

Continuous Quantity.

The phenomena of nature are for the most part manifested in quantities which increase or decrease continuously. When we inquire into the precise meaning of continuous quantity, we find that it can only be described as that which is divisible without limit. We can divide a millemetre into ten, or one hundred, or one thousand, or ten thousand parts, and mentally at any rate we can carry

on the process *ad infinitum*. Any finite space, then, must
be conceived as made up of an infinite number of parts,
each of which must consequently be infinitely small. We
cannot entertain some of the simplest geometrical notions
without allowing this. The conception of a square involves the conception of a side and diagonal, which, as
Euclid admirably proves in the 117th proposition of his
tenth book, have no common measure[f], meaning, as I
apprehend, no finite common measure. Incommensurable
quantities are, in fact, those which have for their only
common measure an infinitely small quantity. It is
somewhat startling to find, too, that in theory incommensurable quantities will be infinitely more frequent than
commensurable. Let any two lines be drawn haphazard ;
it is infinitely unlikely that they will be commensurable,
so that the commensurable quantities, which we are supposed to deal with in practice, are but singular cases
among an infinitely greater number of incommensurable
cases.

Practically, however, we treat all quantities as made up
of the least quantities which our senses, assisted by the
best measuring instruments, can appreciate. So long as
microscopes were uninvented, it was sufficient to regard
an inch as made up of a thousand thousandths of an
inch ; now we must treat it as composed of a million
millionths. We might apparently avoid all mention of
infinitely small quantities, by never carrying our approximations beyond quantities, which the senses can appreciate.
In geometry, as thus treated, we should never assert two
quantities to be equal, but only to be *apparently* equal.
Legendre really adopts this mode of treatment in the
twentieth proposition of the first book of his Geometry ;
and it is practically adopted throughout the physical
sciences, as we shall afterwards see. But though our

[f] See De Morgan, 'Study of Mathematics,' in U. K. S. Library, p. 81.

fingers, and senses, and instruments must stop somewhere, there is no reason why the mind should not go on. We can see that a proof which is only carried through a few steps, in fact, might be carried on without limit, and it is this consciousness of no stopping place, which renders Euclid's proof of his 117th proposition so impressive. Try how we will to circumvent the matter, we cannot really avoid the consideration of the infinitely small and the infinitely great. The same methods of approximation which seem confined to the finite, mentally extend themselves to the infinite[g].

One result which immediately follows from these considerations is, that we cannot possibly adjust any two quantities in absolute equality. The suspension of Mahomet's coffin between two precisely equal magnets, is theoretically conceivable but practically impossible. The story of the 'Merchant of Venice,' turns upon the infinite improbability, that an exact quantity of flesh could be cut. Unstable equilibrium cannot exist in nature, for it is that which is destroyed by an infinitely small displacement. It might be possible to balance an egg on its end practically, because no egg has a surface of perfect curvature. Suppose the egg shell to be perfectly smooth, and the feat would become impossible.

The Fallacious Indications of the Senses.

I may briefly remind the reader how little we can trust to our unassisted senses in estimating the degree, quantity, or magnitude of any phenomenon. The eye cannot correctly estimate the comparative brightness of two luminous bodies which differ much in brilliancy; for we know that the iris is constantly adjusting itself to the intensity

[g] Lacroix, 'Essai sur l'Enseignement ou manière d'étudier les Mathématiques,' 2nd ed. Paris, 1816, pp. 292–294.

of the light received, and thus admits more or less light
according to circumstances. The moon which shines with
almost dazzling brightness by night, is pale and nearly
imperceptible while the eye is yet affected by the vastly
more powerful light of day. Much has been recorded
concerning the comparative brightness of the zodiacal
light at different times [h], but it would be difficult to prove
that these changes are not due to the varying darkness
at the time, or the different acuteness of the observer's
eye. For a like reason it is exceedingly difficult to establish the existence of any change in the form or comparative brightness of nebulæ; the appearance of a nebula
greatly depends upon the keenness of sight of the observer, or the accidental condition of freshness or fatigue
of his eye; the same is true of lunar observations [i]; and
even the use of the best telescope fails to remedy this
difficulty. In judging of colours again, we must remember
that light of any given colour tends to dull the sensibility
of the eye for light of the same colour.

Nor is the eye when unassisted by instruments a much
better judge of magnitude. Our estimates of the size of
minute bright points, such as the fixed stars, are completely falsified by the effects of irradiation. Tycho calculated from the apparent size of the star-discs, that no
one of the principal fixed stars could be contained within
the area of the earth's orbit. Apart, however, from irradiation or other distinct causes of error, our visual estimates
of sizes and shapes are often astonishingly incorrect.
Artists almost invariably draw distant mountains or other
objects in ludicrous disproportion to nearer objects, as a
comparison of a sketch with a photograph at once shows.
The extraordinary apparent difference of size of the sun

[h] 'Cosmos,' Translated by Otté, vol. i. pp. 131–134.

[i] 'Report of the British Association,' 1871, p. 84. Grant's 'History
of Physical Astronomy,' pp. 568–9.

or moon, according as it is high in the heavens or near the horizon, should be sufficient to make us cautious in accepting the plainest indications of our senses, unassisted by instrumental measurement. As to statements concerning the height of the aurora and the distance of meteors, they are to be utterly distrusted. When Captain Parry says that a ray of the aurora shot suddenly downwards between him and the land which was only 3000 yards distant, we must consider him subject to an error of sense[1].

It is true that errors of observation are more usually errors of judgment than of sense. That which is actually seen must be truly seen so far; and if we correctly interpret the meaning of the phenomenon, there would be no error at all. But the weakness of the bare senses as measuring instruments, arises from the fact that they import varying conditions of unknown amount, and we cannot make the requisite corrections and allowances as in the case of a solid and invariable instrument.

Bacon has excellently stated the insufficiency of the senses for estimating the magnitudes of objects, or detecting the degrees in which phenomena present themselves. 'Things escape the senses,' he says[m], 'because the object is not sufficient in quantity to strike the sense: as all minute bodies; because the percussion of the object is too great to be endured by the senses: as the form of the sun when looking directly at it in mid-day; because the time is not proportionate to actuate the sense: as the motion of a bullet in the air, or the quick circular motion of a firebrand, which are too fast, or the hour-hand of a common clock, which is too slow; from the distance of the object as to place: as the size of the celestial bodies, and the size and nature of all distant bodies;

[1] Loomis, 'On the Aurora Borealis.' Smithsonian Transactions, quoting Parry's Third Voyage, p. 61.

[m] 'Novum Organum.'

from prepossession by another object : as one powerful
smell renders other smells in the same room imperceptible ; from the interruption of interposing bodies :
as the internal parts of animals ; and because the object
is unfit to make an impression upon the sense : as the
air or the invisible and untangible spirit which is included in every living body.'

Complexity of Quantitative Questions.

One remark which we may well make in entering
upon quantitative questions, has regard to the great variety
and extent of phenomena presented to our notice.   So
long as we deal only with a simply logical question, that
question is merely, Does a certain event happen ? or, Does
a certain object exist ?   No sooner do we regard the event
or object as capable of more or less, than one question
branches out into many.   We must now ask, How much
is it compared with its cause or necessary condition ?
Does it change when the amount of the cause changes ?
If so, does it change in the same or opposite direction ?   Is
the change in simple proportion to that of the cause ?   If
not, what more complex law of connection holds true ?
This law determined satisfactorily in one series of circumstances may be varied under new conditions, and the
most complex relations of several quantities may ultimately
be established.

In every question of physical science there is thus a
series of steps of progress, the first one or two of which
are usually made with ease, while the succeeding ones
demand more and more careful measurement.   We cannot
lay down any single invariable series of questions which
must be asked from nature.   The exact character of the
questions will vary according to the nature of the case,
but they will usually be of a very evident kind, and we
may readily illustrate them by actual examples.   Suppose,

for instance, that we are investigating the solution of some salt in water. The first is a purely logical question : Is there solution, or is there not ?   Assuming the answer to be in the affirmative, we next inquire, Does the solubility vary with the temperature, or not ?   In all probability some variation will be found to exist, and we shall have at the same time an answer to the further question, Does the quantity dissolved increase, or does it diminish with the temperature ?   In by far the greatest number of cases salts and substances of all kinds dissolve more freely the higher the temperature of the water, but there are a few salts, such as calcium sulphate, which follow the opposite rule.   A considerable number of salts resemble sodium sulphate in becoming more soluble up to a certain temperature, and then varying in the opposite direction. We next require to assign the amount of variation as compared with that of the temperature, assuming at first that the increase of solubility is proportional to the increase of temperature.   Common salt is an instance of very slight variation, and potassium nitrate of very considerable increase with temperature.   Very accurate observations will probably show, however, that the simple law of proportionate variation is only approximately true, and some more complicated law involving the second, third, or higher powers of the temperature may ultimately be established.   All these investigations have to be carried out for each salt separately, since no distinct principles by which we may infer from one substance to another have yet been detected.   There is still an indefinite field for further research open ; for the solubility of salts would probably vary with the pressure under which the medium is placed ; the presence of other salts already dissolved may have effects yet unknown.   The researches already effected as regards the solvent power of water must be repeated as regards alcohol, ether, carbon bisulphide, and other media, so that unless general laws can be detected, this one phenomenon of solution can never be exhaustively treated. The same kind of questions recur as regards the solution or absorption of gases in liquids, the pressure as well as the temperature having then a most decided effect, and Professor Roscoe's researches on the subject present an excellent example of the successive determination of various complicated laws[n].

There is hardly a single branch of scientific research in which similar complications are not ultimately encountered. In the case of gravity, indeed, we arrive at the final law, that the force is invariably the same for all kinds of matter, and depends only on the distance of action. But in other subjects the laws, if simple in their ultimate nature, are disguised and complicated in their apparent results. Thus the effect of heat in expanding solids, or the reverse effect of forcible extension or compression upon the temperature of a body, will vary from one substance to another, will vary as the temperature is already higher or lower, and will probably follow a highly complex law, which in some cases gives negative or exceptional results. In crystalline substances the same researches have to be repeated in each distinct axial direction.

In the sciences of pure observation again, such as those of astronomy, meteorology, and terrestrial magnetism, we meet with many interesting series of quantitative determinations. The so-called fixed stars, as Giordano Bruno divined, are not really fixed, and may be more truly described as vast wandering orbs, each pursuing its own path through space. We must then determine separately for each star the following questions :—

1. Does it move ?
2. In what direction ?

n Watt's 'Dictionary of Chemistry,' vol. ii. p. 790.

3. At what velocity?
4. Is this velocity variable or uniform?
5. If variable, according to what law?
6. Is the direction uniform?
7. If not, what is the form of the apparent path?

The successive answers to such questions in the case of certain binary stars, have afforded a proof that the motions are due to a central force coinciding in law with gravity, and doubtless identical with it. In other cases the motions are usually so small that it is exceedingly difficult to distinguish them with certainty. A coincidence of motions in some constellations has been pointed out by Mr. Proctor, and the parallactic effect due to the sun's proper motion has been surely detected; but the time is yet far off when any general results as regards stellar motions can be established.

The variation in the brightness of stars opens an unlimited field for curious observation. There is not a star in the heavens concerning which we might not have to determine—

1. Does it vary in brightness?
2. Is the brightness increasing or decreasing?
3. Is the variation uniform, that is, simply proportional to time?
4. If not, according to what law does it vary?

In a majority of cases the change will probably be found to have a periodic character, in which case several other questions will arise, such as—

5. What is the length of the period?
6. Are there minor periods within the principal period?
7. What is the form or law of variation within the period?
8. Is there any change in the amount of variation?

9. If so, is it a secular, i. e. a continually growing change, or does it give evidence of a greater period?

Already the periodic changes of a certain number of stars have been determined with accuracy, and the lengths of the periods vary from less than three days up to intervals of time at least 250 times as great. Periods within periods have also been detected[o].

There is, perhaps, no subject in which more complicated quantitative conditions have to be determined than terrestrial magnetism. Since the time when the declination of the compass was first noticed, as some suppose by Columbus, we have had successive discoveries from time to time of the progressive change of declination from century to century; of the periodic character of this change; of the difference of the declination in various parts of the earth's surface; of the varying laws of the change of declination; of the dip or inclination of the needle, and the corresponding laws of its periodic changes; the horizontal and perpendicular intensities have also been the subject of exact measurement, and have been found to vary by place and time, like the directions of the needle; daily and yearly periodic changes have also been detected, and all the elements are found to be subject to occasional storms or abnormal perturbations, in which the eleven year period, now known to be common to many planetary relations, is apparent. The complete solution of these motions of the compass needle involves nothing less than a determination of its position and oscillations in every part of the world at any epoch, the like determination for another epoch, and so on, time after time, until the periods of all changes are ascertained, and the character of the variations determined. This one subject offers to men of science an almost inexhaustible field for

---

[o] Humboldt's 'Cosmos,' translated by Otté, vol. iii. p. 228.

interesting quantitative research [P], in which we shall doubtless at some future time discover the operation of causes now most mysterious and unaccountable.

The Methods of Accurate Measurement.

In studying the modes by which physicists have accomplished very exact measurements, we find that they are very various, but that they may perhaps be reduced under the following three classes :—

1. The increase or decrease of the quantity to be measured in some determinate ratio, so as to bring it within the scope of our senses, and to equate it with the standard unit, or some determinate multiple or sub-multiple of this unit.

2. The discovery of some natural conjunction of events which will enable us to compare directly the multiples of the quantity with those of the unit, or a quantity related in a definite ratio to that unit.

3. Indirect measurement, which gives us not the quantity itself, but some other quantity connected with it by known mathematical relations.

Conditions of Accurate Measurement.

Several conditions are requisite in order that a measurement may be made with great accuracy, and that the result may be closely accordant when several independent measurements are made.

In the first place the magnitude must be exactly defined by sharp terminations, or precise marks of inconsiderable thickness. When a boundary is vague and graduated, like the penumbra in a lunar eclipse, it is impossible to say where the end really is, and different people will come

P Gauss, 'General Theory of Terrestrial Magnetism'; Taylor's 'Scientific Memoirs,' vol. ii. p. 228.

to different results. We may sometimes overcome this difficulty to a certain extent, by observations repeated in a special manner, as we shall afterwards see; but when possible, we should choose opportunities for measurement when precise definition is easy. The moment of occultation of a star by the moon can be observed with great accuracy, because the star disappears with perfect suddenness; but there are many other astronomical conjunctions, eclipses, transits, &c., which occupy a certain length of time in happening, and thus open the way to differences of opinion. It would be impossible to observe with precision the movements of a body possessing no definite points of reference. The spots on the sun, for instance, furnish the only direct criterion of its rotation; and the possibility that these spots have a tendency to move in one direction throws a doubt upon all determinations of the sun's axial movement.

The colours of the complete spectrum shade with perfect continuity into each other, so that their separation is entirely an arbitrary matter. Exact determinations of refractive indices would have been impossible, had we not the fixed dark lines of the solar spectrum as precise points for measurement, or, what comes to the same thing, various kinds of homogeneous light, such as that of sodium, possessing a nearly uniform length of vibration.

In the second place, we cannot measure accurately unless we have the means either of multiplying or dividing a quantity without considerable error, so that we may correctly equate one magnitude with the multiple or submultiple of the other. In some cases we operate upon the quantity to be measured, and bring it into accurate coincidence with the actual standard, as when in photometry we vary the distance of our luminous body, until its illuminating power at a certain point is equal to that of a standard lamp. In other cases we repeat the unit until it equals the object, as in surveying land, or determining a weight by the balance. The requisites of accuracy now are :—(1) That we can repeat unit after unit of exactly equal magnitude ; (2) That these can be joined together so that the aggregate shall really be the sum of the parts. The same conditions apply to subdivision, which may be regarded as a multiplication of subordinate units. In order to measure to the thousandth of an inch, we must be able to add thousandth after thousandth without error in the magnitude of these spaces, or in their conjunction.

The condenser electrometer, as remarked by Thomson and Tait[q], is a good example of an instrument unfitted to give any sure measure of electro-motive force, because the friction between the parts of the condenser often produces more electricity than the original quantity which was to be measured.

Measuring Instruments.

To consider the mechanical construction of scientific instruments, is no part of my purpose in this book. I wish to point out merely the general purpose of such instruments, and the methods adopted to carry out that purpose with great precision. In the first place we must distinguish between the instrument which effects a comparison between two quantities, and the standard magnitude which often forms one of the quantities compared. The astronomer's clock, for instance, is no standard of the efflux of time; it serves but to subdivide, with approximate accuracy, the interval of successive passages of a star across the meridian, which it may effect perhaps to the tenth part of a second, or $\frac{1}{864000}$ part of the whole.

---

[q] 'Elements of Natural Philosophy,' sect. 326, p. 108.

The moving globe itself is the real standard clock, and the transit instrument the finger of the clock, while the stars are the hour, minute, and second marks, none the less useful or accurate because they are disposed at unequal intervals. The photometer is a simple instrument, by which we compare the relative intensity of rays of light falling upon a given spot. The galvanometer shows the comparative intensity of electric currents passing through a wire. The calorimeter guages the quantity of heat passing from a given object. But no such instruments furnish the standard unit in terms of which our results are to be expressed. In one peculiar case alone does the same instrument combine the unit of measurement and the means of comparison. A theodolite, mural circle, sextant, or other instrument for the measurement of angular magnitudes has no need of an additional physical unit; for the very circle itself, or complete revolution, is the natural unit to which all greater or lesser amounts of angular magnitude are referred.

The result of every measurement is to make known the purely numerical ratio existing between the magnitude to be measured, and a certain other magnitude, which should, when possible, be a fixed unit or standard magnitude, or at least an intermediate unit of which the value can be ascertained in terms of the ultimate standard. But though a ratio is the required result, an equation is the mode in which the ratio is determined and expressed. In every measurement we equate some multiple or submultiple of one quantity, with some multiple or submultiple of another, and equality is always the fact which we ascertain by the senses. By the eye, the ear, or the touch, we judge whether there is a discrepancy or not between two lights, two sounds, two intervals of time, two bars of metal. Often indeed we substitute one sense for the other, as when the efflux of time is judged by the marks upon a moving slip of paper, so that equal intervals of time are represented by equal lengths. There is, perhaps, a tendency to reduce all comparisons to the comparison of space magnitudes, but in any case one of the senses must be the ultimate judge of coincidence or non-coincidence.

Since the equation to be established may exist between any multiples or submultiples of the quantities compared, there naturally arise several different modes of comparison adapted to different cases. Let p be the magnitude to be measured, and q that in terms of which it is to be expressed. Then we wish to find such numbers x and y, that the equation $p = \frac{x}{y} q$ may be true. Now this same equation may be presented in four slightly different forms, namely :—

First Form.	Second Form.	Third Form.	Fourth Form.
$p = \frac{x}{y} q$	$p \frac{y}{x} = q$	$py = qx$	$\frac{p}{x} = \frac{q}{y}$

Each of these modes of expressing the same equation corresponds to one mode of effecting a measurement.

When the standard quantity is greater than that to be measured, we often adopt the first mode, and subdivide the unit until we get a magnitude equal to that measured. The angles observed in surveying, in astronomy, or in goniometry are usually smaller than a whole revolution, and the measuring circle is divided by the use of the microscope and screw, until we obtain an angle undistinguishable from that observed. The dimensions of minute objects are determined by subdividing the inch or centimetre, the screw micrometer being the most accurate means of subdivision. Ordinary temperatures are estimated by division of the standard interval between the freezing and boiling points of water, as marked on a thermometer tube.

In a still greater number of cases, perhaps, we multiply

the standard unit until we get a magnitude equal to that to be measured. Ordinary measurement by a foot rule, a surveyor's chain, or the excessively careful measurements of the base line of a trigonometrical survey by standard bars form a sufficient instance of this case.

In the second case, where $p\dfrac{x}{y} = q$, we multiply or divide a magnitude until we get what is equal to the unit, or to some magnitude easily comparable with it. As a general rule the quantities which we desire to measure in physical science are too small rather than too great for easy determination, and the problem consists in multiplying them without introducing error. Thus the expansion of a metallic bar when heated from $0°$ C to $100°$ may be multiplied by a train of levers or cog wheels. In the common thermometer the expansion of the mercury is rendered very apparent, and easily measurable by the fineness of the tube, and many other cases might be quoted. There are some phenomena, on the contrary, which are too great or rapid to come within the easy range of our senses, and our task is then the opposite one of diminution. Galileo found it difficult to measure the velocity of a falling body, owing to the very considerable velocity acquired in a single second. He adopted the elegant device, therefore, of lessening the rapidity by letting the body roll down an inclined plane, which enables us to reduce the accelerating force in any required ratio. The same purpose is effected in the well known experiments performed on Attwood's machine, and the measurement of gravity by the pendulum really depends on the same principle applied in a far more advantageous manner. Sir Charles Wheatstone has invented a beautiful method of galvanometry for strong currents, which consists in drawing off from the main current a certain determinate portion, which is equated by the galvano-

meter to a standard current[r].   In short, he measures not
the current itself but a known fraction of it.

In many electrical and other experiments, we wish to
measure the movements of a needle or other body, which
are not only very slight in themselves, but the manifestations of exceedingly small forces.   We cannot even
approach a delicately balanced needle without disturbing
it.   Under these circumstances the only mode of proceeding with accuracy, is to attach a very small mirror to the
moving body, and employ a ray of light reflected from
the mirror as an index of its movements.   The ray may
be considered quite incapable of affecting the body, and
yet by allowing the ray to pass to a sufficient distance,
the motions of the mirror may be increased to almost any
extent.   A ray of light is in fact a perfectly weightless
finger or index of indefinite length, with the additional
advantage that the angular deviation is by the law of
reflection double that of the mirror.   This method, was
introduced by Gauss, and is now of great importance ;
but in Wollaston's reflecting goniometer a ray of light
had previously been employed as an index finger.   Lavoisier and Laplace had also used a telescope in connection
with the pyrometer.

It is a great advantage in some instruments that they
can be readily made to manifest a phenomenon in a greater
or less degree, by a very slight change in the construction.
Thus either by enlarging the bulb or contracting the tube
of the thermometer, we can make it give more conspicuous
indications of change of temperature.   The barometer, on
the other hand, always gives the variations of pressure
on one scale.   The torsion balance is especially remarkable for the extreme delicacy which may be attained
by increasing the length and lightness of the rod, and the

length and thinness of the supporting thread. Forces so minute as the attraction of gravitation between two balls, or the magnetic and diamagnetic attraction of common liquids and gases, may thus be made apparent, and even measured. The common chemical balance, too, is capable theoretically of indefinite sensibility[s].

The third mode of measurement, which may be called the Method of Repetition, is of such great importance and interest that we must consider it in a separate section. It consists in multiplying both magnitudes to be compared until some multiple of the first is found to coincide very nearly with some multiple of the second. If the multiplication can be effected to an indefinite extent, without the introduction of countervailing errors, the accuracy with which the required ratio can be determined is unlimited, and we thus account for the extraordinary precision with which intervals of time in astronomy are compared together.

The fourth mode of measurement in which we equate submultiples of two magnitudes is comparatively seldom employed, because it does not conduce to accuracy. In the photometer, perhaps, we may be said to use it; we compare the intensity of two sources of light, by placing them both at such distances from a given surface, that the light falling on the surface is tolerable to the eye, and equally intense from each source. Since the intensity of rays diminishes, as the inverse squares of the distances, the relative intensities of the luminous bodies are proportional to the squares of their distances. The equality of intensity of two rays of similarly coloured light, may be most accurately ascertained in the mode suggested by Arago, namely, by causing the rays to pass in opposite directions through two nearly flat lenses pressed together.

---

[s] Watt's ' Dictionary of Chemistry,' art. *Balance*, vol. i. p. 487.

There is an exact equation between the intensities of the beams when Newton's rings disappear, the ring created by one ray being exactly the complement of that created by the other[t].

The Method of Repetition.

The ratio of two quantities can be determined with unlimited accuracy, if we can multiply both the object of measurement and the standard unit without error, and then observe what multiple of the one coincides or nearly coincides with some multiple of the other. Although perfect coincidence can never be really attained, the error thus arising may be indefinitely reduced. For if the equation $py = qx$ be uncertain to the amount e, so that $py = qx \pm e$, then we have $p = q\dfrac{x}{y} \pm \dfrac{e}{y}$, and as we are supposed to be able to make y as great as we like without increasing the error e, it follows that we can approximate as closely as we like to the required ratio $x \div y$.

This method of repetition is naturally employed whenever quantities can be repeated, or repeat themselves without error of juxtaposition, which is especially the case with the motions of the earth and heavenly bodies. In determining the length of the sidereal day, we really determine the ratio between the earth's revolution round the sun, and its rotation on its own axis. We might ascertain the ratio by observing the successive passages of a star across the zenith, and comparing the interval by a good clock with that between two passages of the sun, the difference being due to the angular movement of the earth round the sun. In such observations we should have an error of a considerable part of a second at each

[t] Humboldt's 'Cosmos,' (Bohn), vol. iii. p. 129.

observation, in addition to the irregularities of the clock. But the revolutions of the earth repeat themselves day after day, and year after year, without the slightest interval between the end of one period and the beginning of another. The operation of multiplication is perfectly performed for us by nature. If, then, we can find an observation of the passage of a star across the meridian a hundred years ago, that is of the interval of time between the passage of the sun and the star, the instrumental errors in measuring this interval by a clock and telescope may be greater than in the present day, but will be divided by about 36,524 days, and rendered excessively small. It is thus that astronomers have been able to ascertain the ratio of the mean solar to the sideral day to the 8th place of decimals (1.00273791 to 1), or to the hundred millionth part, probably the most accurate result of measurement in the whole range of science.

The antiquity of this mode of comparison is almost as great as that of astronomy itself. Hipparchus made the first clear application of it, when he compared his own observations with those of Aristarchus, made 145 years previously [u], and thus ascertained the length of the year. This calculation may in fact be regarded as the earliest attempt at an exact determination of the constants of nature. The method is the main resource of astronomers; Tycho, for instance, detected the slow diminution of the obliquity of the earth's axis, by the comparison of observations at long intervals. Living astronomers use the method as much as earlier ones; but so superior in accuracy are all observations taken during the last hundred years to all previous ones, that it is often found preferable to take a shorter interval, rather than incur the risk of greater instrumental errors in the earlier observations.

[u] Montucla, 'Histoire des Mathématiques,' vol. i. p. 258.

It is obvious that many of the slower changes of the heavenly bodies must require the lapse of large intervals of time to render their amount perceptible. Hipparchus could not possibly have discovered many of the smaller inequalities of the heavenly motions, because there were no previous observations of sufficient age or exactness to exhibit them. And just as the observations of Hipparchus formed the starting-point for subsequent comparisons, so a large part of the labour of present astronomers is directed to recording the present state of the heavens so exactly, that future generations of astronomers may detect many changes, which cannot possibly become known in the present age.

The principle of repetition was very ingeniously employed in an instrument first proposed by Mayer in 1767, and carried into practice in the Repeating Circle of Borda [v]. The exact measurement of angles is indispensable, not only in astronomy but also in trigonometrical surveys, and the highest skill in the mechanical execution of the graduated circle and telescope will not prevent terminal errors of considerable amount. If instead of one telescope, the circle be provided with two similar telescopes, these may be alternately directed to two distant points, say the marks in a trigonometrical survey, so that the circle shall be turned through any multiple of the angle subtended by those marks, before the amount of the angular revolution is read off upon the graduated circle. Theoretically speaking, all error arising from imperfect graduation might thus be indefinitely reduced, being divided by the number of repetitions. In practice, however, the advantage of the invention is not found to be great, probably because a certain error is introduced at each observation in the changing or fixing of the telescopes. It is moreover in-

applicable to moving objects like the heavenly bodies, so
that its use is confined to important trigonometrical
surveys.

The pendulum is the most perfect of all instruments,
chiefly because it naturally admits of almost indefinite
repetition. Since the force of gravity never ceases, one
swing of the pendulum is no sooner ended than the other
is begun, so that the juxtaposition of successive units is
absolutely perfect. Provided that the oscillations be equal,
then one thousand oscillations will occupy exactly one
thousand times as great an interval of time as one oscillation. Not only is the subdivision of time entirely de-
pendent on this fact, but in the accurate measurement of
gravity, and many important determination it is of
the greatest service. In the deepest mine, we could not
observe the rapidity of fall of a body for more than a
quarter of a minute, and the measurement of its velocity
would be difficult, and subject to uncertain errors from
resistance of air, &c. In the pendulum, we have a body
which can be kept rising and falling for many hours, in
a medium entirely under our command or if desirable in
a vacuum. Moreover, the comparative force of gravity at
different points, at the top and bottom of a mine for
instance, can be determined with wonderful precision, by
comparing the oscillations of two exactly similar pendulums, with the aid of electric clock signals. To ascertain
the comparative lengths of vibration of two pendulums, it
is only requisite to swing them one in front of the other,
to record by a clock the moment when they coincide in
swing, so that one hides the other, and then count the
number of vibrations until they again come to similar
coincidence. If one pendulum makes m vibrations and
the other n, we at once have our equation $pn = qm$;
which gives the length of vibration of either pendulum in
terms of the other. This method of coincidence, embody-

ing the principle of repetition in perfection, was employed with wonderful skill by Sir George Airy, in his experiments on the Density of the Earth at the Harton Colliery ; the pendulums above and below being compared with clocks, which again were compared with each other by electric signals. So exceedingly accurate was this method of observation, as carried out by Sir George Airy, that he was able to measure a total difference in the vibrations at the top and bottom of the shaft, amounting to only 2·24 seconds in the twenty-four hours, with an error of less than one hundredth part of a second, or one part in 8,640,000 of the whole day [x].

The principle of repetition has been elegantly applied in observing the motion of waves in water. If the canal in which the experiments are made be short, say twenty feet long, the waves will pass through it so rapidly that an observation of one length, as practised by Walker, will be subject to much terminal error, even when the observer is very skilful. But it is a result of the undulatory theory that a wave is quite unaltered, and loses no time by complete reflection, so that it may be allowed to travel backwards and forwards in the same canal, and its motion, say through sixty lengths, or 1200 feet, may be observed with the same accuracy as in a canal 1200 feet long, with the advantage of greater uniformity in the condition of the canal and water [y]. It is always desirable, if possible, to bring an experiment into a small compass, so as to be well under command, and yet we may often by repetition enjoy at the same time the advantage of extensive observation.

One reason of the great accuracy of weighing with a good balance is the fact, that weights placed in the same

[x] 'Philosophical Transactions,' (1856) vol. 146, Part i. p. 297.

[y] Airy, 'On Tides and Waves,' Encyclopædia Metropolitana, p. 345. Scott Russell, 'British Association Report,' 1837, p. 432.

scale are naturally added together without the slightest error. There is no difficulty in the precise juxtaposition of two grammes, but the juxtaposition of two metre measures can only be effected with tolerable accuracy, by the use of microscopes and many precautions. Hence, the extreme trouble and cost attaching to the exact measurement of a base line for a survey, the risk of error entering at every juxtaposition of the measuring bars, and indefatigable attention to all the requisite precautions being necessary throughout the operation[z].

Measurements by Natural Coincidence.

In certain cases a peculiar conjunction of circumstances enables us to dispense more or less with instrumental aids, and to obtain the most exact numerical results in the simplest manner. The mere fact, for instance, that no human being has ever seen a different face of the moon from that familiar to us, conclusively proves that the period of rotation of the moon on its own axis is equal to that of its revolution round the earth. Not only have we the repetition of these movements during 1000 or 2000 years at least, but we have observations made for us at very remote periods, free from instrumental error, no instrument being needed. We learn that the seventh satellite of Saturn is subject to a similar law, because its light undergoes a variation in each revolution, owing to the existence of some dark tract of land ; now this failure of light always occurs while it is in the same position relative to Saturn, clearly proving the equality of the axial and revolutional periods, as Huyghens perceived[a].

---

[z] Herschel's, ' Familiar Lectures on Scientific Subjects,' p. 184.

[a] ' Hugenii Cosmotheoros,' pp. 117–18. Laplace's 'Système,' translated, vol. i. p. 67.

A like peculiarity in the motions of Jupiter's fourth satellite was similarly detected by Maraldi in 1713.

Remarkable conjunctions of the planets may sometimes allow us to compare their periods of revolution, through long intervals of time, with great accuracy. Laplace in explaining the long inequality in the motions of Jupiter and Saturn, was much assisted by a conjunction of these planets, observed by Ibyn Jounis at Cairo, towards the close of the eleventh century. Laplace calculated that such a conjunction must have happened on the 31st of October, A. D. 1087 ; and the discordance between the distances of the planets as recorded, and as assigned by theory, was less than one-fifth of the apparent diameter of the sun. This difference being less than the probable error of the early record, his theory was confirmed as far as facts were available [b].

The ancient astronomers often shewed the highest ingenuity in turning any opportunities of measurement which occurred to good account. Eratosthenes, as early as 250 B. C., happening to hear that the sun at Syene, in Upper Egypt, was visible at the summer solstice at the bottom of a well, proving that it was in the zenith, proposed to determine the dimensions of the earth, by measuring the length of the shadow of a rod at Alexandria on the same day of the year. He thus learnt in a rude manner the difference of latitude between Alexandria and Syene, and finding it to be about one fiftieth part of the whole circumference, he ascertained the dimensions of the earth within about one sixth part of the truth. The use of wells in astronomical observation appears to have been occasionally practised in comparatively recent times, as by Flamsteed in 1679 [c]. Hipparchus employed the moon as an instrument of measurement in several sagacious

[b] Grant's, 'History of Physical Astronomy,' p. 129.

[c] Baily's, 'Account of Flamsteed,' p. lix.

modes. When the moon is exactly half full, the moon, sun, and earth, are at the angles of a right-angled triangle. He proposed therefore at such a time to measure the moon's elongation from the sun, which would give him the two other angles of the triangle, and enable him to judge of the comparative distances of the moon and sun from the earth. His result, though very rude, was far more accurate than any notions previously entertained, and enabled him to form some estimate of the comparative magnitudes of the bodies. Eclipses of the moon were also very useful in ascertaining the longitudes of the stars, which were invisible when the sun was above the horizon. For the moon when eclipsed must be 180° distant from the sun; hence it was only requisite to measure the distance of a fixed star in longitude from the eclipsed moon to obtain with ease its angular distance from the sun.

In later times the eclipses of Jupiter have usefully served to give a measure of an angle; for at the middle moment of the eclipse the satellite must be exactly in the same straight line with the planet and sun, so that we can learn from the known laws of movement of the satellite the longitude of Jupiter as seen from the sun. If at the same time we measure the elongation or apparent angular distance of Jupiter from the sun, as seen from the earth, we have all the angles of the triangle between Jupiter, the sun, and the earth, and can calculate the comparative magnitudes of the sides of the triangle by simple trigonometry.

The transits of Venus over the sun's face are other natural events which seem to give most accurate measurements of the sun's parallax, or apparent difference of position as seen from distant points of the earth's surface. The sun forms a kind of background on which the place of the planet is marked, and serves as a measuring instrument free from all the errors of construction, which affect human instruments. The rotation of the earth, too, by variously affecting the apparent velocity of ingress or egress of Venus, as seen from different places, discloses the amount of the parallax. It has been sufficiently shown that by rightly choosing the moments of observation, the planetary bodies may often be made to reveal their relative distance, to measure their own position, to record their own movements with a high degree of accuracy. With the improvement of astronomical instruments, such conjunctions become less necessary to the progress of the science, but it will always remain advantageous to choose those moments for observation when instrumental errors enter with the least effect.

In other sciences, exact quantitative laws can occasionally be obtained without instrumental measurement, as when we learn the exactly equal velocity of sounds of different pitch, by observing that a peal of bells or a musical performance is heard harmoniously at any distance to which the sound penetrates; this could not be the case, as Newton remarked, if one sound overtook the other. One of the most important principles of the atomic theory, was proved by implication, before the use of the balance was introduced into chemistry. Wenzel observed, before 1777, that when two neutral substances decompose each other, the resulting salts are also neutral. In mixing sodium sulphate and barium nitrate, we obtain insoluble barium sulphate and neutral sodium nitrate. This result could not follow unless the nitric acid, requisite to saturate one atom of sodium, were exactly equal to that required by one atom of barium, so that an exchange could take place without leaving either acid or base in excess [d].

A very important principle of mechanics may also be

[d] Daubeny, 'Atomic Theory,' p. 30.

established by a simple acoustical observation. When
a rod or tongue of metal fixed at one end is set in
vibration, the pitch of the sound may be observed to
be exactly the same, whether the vibrations be small or
great; hence the oscillations are isochronous, or equally
rapid, independently of their magnitude. On the ground
of theory, it can be shown that such a result only
happens when the flexure is proportional to the deflecting
force. Thus the simple observation that the pitch of
the sound of a harmonium, for instance, does not change
with its loudness, establishes an exact law of nature [c].

A closely similar instance is found in the proof that the
intensity of light or heat rays varies inversely as the
square of the distance increases. For the apparent magnitude certainly varies according to this law; hence, if the
intensity of light varied according to any other law, the
brightness of an object would be different at different
distances, which is not observed to be the case. Melloni
applied the same kind of reasoning, in a somewhat different form, to the radiation of heat-rays [f].

Modes of Indirect Measurement.

Some of the most conspicuously beautiful experiments
in the whole range of science, have been devised for the
purpose of indirectly measuring quantities, which in their
extreme greatness or smallness surpass the powers of
sense. All that we need to do, is to discover some
other conveniently measurable phenomenon, which is related in a known ratio or according to a known law,
however complicated, with that to be measured. Having

[e] Jamin, 'Cours de Physique,' vol. i. p. 152.
[f] Balfour Stewart's, 'Elementary Treatise on Heat,' 1st edit. pp. 164,
165.

once obtained experimental data, there is no further difficulty beyond that of arithmetic or algebraic calculation.

Gold is reduced by the gold-beater to leaves so thin, that the most powerful microscope would not detect any measurable thickness. If we laid several hundred leaves upon each other to multiply the thickness, we should still have no more than $\frac{1}{100}$th of an inch at the most to measure, and the errors arising in the superposition and measurement would be considerable. But we can readily obtain an exact result through the connected amount of weight. Faraday weighed 2000 leaves of gold, each $3\frac{3}{4}$ inch square, and found them equal to 384 grains. From the known specific gravity of gold, it was easy to calculate that the average thickness of the leaves was $\frac{1}{282.000}$ of an inch [g].

We must ascribe to Newton the honour of leading the way in methods of minute measurement. He did not call waves of light by their right name, and did not understand their nature; yet he measured their length, though it did not exceed the 2,000,000th part of a metre or the one fifty thousandth part of an inch. He pressed together two lenses of very large but known radii. It was not difficult to calculate the interval between the lenses at any point, by measuring the distance from the central point of contact. Now, with homogeneous rays the successive rings of light and darkness mark the points at which the interval between the lenses is equal to one half, or any multiple of half a vibration of the light, so that the length of the vibration became known. In a similar manner many phenomena of interference of rays of light admit of the measurement of the wave lengths. The fringes of interference arise from rays of light which cross each other at a small angle, and an excessively

[g] Faraday, 'Chemical Researches,' p. 393.

minute difference in the lengths of the waves makes a very perceptible difference in the position of the point at which two rays will interfere and produce darkness.

M. Fizeau has recently employed Newton's rings in an inverse manner, to measure small amounts of motion. By merely counting the number of rings of sodium monochromatic light passing a certain point where two glass plates are in close proximity, he is able to ascertain with the greatest accuracy and ease the change of distance between these glasses, produced, for instance, by the expansion of a metallic bar, connected with one of the glass plates[h].

Nothing excites more admiration than the mode in which scientific observers can occasionally measure quantities, which seem beyond the bounds of human observation. We know the *average* depth of the Pacific Ocean to be 14,190 feet, not by actual sounding, which would be impracticable in sufficient detail, but by noticing the rate of transmission of earthquake waves from the South American to the opposite coasts, the rate of movement being connected by theory with the depth of the water[i]. In the same way the average depth of the Atlantic Ocean is inferred to be no less than 22,157 feet, from the velocity of the ordinary tidal waves. A tidal wave again gives beautiful evidence of an effect of the law of gravity, which we could never in any other way detect. Newton estimated that the moon's force in moving the ocean is only $\frac{1}{2,871,400}$ part of the whole force of gravity, which even the pendulum, used with the utmost skill, would fail to render apparent. Yet the immense extent of the ocean allows the accumulation of the effect into a very palpable amount; and from the comparative

[h] 'Proceedings of the Royal Society,' 30th November, 1866.
[i] Herschel, 'Physical Geography,' § 40.

heights of the lunar and solar tides, Newton roughly
estimated the comparative forces of the moon's and sun's
gravity at the earth[k].

A few years ago it might have seemed impossible that
we should ever measure the velocity with which a star
approaches or recedes from the earth, since the apparent
position of the star is thereby unaltered.   But the spectroscope now enables us to detect and even measure such
motion with considerable accuracy, by the alteration which
it causes in the apparent rapidity of vibration, and consequently in the refrangibility of rays of light of definite
colour.   And while our estimates of the lateral movements of stars depend upon our very uncertain knowledge of their distance, the spectroscope gives the motion
in another direction in absolute quantity, irrespective of
all other quantities known or unknown, excepting the
motion of the earth itself[l].

The rapidity of vibration for each musical tone, having been accurately determined by comparison with the
Syren (p. 12), we can use sounds as indirect indications of
rapid vibrations.   It is now known that the contraction of
a muscle arises from the periodical contractions of each
separate fibre, and from a faint sound or susurrus which
accompanies the action of a muscle, it is inferred that
each contraction lasts for about $\frac{1}{300}$ of a second.   Minute
quantities of radiant heat are now always measured indirectly by the electricity which they produce when falling
upon a thermopile.   The extreme delicacy of the method
seems to be due to the power of multiplication at several
points in the apparatus.   The number of elements or junc-
tions of different metals in the thermopile can be increased

[k] 'Principia,' bk. iii. Prop. 37, 'Corollaries,' 2 and 3.   Motte's translation, vol. ii. p. 310.

[l] Roscoe's, 'Spectrum, Analysis,' 1st ed. p. 296.

so that the tension of the electric current derived from the same intensity of radiation is multiplied; the effect of the current upon the magnetic needle can be multiplied within certain bounds, by passing the current many times round it in a coil; the excursions of the needle can be increased by rendering it astatic and increasing the delicacy of its suspension; lastly, the angular divergence can be observed, with any required accuracy, by the use of an attached mirror and distant scale viewed through a telescope (p. 234). Such is the delicacy of this method of measuring heat, that Dr. Joule succeeded in making a thermopile which would indicate a difference of $\frac{1}{8800}$ part of a degree centigrade [m].

A striking case of indirect measurement is furnished by the revolving mirror of Wheatstone and Foucault, whereby a minute interval of time is estimated in the form of an angular deviation. Wheatstone viewed an electric spark in a mirror rotating so rapidly, that if the duration of the spark had been more than $\frac{1}{72,000}$ of a second, the point of light would have appeared elongated to an angular extent of one-half degree. In the spark, as drawn directly from a Leyden jar, no elongation was apparent, so that the duration of the spark was immeasurably small; but when the discharge took place through a bad conductor, the elongation of the spark denoted a sensible duration[n]. In the hands of Foucault the rotating mirror gave a measure of the time occupied by light in passing through a few metres of space.

Comparative Use of Measuring Instruments.

In almost every case a measuring instrument serves,

[m] 'Philosophical Transactions' (1859), vol. cxlix. p. 94.
[n] Watts' 'Dictionary of Chemistry,' vol. ii. p. 393.

and should serve only as a means of comparison between two or more magnitudes. As a general rule, we should not even attempt to make the divisions of the measuring scale exact multiples or submultiples of the unit, but, regarding them as arbitrary marks, should determine their values by comparison with the standard itself. Thus the perpendicular wires in the field of a transit telescope, are fixed at nearly equal but arbitrary distances, and those distances are afterwards determined, as first suggested by Malvasia, by watching the passage of star after star across them, and noting the intervals of time by the clock. Owing to the perfectly regular motion of the earth, these time intervals give an exact determination of the angular intervals. In the same way, the angular value of each turn of the screw micrometer attached to a telescope, can be easily and accurately ascertained.

When a thermopile is used to observe radiant heat, it would be almost impossible to calculate on à priori grounds what is the value of each division of the galvanometer circle, and still more difficult to construct a galvanometer, so that each division should have a given value. But this is quite unnecessary, because by placing the thermopile before a body of known dimensions, at a known distance, with a known temperature, and radiating power, we measure a known amount of radiant heat, and inversely measure the value of the indications of the thermopile. In a similar way Mr. Joule ascertained the actual temperature produced by the compression of bars of metal. For having inserted a simple thermopile composed of a single junction of copper and iron wire, and noted the deflections of the galvanometer, he had only to dip the bars into water of different temperatures, until he produced a like deflection, in order to ascertain the temperature developed by pressure[o].

[o] 'Philosophical Transactions' (1859), vol. cxlix. p. 119, &c.

In many instances we are indeed obliged to accept a very carefully constructed instrument as a standard, as in the case of a standard barometer. But it is then best to treat all inferior instruments comparatively only, and determine the values of their scales by comparison with the assumed standard.

Systematic Performance of Measurements.

When a large number of accurate measurements have to be effected, it is usually desirable to make a certain number of determinations with scrupulous care, and afterwards use them as points of reference for the remaining determinations. In the trigonometrical survey of a country, the principal triangulation fixes the relative positions and distances of a few points with rigid accuracy. A minor triangulation refers every prominent hill or village to one of the principal points, and then the details are filled in by reference to the secondary points. The survey of the heavens is effected in a like manner. The ancient astronomers compared the right ascensions of a few principal stars with the moon, and thus ascertained their positions with regard to the sun; the minor stars were afterwards referred to the principal stars. Tycho followed the same method, except that he used the more slowly moving planet Venus instead of the moon. Flamsteed was in the habit of using about seven stars, favourably situated at points all round the heavens. The distances of the other stars from these standard points, were determined in his early observations by the use of the quadrant P. Even since the introduction of the transit telescope and mural circle, tables of standard stars are formed at Greenwich, the positions being determined with every

P Baily's 'Account of Flamsteed,' pp. 378-380.

possible accuracy, so that they can be employed for purposes of reference by all astronomers.

In ascertaining the specific gravities of substances, all gases are referred to atmospheric air at a given temperature and pressure; all liquids and solids are referred to water. We require to compare the densities of water and air with great care, and the comparative densities of any two substances whatever can then be with ease ascertained.

In comparing a very great with a very small magnitude, it is usually desirable to break up the process into several steps, using intermediate terms of comparison. We should never think of measuring the distance from London to Edinburgh by laying down measuring rods throughout the whole distance. A base of several miles in length is selected on level ground, and compared on the one hand with the standard yard, and on the other with the distance of London and Edinburgh, or any other two points, by trigonometrical survey. It would be exceedingly difficult to compare the light of a star with that of the sun, which would be about thirty thousand million times greater; but Sir J. Herschel[q] effected the comparison by using the full moon as an intermediate unit. Wollaston ascertained that the sun gave 801,072 times as much light as the full moon, and Herschel determined that the light of the latter exceeded that of a Centauri 27,408 times, so that we find the ratio between the light of the sun and star to be that of about 22,000,000,000 to 1.

The Pendulum.

By far the most perfect and beautiful of all instruments of measurement is the pendulum. Consisting

[q] Herschel's 'Astronomy,' § 817, 4th. ed. p. 553.

merely of a heavy body suspended freely at an invariable distance from a fixed point, it is the most simple in construction; and yet all the highest problems of physical measurement depend upon its careful use. Its excessive value arises from two circumstances, which render it at once most accurate and indispensable.

(1) The method of repetition is eminently applicable to it, as already described (p. 339.)

(2) Unlike any other instrument, it connects together three different variable quantities, those of space, time, and force.

In most works on natural philosophy it is shown, that when the oscillations of the pendulum are infinitely small, the square of the time occupied by an oscillation is directly proportional to the length of the pendulum, and indirectly proportional to the force affecting it, of whatever kind. The whole theory of the pendulum is contained in the formula, first given by Huyghens in his Horologium Oscillatorium,

$$\text{time of oscillation} = 3\cdot14159 \ldots \times \sqrt{\frac{\text{length of pendulum}}{\text{force.}}}$$

The quantity $3\cdot14159$ is the constant ratio of the circumference and radius of a circle, and is of course known with accuracy. Hence, any two of the three quantities concerned being given, the third may be found; or any two being maintained invariable, the third will be invariable. Thus a pendulum of invariable length suspended at the same place, where the force of gravity may be considered uniform, furnishes a theoretically perfect measure of time. The same invariable pendulum being made to vibrate at different points of the earth's surface, and the time of vibration being astronomically determined, the force of gravity becomes accurately known. Finally, with a known force of gravity, and time of vibration ascertained by reference to the stars, the length is determinate.

In the first use all astronomical observations depend upon it. In the second employment it has been almost equally indispensable. The primary principle that gravity is equal in all matter was proved by Newton's and Gauss' pendulum experiments. The torsion pendulum of Michell, Cavendish, and Baily, depending upon exactly the same principles as the ordinary pendulum, gave the density of the earth, one of the foremost natural constants. Kater and Sabine, by pendulum observations in different parts of the earth, ascertained the variation of gravity, whence comes a determination of the earth's ellipticity. The laws of electric and magnetic attraction have also been determined by the method of vibrations, which is in constant use in the measurement of the horizontal force of terrestrial magnetism.

We must not confuse with the ordinary use of the pendulum its application by Newton, to show the absence of internal friction against space[r], or to ascertain the laws of motion and elasticity[s]. In such cases the extent of vibration is the quantity measured, and the principles of the instrument are different.

Attainable Accuracy of Measurement.

It is a matter of some interest to compare the degrees of accuracy, which can be attained in the measurement of different kinds of magnitude. Few measurements of any kind are exact to more than six significant figures[t], but it is seldom that such a point of accuracy can be hoped for. Time is the magnitude which seems to be capable of the most exact discrimination, owing to the properties of the

---

[r] 'Principia,' bk. ii. Sect. 6. Prop. 31. Motte's Translation, vol. ii p. 107.

[s] Ibid. bk. i. Law iii. Corollary 6. Motte's Translation, vol. i. p. 33.

[t] Thomson and Tait's 'Natural Philosophy,' vol. i. p. 333.

pendulum, and the principle of repetition already described
(pp. 339, 353). As regards short intervals of time, it has
already been stated that Sir George Airy was able to
estimate a difference of $2\frac{1}{4}$ seconds per day, between two
pendulums with an uncertainty of less than ·01 of a second,
or one part in 8,640,000, an exactness, as he truly remarks,
'almost beyond conception[u]'. The ratio between the mean
solar and the sidereal day, too, is known to about one part
in one hundred millions, or to the eighth place of decimals
(p. 337).

Determinations of weight seem to come next in exactness, owing to the fact that repetition without error is
applicable to them (p. 340). An ordinary good balance
should show about one part in 500,000 of the load[x]. The
finest balance employed by M. Stas, turned with $\frac{1}{33}$ of a
milligramme, when loaded with 25 grammes in each pan,
that is, with one part in 825,000 of the load[y]. But balances
have certainly been constructed to show one part in a
million[z], and Ramsden is commonly said to have constructed a balance for the Royal Society, to indicate one
part in seven millions, though this is hardly credible.
Professor Clerk Maxwell takes it for granted that one
part in five millions can be detected, but we ought to
discriminate between what a balance can do when first
constructed, and when in continuous use.

Determinations of lengths, unless performed with extraordinary care, are open to much error in the junction of
the measuring bars. Even in measuring the base line of
a trigonometrical survey, the accuracy generally attained
is only that of about one part in 60,000, or an inch in the

[u] 'Philosophical Transactions,' (1856), vol. cxlvi pp. 330–1.

[x] Thomson and Tait, 'Natural Philosophy,' vol. i. p. 333.

[y] 'First Annual Report of the Mint,' p. 106.

[z] Jevons, in Watts' 'Dictionary of Chemistry,' vol. i. p. 483.

mile[a]; but it is said that in four measurements of a
base line carried out very recently at Cape Comorin, the
greatest error was 0·077 inch in 1·68 mile, or one part in
1,382,400, an almost incredible degree of accuracy[b].. Sir J.
Whitworth has shown that touch is even a more delicate
mode of measuring lengths than sight, and by means of a
splendidly executed screw, and a small cube of iron placed
between two flat-ended iron bars, so as to be suspended
when touching them, he can detect a change of dimension
in a bar, amounting to no more than one-millionth of an
inch[c].

[a] Thomson and Tait, 'Natural Philosophy,' vol. i. p. 333.

[b] 'Athenæum,' February 28, 1870, p. 295.

[c] British Association, Glasgow, 1856. 'Address of the President of
the Mechanical Section.'

CHAPTER XIV.

UNITS AND STANDARDS OF MEASUREMENT.

INSTRUMENTS of measurement are, as we have seen, only means of comparison between one magnitude and another, and as a general rule we must assume some one arbitrary magnitude, in terms of which all results of measurement are to be expressed. Mere ratios between any series of objects will never tell us their absolute magnitudes; we must have at least one ratio for each, and we must have one absolute quantity. The number of ratios n are expressible in n equations, which will contain at least $n + 1$ quantities, so that if we employ them to make known n magnitudes, we must have one magnitude known. Hence, whether we are measuring time, space, density, weight, mass, energy, or any other physical quantity, we must refer to some concrete standard, some actual object, which if once lost and irrecoverable, all our measures lose their absolute meaning. This concrete standard is in all, except two, cases absolutely arbitrary in point of theory, and its selection a question of practical convenience.

Of the two cases in which a natural standard unit is ready made for us, one case is that of number itself. Abstract number needs no special unit; for any object by existing or being thought of as separate from other objects (p. 176), furnishes us with a unit, and is the only standard required.

Angular magnitude is the second case in which we have a natural and almost necessary unit of reference, namely, the whole revolution or *perigon*, as it has been called by Mr. Sandeman[a].

It is a necessary result of the uniform properties of space, that all complete revolutions are equal to each other, so that we need not select any one, and can always refer anew to space itself. Whether we take the whole perigon, its half, or its quarter, is really immaterial; Euclid took the right angle, because the Greek geometers had never generalized their notions of angular magnitude sufficiently to conceive clearly angles of all magnitude, or of unlimited *quantity of revolution.* But Euclid defines a right angle as half that made by a line with its own continuation, not called by him an angle, and which is of course equal to half a revolution. In mathematical analysis, again, a different fraction of the perigon is taken, namely, such a fraction that the arc or portion of the circumference included within it is equal to the radius of the circle. This angle, called by De Morgan the *arcual unit*, is equal to about $57°, 17', 44''\cdot8$, or decimally $57°\cdot295779513\ldots$, and is such that the half revolution contains $3\cdot14159265\ldots$ such units[b]. Though this standard angle is naturally employed in mathematical analysis, and any other unit would introduce needless complexity, we must not look upon it as a distinct unit, since its amount is connected with that of the half perigon, by a natural constant $3\cdot14159\ldots$ usually signified by the letter π.

When we pass to other species of quantity, the choice of unit is found to be entirely arbitrary. There is abso-

[a] 'Pelicotetics, or the Science of Quantity; an Elementary Treatise on Algebra, and its groundwork Arithmetic.' By Archibald Sandeman, M.A. Cambridge, (Deighton, Bell, and Co.) 1868, p. 304.

[b] De Morgan's 'Trigonometry and Double Algebra,' p. 5.

lutely no mode of defining a length, but by selecting some physical object exhibiting that length between certain obvious points—as, for instance, the extremities of a bar, or marks made upon its surface.

Standard Unit of Time.

Time is the great independent variable of all change, that which itself flows on uninterruptedly, and brings the variety which we call life and motion. When we reflect upon its intimate nature, Time, like every other element of existence, proves to be an inscrutable mystery. We can only say with St. Augustin, to one who asks us what is time, 'I know when you do not ask me.' The mind of man will ask what can never be answered, but one result of a true and rigorous logical philosophy must be to convince us, that scientific explanation can only take place between phenomena which have something in common, and that when we get down to primary notions, like those of time and space, the mind must meet a point of mystery beyond which it cannot penetrate. A definition of time must not be looked for; if we say with Hobbes[c], that it is 'the phantasm of before and after in motion,' or with Aristotle that it is 'the number of motion according to former and latter;' we obviously gain nothing, because the notion of time is involved in the expressions *before and after, former and latter*. Time is undoubtedly one of those primary notions which can only be defined physically, or by observation of phenomena which proceed in time.

If we have not advanced a step beyond Augustin's acute reflections on this subject[d], it is curious to observe the

[c] 'English Works of Thos. Hobbes,' Edit. by Molesworth, vol. i. p. 95.
[d] 'Confessions,' bk. xi. chapters 20–28.

wonderful advances which have been made in the practical measurement of its efflux. The rude sun-dial or the rising of a conspicuous star, gave points of reference, while the flow of water from the clepsydra, the burning of a candle, or, in the monastic ages, even the continuous equable chanting of psalms, gave the means of roughly subdividing periods, and marking the hours of the day and night[e]. The sun and stars still furnish the standard of time, but means of accurate subdivision have become requisite, and this has been furnished by the pendulum and the chronoscope. By the pendulum we can accurately divide the day into seconds of time. By the chronograph we can subdivide the second into a hundred, a thousand, or even a million parts. Wheatstone measured the duration of an electric spark, and found it to be no more than $\frac{1}{115,200}$ part of a second, while more recently Captain Noble has been able to appreciate intervals of time, not exceeding the millionth part of a second.

When we come to inquire precisely what phenomenon it is that we thus so minutely measure, we meet insurmountable difficulties. Newton distinguished time according as it was *absolute* or *apparent* time, in the following words :—

'Absolute, true, and mathematical time of itself and from its own nature, flows equably without regard to anything external, and by another name is called *duration;* relative, apparent and common time, is some sensible and external measure of duration by the means of motion[f].' Though we are perhaps obliged to assume the existence of a uniformly increasing quantity which we call time,

---

[e] Sir G. C. Lewis gives many curious particulars concerning the measurement of time, 'Astronomy of the Ancients,' pp. 241, &c.

[f] 'Principia,' bk. i. 'Scholium to Definitions.' Translated by Motte, vol. i. p. ix. See also, p. 11.

yet we cannot feel or know abstract and absolute time.
Duration must be made manifest to us by the recurrence
of some phenomenon. The succession and change of our
own thoughts is no.doubt the first and simplest measure
of time, but a very rude one, because in some persons and
circumstances the thoughts evidently flow with much
greater rapidity than in other persons and circumstances.
In the absence of all other phenomena, the interval between one thought and another, would necessarily become
the unit of time. The earth, as I have already said, is
the real clock of the astronomer, and is practically assumed
as invariable in its movements. But on what ground is
it so assumed ? According to the first law of motion, every
body perseveres in its state of rest or of uniform motion
in a right line, unless it is compelled to change that state
by forces impressed thereon. Rotatory motion is subject
to a like condition, namely, that it perseveres uniformly
unless disturbed by extrinsic forces. Now uniform motion means motion through equal spaces in equal times,
so that if we have a body entirely free from all resistance
or perturbation, and can measure equal spaces of its path,
we have a perfect measure of time. But let it be remembered at the same time, that this law has never been
absolutely proved by experience ; for we cannot point to
any body, and say that it is wholly unresisted or undisturbed ; and even if we had such a body, we should need
some entirely independent standard of time to ascertain
whether its motion was really uniform. As it is in moving
bodies that we find the best standard of time, we cannot
theoretically speaking use them to prove the uniformity
of their own movements, which would amount to a *petitio
principii*. Our experience amounts to this, that when
we examine and compare the movements of bodies which
seem to us nearly free from disturbance, we find them
give nearly harmonious measures of time. If any one

body which seems to us to move uniformly is not doing so, but is subject to fits and starts unknown to us, because we have no absolute standard of time, then all other bodies must be subject to exactly the same arbitrary fits and starts, otherwise there would be a discrepancy between them disclosing the irregularities. Just as in comparing together a number of chronometers, we should soon detect bad ones by their irregular going, as measured by the others, so in nature we detect disturbed movement by its discrepancy from that of other bodies, which we believe to be undisturbed, and which agree very nearly among themselves. But inasmuch as the measure of motion involves time, and the measure of time involves motion, there must be ultimately an assumption. We may define equal times, as times during which a moving body under the influence of no force describes equal spaces[g], but all we can say in its support is, that it leads us into no known difficulties, and that to the best of our experience, one freely moving body gives exactly the same results as any other.

When we inquire where the freely moving body is, no satisfactory answer can be given. Practically the rotating globe is sufficiently accurate, and Thomson and Tait say: 'Equal times are times during which the earth turns through equal angles[h]'. No long time has passed since astronomers thought it impossible to detect any inequality in its movement. Poisson was supposed to have proved that a change in the length of the sidereal day, amounting to one ten-millionth part in 2500 years, was incompatible with an ancient eclipse recorded by the Chaldæans, and similar calculations were made by Laplace. But it is now known that these calculations were somewhat in error,

---

[g] Rankine, 'Philosophical Magazine,' Feb. 1867, vol. xxxiii. p. 91.

[h] 'Treatise on Natural Philosophy,' vol. i. p. 179.

and that the dissipation of energy arising out of the friction of tidal waves, and the radiation of the heat into space, has slightly decreased the rapidity of the earth's rotatory motion. The sidereal day is now longer by one part in 2,700,000, than it was in 720 B.C. Even before this discovery, it was certain that the invariable rotation depended upon the perfect maintenance of the earth's internal heat, which is requisite in order that the earth's dimensions shall be unaltered. Now the earth being far superior in temperature to empty space, must cool more or less rapidly, so that it cannot furnish an absolute measure of time. Similar objections could be raised to all other rotating bodies within our cognizance.

The moon's motion round the earth, and the earth's motion round the sun, form the next best measure of time. They are subject, indeed, to all kinds of disturbance from other planets, but it is believed that these must in the course of time run through their rhythmical courses, and leave the mean distances unaffected, and consequently, by the third Law of Kepler, the periodic times unchanged. But there is more reason than not to believe that the earth encounters a certain slight resistance in passing through space, like that which is so apparent in Encke's comet. There may also be a certain dissipation of energy in the electrical relations of the earth to the sun, possibly identical with that which is manifested in the retardation of comets[i]. It is probably an untrue assumption then, that the earth's orbit remains quite invariable, and if so our last hope of getting a really uniform measure of time disappears, and we are reduced to accepting such as are sufficient for all practical purposes.

[i] 'Proceedings of the Manchester Philosophical Society,' 28th Nov. 1871, vol. xi. p. 33.

It is just possible that in the course of time, some other body may be found to furnish a better standard of time than the earth in its annual motion. The greatly superior mass of Jupiter and its satellites, and their greater distance from the sun, may render the electrical dissipation of energy less considerable even than in the case of the earth. But the choice of the best measure will always be an open one, and whatever moving body we assume, may ultimately be shown to be subject to disturbing forces.

The pendulum, although so admirable an instrument for subdivision of time, entirely fails as a standard; for though the same pendulum affected by the same force of gravity would perform equal vibrations in equal times, yet the slightest change in the form or weight of the pendulum, the slightest corrosion of any part, or the most minute displacement of the point of suspension, would falsify the results, and there enter many other difficult questions of temperature, resistance, length of vibration, &c.

Thomson and Tait are of opinion[k] that the ultimate standard of chronometry must be founded on the physical properties of some body of more constant character than the earth; for instance, a carefully arranged metallic spring, hermetically sealed in an exhausted glass vessel. Although their suggestion is no doubt theoretically correct, it is hard to see how we can be sure that the dimensions and elasticity of a piece of wrought metal will remain perfectly unchanged for the few millions of years contemplated by them. A nearly perfect gas, like hydrogen, is perhaps the only kind of substance in the unchanged elasticity of which we could have confidence. Moreover, it is difficult to perceive how the undulations of such a

---

[k] 'The Elements of Natural Philosophy,' part i. p. 119.

spring could be observed with the requisite accuracy. We thus appear to be devoid of any hope of establishing a sure standard of the efflux of time.

The Unit of Space and the Bar Standard.

Next in importance after the measurement of time is that of space. Time comes first in theory, because phenomena, our internal thoughts for instance, may change in time without regard to space magnitude. As to the phenomena of outward nature, they tend more and more to resolve themselves into the motion of molecules, and motion cannot be conceived or measured without reference both to time and space.

Turning now to space measurements, we find it almost equally difficult to fix and define once and for ever, a unit magnitude. There are three different modes in which it has been proposed to attempt the perpetuation of a standard length.

(1) By constructing an actual specimen of the standard yard or metre, in the form of a bar.

(2) By assuming the globe itself to be the ultimate standard of magnitude, the practical unit being a submultiple of some dimension of the globe.

(3) By adopting the length of a simple pendulum, beating seconds as a standard of reference.

At first sight it might seem that there was no great difficulty in this matter, and that any one of these methods might serve well enough ; but the more minutely we inquire into the details, the more hopeless appears to be the attempt to establish an invariable standard. We must in the first place point out a principle not of an obvious character, namely, that *the standard length must be defined by one single object*[1]. To make two bars of exactly the

---

[1] See Harris' ' Essay upon Money and Coins,' part ii. [1758] p. 127.

same length, or even two bars bearing a perfectly defined ratio to each other, is beyond the power of human art. If two copies of the standard metre be made and declared equally correct, future investigators will certainly discover some discrepancy between them, proving of course that they cannot both be the standard, and giving cause for dispute as to what magnitude should then be taken as correct.

If one invariable bar could be constructed and maintained as the absolute standard, no such inconvenience could arise. Each successive generation as it acquired higher powers of measurement, would detect errors in the copies of the standard, but the standard itself would be unimpeached, and would, as it were, become by degrees more and more accurately known. Unfortunately to construct and preserve a metre or yard is also a task which is either impossible, or what comes nearly to the same thing, cannot be shown to be possible. Passing over the practical difficulty of defining the ends of the standard length with complete accuracy, whether by dots or lines on the surface, or by the terminal points of the bar, we have no means of proving that substances remain of invariable dimensions. Just as we cannot tell whether the rotation of the earth is uniform, except by comparing it with other moving bodies, believed to be more uniform in motion, so we cannot detect the change of length in a bar, except by comparing it with some other bar supposed to be invariable. But how are we to know which is the invariable bar? It is certain that many rigid and apparently invariable substances do change in dimensions. The bulb of a thermometer certainly contracts by age, besides undergoing rapid changes of dimensions when warmed or cooled through 100° Cent.[m] Can we

---

[m] Watts' 'Dictionary of Chemistry,' vol. v. pp. 766, 767. Dr. Joule has recently confirmed the statements concerning the contraction of a thermometer-bulb.

be sure that even the most solid metallic bars do not slightly contract by age, or undergo variations in their structure by change of temperature. M. Fizeau was induced to try whether a quartz crystal, subjected to several hundred alternations of temperature, would be modified in its physical properties, and he was unable to detect any change in the coefficients of expansion[n]. It does not follow however, that, because no apparent change was discovered in a quartz crystal, newly-constructed bars of metal would undergo no change.

The only principle, as it seems to me, upon which the perpetuation of a standard of length can be ultimately rested, is that, if a variation of length occurs, it will in all probability be of different amount in different substances. If then a great number of standard metres were constructed of all kinds of different metals, alloys; hard rocks, such as granite, serpentine, slate, quartz, limestone ; artificial substances, such as porcelain, glass, &c., &c., careful comparison would show from time to time the comparative variations of length of these different substances. The most variable substances would be the most divergent, and the true standard would be furnished by the mean length of those which agreed most closely with each other, just as uniform motion is that of those bodies which agree most closely in indicating the efflux of time.

The Terrestrial Standard.

The second method assumes that the globe itself is a body of invariable dimensions. The founders of the metrical system selected the ten-millionth part of the distance from the equator to the pole as the definition of the metre, and the late Sir John Herschel proposed[o] that

[n] 'Philosophical Magazine,' (1868), 4th Series, vol. xxxvi. p. 32.
[o] 'Familiar Lectures on Scientific Subjects,' (1866) p. 191.

the English inch, which is now almost exactly the
500,500,000th part of the polar axis of the earth, should
be made exactly equal to the 500,000,000th part, and be
adopted as our standard.   The first imperfection in such
a method is that the earth is certainly not invariable in
size; for we know that it is superior in temperature to surrounding space, and must be slowly cooling and contracting. There is much reason to believe that all earthquakes,
volcanoes, mountain elevations, and changes of sea level,
are evidences of this contraction as asserted by Mr. Mallet[p].
But such is the vast bulk of the earth and the duration
of its past existence, that this contraction is perhaps less
rapid in proportion than that of any bar or other material
standard which we can construct.

The second and chief difficulty of this method arises
from the vast size of the earth, which prevents us from
making any comparison with the ultimate standard, except by a trigonometrical survey of a most elaborate and
costly kind. The French physicists, who first proposed
the method, attempted to obviate this inconvenience by
carrying out the survey once for all, and then constructing
a standard metre, which should be exactly the one ten
millionth part of the distance from the pole to the
equator.   But since all measuring operations are merely
approximate, as so often stated in previous pages, it was
impossible that this operation could be perfectly achieved.
Accordingly it was shown by Colonel Puissant in 1838,
that the supposed French metre was erroneous to the considerable extent of one part in 5527, the quadrant of the
earth's circumference measuring 10,001,789 instead of
10,000,000 of such metres. It then became necessary
either to alter the length of the assumed metre, or
otherwise to abandon its supposed relation to the earth's
dimensions.

[p] 'Proceedings of the Royal Society,' 20th June, 1872, vol. xx. p. 438.

The French Government and the present International Metrical Commission have for obvious reasons decided in favour of the latter course, and have thus reverted to the first method of defining the metre by a given bar. As from time to time the ratio between this assumed standard metre and the dimensions of the earth becomes more and more accurately known, we have the better means of restoring that metre by actual reference to the globe if required. But until lost, destroyed, or for some clear reason discredited, the bar metre and not the globe is the standard. Any of the more accurate measurements of the English trigonometrical survey might in like manner be employed to restore our standard yard, in terms of which the results are recorded٩.

The Pendulum Standard.

The third method of defining a standard length, by reference to the seconds' pendulum, was first proposed by Huyghens, and was at one time adopted by the English Government. From the principle of the pendulum (p. 353) it clearly appears that if the time of oscillation and the force actuating the pendulum be the same, the length must be the same. We do not get rid of theoretical difficulties, for we must practically assume the attraction of gravity at some point of the earth's surface, say London, to be unchanged from time to time, and the sidereal day to be invariable, neither assumption being absolutely correct so far as we can judge. The pendulum, in short, is only an indirect means of making one physical quantity of space depend upon two other physical quantities of time and force.

The practical difficulties are, however, of a far more

٩ Thomson and Tait's 'Elements of Natural Philosophy,' Part 1. p. 119.

serious character than the theoretical ones. The length
of a pendulum is not the ordinary length of the instrument, which might be greatly varied, without affecting
the duration of a vibration, but the distance from the
centre of suspension to the centre of oscillation. There is
no direct means of determining this centre, which depends
upon the average momentum of all the particles of the
pendulum as regards the centre of suspension. Huyghens
discovered that the centres of suspension and oscillation
are interchangeable, and Captain Kater pointed out that
if a pendulum vibrates with exactly the same rapidity
when suspended from two different points, the distance
between these points is the true length of the equivalent
simple pendulum[r]. But the practical difficulties in employing Kater's reversible pendulum are considerable, and
questions regarding the disturbance of the air, the force
of gravity or even the interference of electrical attractions
have to be entertained. It has been shown that all the
experiments made under the authority of government for
establishing the ratio between the standard yard and the
seconds' pendulum, were vitiated by an error in the corrections for the resisting, adherent or buoyant power of the
air in which the pendulum swung. Even if such corrections were rendered unnecessary by operating in a vacuum,
other difficult questions remain[s]. Gauss' mode of comparing the vibrations of a wire pendulum when suspended
at two different lengths is open to equal or greater practical difficulties. Thus it is found that the pendulum
standard cannot compete in accuracy and certainty with
the simple bar standard, and the method would only be
useful as an accessory mode of restoring the bar standards
if at any time again destroyed.

[r] Kater's 'Treatise on Mechanics,' Cabinet Cyclopædia, p. 154.
[s] Grant's 'History of Physical Astronomy,' p. 156.

Unit of Density.

Before we can measure and define the phenomena of nature, we require a third independent unit, which shall enable us to define the quantity of matter which occupies any given space. All the motions and changes of nature, as we shall see, are probably so many manifestations of energy ; but energy requires some substratum or material machinery of molecules, in and by which it may be exerted. Very little observation shows that, as regards force, there may be two modes of variation of matter. The force required to set a body in motion, varies in simple proportion to the bulk or cubic dimensions of the matter, but also according to its quality. Two cubic inches of iron of uniform quality, will require twice as much force to produce a certain velocity in a given time as one cubic inch ; but one cubic inch of gold will require more force than one cubic inch of iron. There is then some new measurable quality in matter apart from its bulk, which we may call *density*, and which is, strictly speaking, indicated by its capacity to resist and absorb the action of force. For the unit of density we may assume that of any substance which is uniform in quality, and can readily be referred to from time to time. Pure water at any definite temperature, for instance that of snow melting under an inappreciable pressure, furnishes a natural and invariable standard of density, and by testing equal bulks of various substances compared with a like bulk of ice-cold water, as regards the velocity produced in a unit of time by the same force, we should ascertain the densities of those substances as expressed in that of water.

Practically the force of gravity is used to measure density ; for a simple and beautiful experiment with the pendulum, performed by Newton and Gauss, shows that all kinds of matter equally gravitate, that is, the attractive power of a substance is exactly proportional to its density. Two portions of matter then which are in equilibrium in the balance, may be assumed to possess equal inertia, and their densities will therefore be inversely as their cubic dimensions.

Unit of Mass.

Multiplying the number of units of density of a portion of matter, by the number of units of space occupied by it, we arrive at the quantity of matter, or, as it is usually called, the units of *mass*, as indicated by the inertia and gravity it possesses. To proceed in the most simple and logical manner, the unit of mass ought to be that of a cubic unit of matter of the standard density. The founders of the French metrical system took as their unit of mass, the cubic centimetre of water, at the temperature of maximum density (about 4° Centigrade). They called this unit of mass the *gramme*, and constructed standard specimens of the kilogram, which might be readily referred to by all who required to employ accurate weights. Unfortunately, however, the determination of the bulk of a given weight of water at a certain temperature is an operation involving many practical and theoretical difficulties, and it can not be performed in the present day with a greater exactness than that of about one part in 5000, the results of careful observers being sometimes found to differ as much as one part in 1000 [t].

Weights, on the other hand, can be compared with each other to at least one part in a million. Hence if different specimens of the kilogram be prepared by direct

[t] Clerk Maxwell's 'Theory of Heat,' p. 79.

weighing against water, they will not agree very closely
with each other ; and, as a matter of fact, the two principal
standard kilograms neither agree with each other, nor
with their true definition ". The so-called Kilogram des
Archives weighs 15432·34874 grains according to Prof.
W H. Miller, while the kilogram deposited at the
Ministry of the Interior in Paris, as the standard for
commercial purposes, weighs 15432·344 grains [x].

Now since a standard weight constructed of platinum,
or platinum and iridium, can be preserved in all probability free from any appreciable alteration, and since it
can be very accurately compared with other weights, we
shall ultimately attain the greatest exactness in our
recorded measurements of weight and mass, by assuming
some single standard kilogram as a provisional standard,
leaving the determination of its actual mass in units of
space and density for future investigation. This is what
is practically done at the present day, and thus a unit of
mass takes the place of the unit of density, both in the
French and the present English systems. The English
pound is defined by a certain lump of platinum, carefully
preserved at Westminster, and is an entirely arbitrary
mass, made to agree as nearly as possible with old English
pounds. The gallon, the old English unit of cubic measurement, is defined by the condition that it shall contain exactly ten pounds weight of water at 62° Fahr.; and
although it is stated that it has the capacity of about
277·274 cubic inches, this ratio between the cubic and
linear system of measurement is not legally enacted, but
is left open to investigation from time to time. While
the French metric system as originally designed was
theoretically perfect, it does not seem to differ practically
in this point from the English system.

" Thomson and Tait's 'Treatise on Natural Philosophy,' vol. i.
p. 325.   [x] Ibid.

Subsidiary Units.

Having once established the standard units of time, space, and density or mass, we might employ them for the expression of all quantities of such nature. But it is often found convenient in particular branches of science, to use multiples or submultiples of the original units, for the expression of quantities, in a clear and simple manner. We use the mile rather than the yard when treating of the magnitude of the globe, and the mean distance of the earth and sun is not too large a unit when we have to describe the distances of the stars. On the other hand, when we are occupied with microscopic objects, the inch, the line or the millimetre, become the most convenient terms of expression.

It is allowable for a scientific man to introduce a new unit in any branch of knowledge, provided that it assists precise expression, and is carefully brought into relation with the primary units. Thus Prof. A. W. Williamson has proposed as a convenient unit in chemical science, an absolute volume equal to about 11·2 litres, representing the bulk of one gramme of hydrogen gas at standard temperature and pressure, or the *equivalent* weight of any other gas, such as 16 grammes of oxygen, 14 grammes of nitrogen, &c.; in short, the bulk of that quantity of any one of those gases which weighs as many grammes as there are units in the number expressing its atomic weight[y]. Professor Hofmann has also proposed a new concrete unit for chemists, called a *crith*, to be defined by the weight of one cubic decimetre or litre of hydrogen gas at 0° C. and 0°·76mm., weighing about 0·0896 grammes[z]. Both these units if adopted must be regarded as purely subordinate units, ultimately defined by reference to the primary units, and not involving any new assumption.

[y] 'Chemistry for Students,' by A. W. Williamson. Clarendon Press Series, 2nd ed. Preface p. vi.   [z] 'Introd. to Chemistry,' p. 131.

Derived Units.

The standard units of time, space, and mass having been once fixed, it becomes obvious that many kinds of magnitude are naturally measured by units immediately derived from one or more of the three principal ones. From the standard metre of linear magnitude follows in the most obvious manner the centaire or square metre, the unit of superficial magnitude, and the litre or cube of the tenth part of a metre, the standard of capacity or volume. Velocity of motion, again, is expressed by the ratio of the space passed over, when the motion is uniform, to the time occupied; hence the unit velocity will be that of a body which passes over a unit of space in a unit of time, say one metre per second. Momentum is measured by the mass moving, regard being paid both to the amount of matter and the velocity at which it is moving. Hence the unit of momentum will be that of a unit volume of matter of the unit density moving with the unit velocity, or in the French system, a cubic centimetre of water of the maximum density moving one metre per second.

An accelerating force is measured by the ratio of the momentum generated to the time occupied, the force being supposed to act uniformly. The unit of force will therefore be that which generates a unit of momentum in a unit of time, or which causes, in the French system, one cubic centimetre of water at maximum density to acquire in one second a velocity of one metre per second. The force of gravity is the most familiar kind of force, and as when acting unimpeded upon any substance it produces in a second a velocity of 9·80868 .... metres per second in Paris, it follows that the absolute unit of force is about the tenth part of the force of gravity. If we employ British weights and measures, the absolute unit of force is represented by the gravity of about half an ounce, since the force of gravity of any portion of matter acting upon that matter during one second, produces a final velocity of 32·1889 feet per second or about 32 units of velocity. Although from its perpetual presence and approximate uniformity we find in gravity the most convenient force for reference, and thus habitually employ it to estimate quantities of matter or mass, we must remember that it is only one of many instances of force. Strictly speaking, we should express weight in terms of force, but practically we express all forces in terms of weight.

We still require the unit of energy, a more complex notion. The momentum of a body expresses the quantity of motion which belongs or would belong to the aggregate of the particles, but when we consider how this motion is related to the action of a force producing or removing it, we find that the effect of a force is proportional to the mass multiplied by the square of the velocity and it is most convenient to take half this product as the expression required. But it is shown in books upon Dynamics that it will be exactly the same thing if we define energy by a force acting through a certain space. The natural unit of energy will then be that which overcomes a unit of force acting through a unit of space ; when we lift one kilogram through one metre, against gravity, we therefore accomplish 9·80868 . . . . units of work, that is, we turn so many units of potential energy existing in the muscles, into potential energy of gravitation. In lifting one pound through one foot there is in like manner a conversion of 32·1889 units of energy. Accordingly the unit of energy will be that required to lift a kilogram through about one tenth part of a metre against gravity, or, in the English system, to lift one pound through the thirty-second part of a foot.

Every person is at perfect liberty to measure and record

quantities in terms of any unit which he likes to adopt. He may use the yard for linear measurement and the litre for cubic measurement, only there will then be a complicated relation between his different results. The system of derived units which we have been briefly considering, is that which gives the most simple and natural relation between quantitative expressions of different kinds, and therefore conduces to ease of comprehension and saving of laborious calculation.

Provisionally Independent Units.

Ultimately, as we can hardly doubt, all phenomena will be recognised as so many manifestations of energy ; and, being expressed in terms of the unit of energy, will be referable to the primary units of space, time, and mass. To effect this reduction, however, in any particular case, we must not only be able to compare different quantities of the phenomenon, but to trace the whole series of steps by which it is connected with the primary notions. We can readily observe that the intensity of one source of light is greater than that of another ; and, knowing that the intensity of light decreases as the square of the distance, we can easily determine their comparative brilliance. Hence we can express the intensity of light falling upon any surface, if we have a unit in which to make the expression. Light is undoubtedly one form of energy, and the unit ought therefore to be the unit of energy. But at present it is quite impossible to say how much energy there is in any particular amount of light. The question then arises,—Are we to defer the measurement of light until we can fully and accurately assign its relation to other forms of energy ? If we answer Yes, it is equivalent to saying that the science of light must stand still perhaps for a generation ;

and not only this science but almost every other. The true course evidently is to select, as the provisional unit of light, some light of convenient intensity, which can be reproduced from time to time in exactly the same intensity, and which is defined by physical circumstances. All the phenomena of light may be experimentally investigated relatively to this unit, for instance that obtained after much labour by Bunsen and Roscoe[a]. In after years it will become a matter of inquiry what is the energy exerted in such unit of light; but it may be long before the relation is exactly determined.

A provisionally independent unit, then, means one which is assumed and physically defined in a safe and reproducible manner, in order that particular quantities may be compared *inter se* more accurately than they can yet be referred to the primary units. In reality almost all our measurements are made by such independent units. Even the unit of mass is practically an independent one, as we have seen (p. 373).

Similarly the unit of heat ought to be simply the unit of energy, already described. But a weight can be measured to the one-millionth part, and temperature to less than the thousandth part of a degree Fahrenheit, and to less therefore than the five-hundredth thousandth part of the absolute temperature, whereas the mechanical equivalent of heat is probably not known to the thousandth part. Hence the need of a provisional unit of heat, which is often taken as that requisite to raise a unit weight of water (say one gramme) through one degree Centigrade of temperature, that is from 0° to 1°. This quantity of heat is capable of approximate expression in terms of time, space, and mass; for by the natural constant, determined by Dr. Joule, and called the mechanical

[a] ' Philosophical Transactions ' (1859), vol. cxlix. p. 884, &c.

equivalent of heat, we know that the assumed unit of heat is equal to the energy of 423·55 gramme-metres, or that energy which will raise the mass of 423·55 grammes through one metre against 9·80868 absolute units of force. Heat may also be expressed in terms of the quantity of ice at 0° Cent., which it is capable of converting into water under an inappreciable pressure.

The science of electricity has lately become so much a matter of quantity, that it is necessary to have some means of accurate expression. When we know exactly the mechanical equivalent of electricity, we can express quantities of electricity in terms of energy, but in the meantime we need some easy available unit. The British Association accordingly have selected as the unit of electrical force that which can just overcome the resistance offered by a piece of pure silver wire 1 metre in length, and 1 millemetre in diameter. This unit must be regarded as merely a convenient provision for working purposes, to be employed for the easy expression of quantities not yet brought into precise relation with the ultimate standards of time, space, and mass. There may also be other provisionally independent units employed in electrical science, such as the voltametric unit of current strength, namely, that current which by decomposing water produces one cubic centimetre of detonating gas at 0° Cent. and 760 mm. of pressure in one minute. The unit of electrical quantity, again, is that quantity which when concentrated in a point and acting on an equal quantity also concentrated in a point at a unit of distance, exerts a repulsion equal to the unit of force. There must also be a unit of electro-magnetic force. All these electrical units must, however, be definitely related to each other, and to the fundamental units, and it is a matter for continual investigation to determine such relations more and more accurately.

Natural Constants and Numbers.

Having acquired accurate measuring instruments, and decided upon the units in which the results shall be estimated and expressed, there remains the question, What use shall be made of our powers of measurement? Our principal object must be to discover general quantitative laws of nature; but a very large amount of preliminary labour is employed in the accurate determination of the dimensions of existing objects, and the numerical relations between diverse forces and phenomena. Step by step every part of the material universe is surveyed and brought into known relations with other parts. Each manifestation of energy is correlated with each other kind of manifestation. Professor Tyndall has described the care with which such operations are conducted [b].

'Those who are unacquainted with the details of scientific investigation, have no idea of the amount of labour expended on the determination of those numbers on which important calculations or inferences depend. They have no idea of the patience shown by a Berzelius in determining atomic weights; by a Regnault in determining coefficients of expansion; or by a Joule in determining the mechanical equivalent of heat. There is a morality brought to bear upon such matters which, in point of severity, is probably without a parallel in any other domain of intellectual action.'

Every new natural constant which is recorded brings many fresh inferences within our power. For if n be the number of such constants known, then $\frac{1}{2}(n^2-n)$ is the number of ratios which are within our powers of calculation, and this increases with the square of n. We thus gradually piece together a map of nature, in which the lines of inference from one phenomenon to another

[b] Tyndall's 'Sound,' 1st ed. p. 26.

rapidly grow in complexity, and the powers of scientific prediction are correspondingly augmented.

The late Mr. Babbage[c] proposed the formation of a complete collection of all the constant numbers of nature; but such a collection would be almost coextensive with the whole mass of scientific literature. Almost all numbers occurring in works on Chemistry, Mineralogy, Physics, Astronomy, &c. are natural constants, and it would be impracticable to give in any one work more than a selection of the more important numbers.

Our present object will be to classify these constant numbers roughly, according to their comparative generality and importance, under the following heads :—

(1) Mathematical constants.
(2) Physical constants.
(3) Astronomical constants.
(4) Terrestrial numbers.
(5) Organic numbers.
(6) Social numbers.

Mathematical Constants.

At the head of the list of natural constants must come those which express the necessary relations of numbers to each other. The ordinary Multiplication Table is the most familiar and the most important of such series of constants, and is, theoretically speaking, infinite in extent. Next we must place the Arithmetical Triangle, the significance of which has already been pointed out (p. 206.) Tables of logarithms also contain vast series of natural constants, arising out of the relations of pure numbers. At the base of all logarithmic theory is the mysterious natural constant commonly denoted by E, *e*, or ϵ, being equal to the infinite series $1 + \frac{1}{1} + \frac{1}{1.2} + \frac{1}{1.2.3} + \frac{1}{1.2.3.4} + \ldots$,

[c] British Association, Cambridge, 1833. Report, pp. 484–490.

and thus consisting of the sum of the ratios between the numbers of permutations and combinations of o, 1, 2, 3, 4, &c. things.

Tables of prime numbers and of the factors of composite numbers must not be forgotten.

Another vast and in fact infinite series of numerical constants contains those connected with the measurement of angles, and embodied in trigonometrical tables, whether as natural or logarithmic sines, cosines, and tangents. It should never be forgotten that though these numbers find their chief employment in connexion with trigonometry, or the measurement of the sides of a right-angled triangle, yet the numbers themselves arise out of simple numerical relations bearing no special relation to space.

Foremost among trigonometrical constants is the well known number π, usually employed as expressing the ratio of the circumference and the diameter of a circle; from π follows the value of the arcual or natural unit of angular value as expressed in ordinary degrees (see p. 358).

Among other mathematical constants not uncommonly used may be mentioned tables of factorials (p. 202), tables of Bernouilli's numbers, tables of the error function [d], which latter are indispensable not only in the theory of probability but also in several other branches of science.

It should also be clearly understood that the mathematical constants and tables of reference already in our possession, although very extensive, are only an infinitely small part of what might be formed. With the progress of science the tabulation of new functions will be continually demanded, and it is worthy of consideration whether public money should not be constantly available

[d] See J. W. L. Glaisher, 'Philosophical Magazine,' 4th Series, vol. xlii. p. 421.

to reward the enormous labour which must be undertaken in these calculations. Such labours once successfully completed must benefit the whole human race as long as it shall exist. A valuable account of all the chief mathematical tables yet published will be found in De Morgan's article on *Tables,* in the 'English Cyclopædia,' Division of Arts and Sciences, vol. vii. p. 976.

Physical Constants.

The second class of constants contains those which refer to the actual constitution of matter. For the most part they depend upon the peculiarities of the chemical substance in question, but we may begin with those which are of the most general character. In a first sub-class we may place the velocity of light or heat undulations, the numbers expressing the relation between the lengths of the undulations, and the rapidity of the undulations, these numbers depending only on the properties of the ethereal medium, and being probably the same in all parts of the universe. The theory of heat gives rise to several numbers of the highest importance, especially Joule's mechanical equivalent of heat, the absolute zero of temperature, the mean temperature of empty space, &c.

Taking into account the diverse properties of the elements we must have tables of the atomic weights, the specific heats, the specific gravities, the refractive powers, not only of the elements, but their almost infinitely numerous compounds. The properties of hardness, elasticity, viscosity, expansion by heat, conducting powers for heat and electricity, must also be determined in immense detail. There are, however, certain of these numbers which stand out prominently because they serve as intermediate units or terms of comparison. Such are, for instance, the absolute coefficients of expansion of air, water, and mercury, the temperature of the maximum density of water ($39°\cdot101$ Fahr. or $4°\cdot0$ Cent.), the latent heats of water and steam, the boiling-point of water under standard pressure, the melting and boiling-points of mercury, and so on.

Astronomical Constants.

The third great class consists of numbers possessing far less generality because they refer, not to the universal properties of matter, but to the special forms and distances in which matter has been disposed in the part of the universe open to our examination. We have, first of all, to define the magnitude and form of the earth, its mean density, the constant of aberration of light expressing the relation between the earth's mean velocity in space and the velocity of light. From the earth, as our observatory, we then proceed to lay down the mean distances of the sun, and of the planets from the same centre ; all the elements of the planetary orbits, the magnitudes, densities, masses, periods of axial rotation of the several planets are by degrees determined with growing accuracy. The same labours must be gone through for the satellites. Catalogues of comets with the elements of their orbits, as far as ascertainable, must not be omitted.

From the earth's orbit as a new base of observations, we next proceed to survey the heavens and lay down the apparent positions, magnitudes, motions, distances, periods of variation, &c. of the stars. All catalogues of stars from those of Hipparchus and Tycho, are full of numbers expressing rudely the conformation of the visible universe. But there is obviously no limit to the labours of astronomers ; not only are millions of distant stars awaiting their first measurements, but those already registered require endless scrutiny as regards their movements in the three dimensions of space, their periods of revolution, their changes of brilliancy and colours. It is obvious that though astronomical numbers are conventionally called *constant*, they are in all cases probably subject to more or less rapid variation.

Terrestrial Numbers.

Our knowledge of the globe we inhabit involves many numerical determinations, which have little or no connexion with astronomical theory. The extreme heights of the principal mountains, the mean elevation of continents, the mean or extreme depths of the oceans, the specific gravities of rocks, the temperature of mines, all the host of numbers expressing the meteorological or magnetic conditions of every part of the surface must fall into this class. Many of such numbers are hardly to be called constant, being subject to periodic or even secular changes, but they are no more variable in fact than many which in astronomical science are set down as constant. In many cases quantities which seem most variable may go through rhythmical changes resulting in a nearly uniform average, and it is only in the long progress of physical investigation that we can hope to discriminate successfully between those elemental numbers which are absolutely fixed and those which vary. In the latter case the law of variation becomes the constant relation which is the object of our search.

Organic Numbers.

All the forms and properties of brute nature having been sufficiently defined by the previous classes of numbers, the organic world, both vegetable and animal, remains outstanding, and offers a higher series of phenomena for our investigation. All exact knowledge relating to the forms and sizes of living things, their numbers, the quantities of various compounds which they consume, contain, or excrete, their muscular or nervous energy, &c. must be placed apart in a class by themselves. All such numbers are doubtless more or less subject to variation, and but in a minor degree capable of exact determination. Man, so far as he is an animal, and as regards his physical form, must also be treated in this class.

Social Numbers.

Little or no allusion has hitherto been made in this work to the fact that man in his economical, sanitary, intellectual, æsthetic, or moral relations may become the subject of exact sciences, the highest and most useful of all sciences. Every one who is in any degree engaged in statistical inquiry or study must so far acknowledge the possibility of natural laws governing such statistical facts. Hence we must certainly allot a distinct place to all numerical information relating to the numbers, ages, physical and sanitary condition, mortality, of all different peoples, in short, to vital statistics. Economic statistics, comprehending the quantities of commodities produced, existing, exchanged, and consumed, constitute another most extensive body of science. In the progress of reason exact investigation may possibly subdue regions of phenomena which at present defy all analysis and scientific treatment. That scientific method can ever exhaust the phenomena of the human mind is on the other hand incredible.

CHAPTER XV.

IN the two preceding chapters we have been engaged in considering how a phenomenon may be accurately measured and expressed. So delicate and complex an operation is a measurement which pretends to any considerable degree of exactness, that no small part of the skill and patience of physicists is usually spent upon this operation. Much of this difficulty arises from the fact that it is scarcely ever possible to measure one simple phenomenon at a time. The ultimate object must be to discover the mathematical equation or law connecting a quantitative cause with its quantitative effect; this purpose usually involves, as we shall see, the varying of one condition at a time, the other conditions being maintained constant. The labours of the experimentalist would be comparatively light if he could carry out this rule of varying one circumstance at a time. He would then obtain a series of corresponding values of the variable quantities concerned, from which he might by proper hypothetical treatment obtain the required law of connexion. But in reality it is seldom possible to carry out this direction except in an approximate manner. Before then we proceed to the consideration of the actual process of quantitative induction, it is necessary to review the several devices by which the complication of effects can be disentangled. Every phenomenon measured will usually be the sum difference or product of two or more different effects, and these must be in some way analysed and separately measured before we possess the materials for a true inductive treatment.

Illustrations of the Complication of Effects.

It is easy to bring forward a multitude of instances to show that a phenomenon is seldom to be observed simple and alone. A more or less elaborate process of analysis is almost always necessary. Thus if an experimentalist wishes to observe and measure the expansion of a liquid by heat, he places it in a thermometer tube and registers the rise of the column of liquid in the narrow tube. But he cannot heat the liquid without also heating the glass, so that the change observed is really the difference between the expansions of the liquid and the glass. More minute investigation will show the necessity perhaps of allowing for further effects, namely the compression of the liquid or the expansion of the bulb due to the increased pressure of the column as it becomes lengthened.

In a great many cases an observed effect will be apparently at least the simple sum of two separate and independent effects. The heat evolved in the combustion of oil is partly due to the carbon and partly to the hydrogen. A measurement of the heat yielded by the two jointly, cannot inform us how much proceeds from the one and how much from the other. If by some separate determination we can ascertain how much the hydrogen yields, then by mere subtraction we learn what is due to the carbon; and *vice versâ*. The heat conveyed by a liquid, may be partly conveyed by true conduction, partly by convection. The light dispersed in the interior of a liquid consists both of what is reflected by floating particles and what is due to true fluorescence [a]; and we must find some mode of determining one portion before we can learn the other.

[a] Stokes, 'Philosophical Transactions' (1852), vol. cxlii. p. 529.

The apparent motion of the spots on the sun, is the algebraic sum of the sun's axial rotation, and of the proper motion of the spots upon the sun's face; hence the difficulty of ascertaining by direct observations the period of the sun's rotation.

We cannot obtain the weight of a portion of liquid in a chemical balance without weighing it with the containing vessel. Hence to have the real weight of the liquid operated upon in an experiment, we must have a separate weighing of the vessel, with or without the adhering film of liquid according to circumstances. This is likewise the mode in which a cart and its load are weighed together, the tare or weight of the cart previously ascertained being deducted. The variation in the height of the barometer is a joint effect, partly due to the real variation of the atmospheric pressure, partly due to the expansion of the mercurial column by heat. The effects may be discriminated, if, instead of one barometer tube we have two tubes placed closely side by side, so as to have exactly the same temperature. If one of them be closed at the bottom so as to be unaffected by the atmospheric pressure, it will show the changes due to temperature only, and, by subtracting these changes from those shown in the other tube, we get the real oscillations of atmospheric pressure. But this correction, as it is called, of the barometric reading, is better effected by calculation from the readings of an ordinary thermometer.

In a great many other cases a quantitative effect will be the difference of two causes acting in opposite directions. The late Sir John Herschel invented an instrument like a large thermometer which he called the Actinometer [b], and M. Pouillet constructed a somewhat similar instrument

[b] 'Admiralty Manual of Scientific Enquiry,' edited by Sir John Herschel, 2nd ed. p. 299.

called the Pyrheliometer, for ascertaining the heating power of the sun's rays. In both instruments the heat of the sun was absorbed by a reservoir containing water, and the rise of temperature of the water was exactly observed, either by its own expansion or by the readings of a delicate thermometer immersed in it. The details of the construction and use of these instruments are immaterial to our immediate purpose. Now in exposing the actinometer to the sun, we do not obtain the full effect of the heat absorbed, because the receiving surface is at the same time radiating heat into empty space. The observed increment of temperature is in short the difference between what is received from the sun and lost by radiation. But the latter quantity is capable of ready determination; we have only to shade the instrument from the direct rays of the sun, while leaving it exposed to the rest of the open sky, and we can observe how much it cools in a certain time. The total effect of the sun's rays will obviously be the apparent effect *plus* the cooling effect in an equal time. By alternate exposure in sun and shade during equal intervals the desired result may be obtained with considerable accuracy [c].

Two quantitative effects were beautifully distinguished in an experiment of John Canton, devised in 1761 for the purpose of demonstrating the compressibility of water [d]. He constructed a thermometer with a large bulb full of water and a short capillary tube, the part of which above the water was freed from air. Under these circumstances the water was relieved from the pressure of the atmosphere, but the glass bulb in bearing that pressure was somewhat contracted. He next placed the instrument under the receiver of an airpump, and on exhausting the air, observed the water sink in the tube. Having thus

[c] Pouillet, 'Taylor's Scientific Memoirs,' vol. iv. p. 45.

[d] Jamin, 'Cours de Physique,' vol. i. p. 158.

obtained a measure of the effect of atmospheric pressure on the bulb, he opened the top of the thermometer tube and admitted the air. The level of the water now sank still more, partly from the pressure on the bulb being now compensated, and partly from the compression of the water by the atmospheric pressure. It is obvious that the amount of the latter effect was approximately the difference of the two observed depressions.

Not uncommonly indeed the actual phenomenon which we wish to measure is considerably less than various disturbing effects which enter into the question. Thus the compressibility of mercury is considerably less than the expansion of the vessels in which it is measured under pressure, so that the attention of the experimentalist has chiefly to be concentrated on the change of magnitude of the vessels. Many astronomical phenomena, such as the parallax or proper motions of the fixed stars, are far less than the instrumental imperfections, and the other phenomena of precession, nutation, aberration, &c. Even Flamsteed imagined he had discovered the parallax of the pole star[c], and time after time astronomers mistook various other phenomena for that minute motion which they were so desirous to discover.

Methods of Eliminating Error.

In any particular experiment it is the object of the experimentalist to measure a single effect only, and he endeavours to obtain that effect free from any interfering effects. If this cannot be, as it seldom or never can really be, he makes the effect as considerable as possible compared with the other effects, which he reduces to a minimum, and treats as noxious errors. Those quantities,

[c] Baily's ' Account of the Rev. John Flamsteed,' p. 58.

which are called *errors* in one case, may really be most important and interesting phenomena in another investigation. When we speak of eliminating error we really mean disentangling the complicated phenomena of nature. The physicist rightly wishes to treat one thing at a time, but as this object can seldom be rigorously carried into practice, he has to seek some mode of counteracting the tendency to error.

The general principle of the subject is that a single observation can render known only a single quantity. Hence if several different quantitative effects are known to enter into any investigation, we must have at least as many distinct results of observation as there are quantities to be determined. Every complete experiment will therefore consist in general of several operations. Guided if possible by previous knowledge of the causes in action, we must arrange these determinations, so that by a simple mathematical process we may distinguish the separate quantities. There appear to be five principal methods in which we may accomplish this object; these methods are specified below and illustrated in the succeeding sections.

(1) *The Method of Avoidance.* The physicist may seek for some special mode of experiment or opportunity of observation, in which the error is non-existent or inappreciable.

(2) *The Differential Method.* He may find opportunities of observation when all interfering phenomena remain constant, and only the subject of observation is at one time present and another time absent; the difference between two exact observations then gives its amount.

(3) *The Method of Correction.* He may endeavour to estimate the amount of the interfering force by the best available mode, and then make a corresponding correction in the results of observation.

(4) *The Method of Compensation.* He may invent some mode of neutralizing the interfering force by balancing against it an exactly equal and opposite force of unknown amount.

(5) *The Method of Reversal.* He may so conduct the experiment that the interfering force may act in opposite directions, in alternate observations, the mean result being free from interference.

1. *Method of Avoidance of Error.*

Astronomers always seek opportunities of observation when errors will have the smallest effect. In spite of elaborate observations and long continued theoretical investigation, it is not found possible to assign any satisfactory law to the refractive power of the atmosphere. Although the apparent change of place of a heavenly body thus produced, may be more or less accurately calculated, yet the error depends upon the temperature and pressure of the atmosphere, and, when a ray is highly inclined to the perpendicular, the uncertainty in the refraction becomes very considerable. Hence astronomers always make their observations, if possible, when the object is at the highest point of its daily course, i.e. on the meridian. In some kinds of investigation, as, for instance, in the determination of the latitude of an observatory, the astronomer is at liberty to select one or more stars out of the countless number visible. There is an evident advantage in such a case, in selecting a star which passes close to the zenith, so that it may be observed almost entirely free from atmospheric refraction, as was done by Hooke. It was ingeniously suggested by Wallis that the parallax of the fixed stars might perhaps be detected by observations of the greatest azimuth east and west of some circumpolar star, since the refractive power of the atmosphere which affects only the altitude would thus be entirely avoided [f].

Astronomers also endeavour to render their clocks as accurate as possible, by removing the source of variation. The pendulum is perfectly isochronous so long as its length remains invariable, and the vibrations are exactly of equal length. They render it nearly invariable in length, that is in the distance between the centres of suspension and oscillation, by a compensatory arrangement for the change of temperature. But as this compensation may not be perfectly accomplished, some astronomers place their chief controlling clocks in a cellar, or other apartment, where the changes of temperature may be as slight as possible. At the Paris Observatory a clock has been placed in the caves beneath the building, where there is no appreciable difference between the summer and winter temperature.

To avoid the effect of unequal oscillations Huyghens made his beautiful investigations, which resulted in the discovery that a pendulum, of which the centre of oscillation moved upon a cycloidal path, would be perfectly isochronous, whatever the variation in the length of oscillations. But though a pendulum may be rendered in some degree cycloidal by the use of a steel suspension spring, it is found that the mechanical arrangements requisite to produce a truly cycloidal motion introduce more error than they avoid. Hence astronomers seek to reduce the error to the smallest amount by maintaining their clock pendulums in uniform movement [g]; and in fact while a clock is in good order and has the same weights, there need be little change in the length of oscillation.

[f] Grant, 'History of Physical Astronomy,' p. 548.
[g] Montucla, 'Histoire des Mathématiques,' vol. ii. p. 420.

When a pendulum cannot be made to swing uniformly, as in experiments upon the force of gravity, it becomes requisite to resort to the third method, and a correction is introduced, calculated on theoretical grounds from the amount of the observed change in the length of vibration.

It has been mentioned that the apparent expansion of a liquid by heat, when contained in a thermometer tube or other vessel, is the difference between the real expansion of the liquid and that of the containing vessel. The effects can be accurately distinguished provided that we can learn the real expansion by heat of any one convenient liquid; for by observing the apparent expansion of the same liquid in any required vessel we can by difference learn the amount of expansion of the vessel due to any given change of temperature. When we once know the change of dimensions of the vessel, we can of course determine the absolute expansion of any other liquid tested in it. Thus it became an all-important object in scientific research to measure with accuracy the absolute dilatation by heat of some one liquid, and mercury owing to several circumstances was by far the most suitable. Dulong and Petit devised a beautiful mode of effecting this by simply avoiding altogether the effect of the change of size of the vessel. Two upright tubes full of mercury were connected by a fine tube at the bottom, and were maintained at two different temperatures. As mercury was free to flow from one tube to the other by the connecting tube, the two columns necessarily exerted equal pressures by the principles of hydrostatics. Hence it was only necessary to measure very accurately by a cathetometer the difference of level of the surfaces of the two columns of mercury, to learn the difference of length of columns of equal hydrostatic pressure, which at once gives the difference of density of the mercury, and the dilatation by heat. The changes of dimension in the containing tubes now became a matter of entire indifference, and the length of a column of mercury at different temperatures was measured as easily as if it had formed a solid bar. The experiment was carried out by Regnault with many improvements of detail, and the absolute dilatation of mercury, at temperatures between 0° Cent. and 350°, was determined almost as accurately as was needful [h].

The presence of a large and uncertain amount of error may often render a method of experiment valueless. Foucault's beautiful mode of demonstrating the rotation of the earth by the motion of a pendulum was thus frustrated. The slightest lateral disturbance of the pendulum gave it an elliptical path with a progressive motion of the axis of the ellipse, and this motion of an unknown amount disguised and overpowered that due to the rotation of the earth [i]. Faraday's laborious experiments on the relation of gravity and electricity were much obstructed, too, by the fact that it is almost impossible to move a large weight of iron or even lead without generating currents of electricity, either by friction or induction. To distinguish the electricity directly due to the action of gravity from the greater quantities indirectly produced would have been a problem of excessive difficulty. Baily in his experiments on the density of the earth was aware of the existence of inexplicable disturbances which have since been referred to the action of electricity with much probability [k]. The skill and ingenuity of the experimentalist are often exhausted in devising a form of apparatus in which such causes of error shall be reduced to a minimum.

[h] Jamin, 'Cours de Physique,' vol. ii. pp. 15–28.
[i] 'Philosophical Magazine,' 1851, 4th Series, vol. ii. *passim.*
[k] Hearn, 'Philosophical Transactions,' 1847, vol. cxxxvii. pp. 217–221.

In some rudimentary experiments we may wish merely to establish the existence of a quantitative effect without precisely measuring its amount; if there exist causes of error of which we can neither render the amount known or inappreciable, the best way will be to make them all negative so that the quantitative effects will be less than the truth rather than greater. Mr. Grove, for instance, in proving that the magnetization or demagnetization of a piece of iron raises its temperature, took care to maintain the electro-magnet by which the iron was acted upon at a lower temperature, so that it would cool rather than warm the iron by radiation or conduction[1].

Rumford's celebrated experiment to prove that heat was generated out of mechanical force in the boring of a cannon was subject to the difficulty that heat might be brought to the cannon by conduction from neighbouring bodies. It was an ingenious device of Davy to produce friction by a piece of clock-work resting upon a block of ice in an exhausted receiver; as the machine rose in temperature above 32°, it was certain that no heat was received by conduction from the support[m]. In many other experiments ice may be employed to prevent the access of heat by conduction, and this device, first put in practice by Murray[n], is beautifully employed in Bunsen's calorimeter.

To obtain the true temperature of the air, though apparently so easy, is really a very difficult matter, because the thermometer employed is sure to be affected either by the sun's rays, the radiation from neighbouring objects, or the escape of heat into space. These sources

---

[1] 'The Correlation of Physical Forces,' 3rd ed. p. 159.

[m] 'Collected works of Sir H. Davy,' vol. ii. pp. 12–14. 'Elements of Chemical Philosophy,' p. 94.

[n] 'Nicholson's Journal,' vol. i. p. 241; quoted in 'Treatise on Heat,' Useful Knowledge Society, p. 24.

of error are too fluctuating to allow of correction, so that the only accurate mode of procedure is that devised by Dr. Joule, of surrounding the thermometer with a copper cylinder ingeniously adjusted to the temperature of the air, as described by him, so that the effect of radiation shall be nullified [o].

When the avoidance of error cannot be carried into effect, it will yet be desirable to reduce the absolute amount of the interfering error as much as possible before employing the succeeding methods to correct the result. As a general rule we can determine a quantity with less inaccuracy as it is smaller, so that if the error itself be small the error in determining that error will be of a still lower order of magnitude. But in some cases the absolute amount of an error is of no consequence, as in the index error of a divided circle, or the difference between a chronometer and astronomical time. Even the rate at which a clock gains or loses is a matter of little importance provided it remains constant, so that a sure calculation of its amount can be made.

2. *Differential Method.*

When we cannot avoid the entrance of error, we can often resort with great success to the second mode of measuring phenomena under such circumstances that the error shall remain nearly or quite the same in all the observations, and neutralize itself as regards the purposes in view. This mode is available whenever we want a difference between quantities and not the absolute quantity of either. The determination of the parallax of the fixed stars is exceedingly difficult, because the amount of parallax is far less than most of the corrections

[o] Clerk Maxwell, 'Theory of Heat,' p. 228. 'Proceedings of the Manchester Philosophical Society,' Nov. 26, 1867, vol. vii. p. 35.

for atmospheric refraction, nutation, aberration, precession, instrumental irregularities, &c., and can with difficulty be detected among these phenomena of various magnitude. But, as Galileo long ago suggested, all such difficulties would be avoided by the differential observation of stars, which though apparently close together are really far separated on the line of sight. Two such stars in close apparent proximity will be subject to almost exactly equal errors, so that all we need do is to observe the apparent change of place of the nearer star as referred to the more distant one. A good telescope furnished with an accurate micrometer is alone needed for the application of the method. Huyghens appears to have been the first observer who actually tried to employ the method practically P, but it was not until 1835 that the improvement of telescopes and micrometers enabled Struve to detect in this way the parallax of the star α Lyræ.

It is one of the many advantages of the observation of transits of Venus for the determination of the solar parallax that the refraction of the atmosphere affects in an exactly equal degree the planet and the portion of the sun's face over which it is passing. Thus the observations are strictly of a differential nature.

By the process of substitutive weighing it is possible to ascertain the equality or inequality of two weights with almost perfect freedom from error. If two weights A and B be placed in the scales of the best balance we cannot be sure that the equilibrium of the beam indicates exact equality, because the arms of the beam may be unequal or unbalanced. But if we take B out and put another weight C in, and equilibrium still exists, it is apparent that the same causes of erroneous

P History of ' Physical Astronomy,' p. 549. Herschel's ' Outlines of Astronomy,' 4th ed. p. 550.

weighing exist in both cases, supposing that the balance has not been disarranged, and that B must be exactly equal to C, since it has exactly the same effect under the same circumstances. In like manner it is a general rule that, if by any uniform mechanical process we get a copy of an object, it is unlikely that this copy will be precisely the same in magnitude and form as the original, but two copies will equally diverge from the original, and will therefore almost exactly resemble each other.

Leslie's Differential Thermometer [q] was well adapted to the experiments for which it was invented. Having two equal bulbs any alteration in the temperature of the air will act equally by conduction on each and produce no change in the indications of the instrument. Only that radiant heat which is purposely thrown upon one of the bulbs will produce any effect. This thermometer in short carries out the principle of the differential method in a mechanical manner.

3. *Method of Correction.*

Whenever the result of an experiment is affected by an interfering cause to an amount either invariable or exactly calculable, it is sufficient simply to add or subtract this calculated amount. We are said to correct observations when we thus eliminate what is due to extraneous causes, although of course we are only separating the correct effects of several agents. Thus the variation in the height of the barometrical column is partly due to the change of temperature, and since the coefficient of absolute dilatation of mercury has been exactly determined, as already described (p. 395), we have only to make cal-

---

[q] Leslie's 'Inquiry into the Nature of Heat,' p. 10.

culations of a simple character, or, what is better still, tabulate a series of such calculations for general use, and the correction for temperature can be made with all desired accuracy. The height of the mercury in the barometer is also affected by capillary attraction, which depresses it by a constant amount depending on the diameter of the tube. The requisite corrections can be estimated with accuracy sufficient for most purposes, more especially as we can check the correctness of the reading of a barometer by comparison with a perfect standard barometer, and introduce if need be an index error including both the error in the affixing of the scale and the effect due to capillarity. But in constructing the standard barometer itself we must take greater precautions; the capillary depression depends somewhat upon the quality of the glass, the absence of air, and the perfect cleanliness of the mercury, so that we cannot with confidence assign the exact amount of the effect. Hence a standard barometer is constructed with a wide tube, sometimes even an inch in diameter, so that the capillary effect may be rendered of little account[r]. Gay Lussac made barometers in the form of a siphon so that the capillary forces acting equally at the upper and lower surfaces should balance and destroy each other, but the method fails in practice because the lower surface, being open to the air, becomes sullied and subject to a different force of capillarity.

In a great many mechanical experiments friction is an interfering condition, and drains away a portion of the energy intended to be operated upon in a definite manner. We should of course reduce the friction in the first place to the lowest possible amount, but as it cannot be altogether prevented, and is not calculable with certainty from any general laws, we must determine it

---

[r] Watts' ‘Dictionary of Chemistry,’ vol. i. pp. 513–15.

separately for each apparatus by suitable experiments. Thus Smeaton, in his admirable but now almost forgotten researches concerning water-wheels, eliminated friction in the most simple manner by determining by trial what weight, acting by a cord and roller upon his model water-wheel, would make it turn without water as rapidly as the water made it turn. In short, he ascertained what weight concurring with the water would exactly compensate for the friction[s]. In Dr. Joule's experiments to determine the mechanical equivalent of heat by the condensation of air, a considerable amount of heat was produced by friction of the condensing pump, and a small portion by stirring the water employed to measure the heat. This heat of friction was ascertained by simply repeating the experiment in an exactly similar manner except that no condensation was effected, and observing the change of temperature then produced[t].

We may describe as *test experiments* any in which we perform operations not intended to give the quantity of the principal phenomenon, but some quantity which would otherwise remain as an error in the result. Thus in astronomical observations almost every source of error may be avoided by increasing the number of observations and distributing them in such a manner as to produce in the final mean as much error in one way as in the other. But there is one source of error, first discovered by Maskelyne, which cannot be avoided, because it affects all observations in the same direction and to the same average amount, namely the Personal Error of the observer or the inclination to record the passage of a star across the wires of the telescope a little too soon or a little too late. This personal error was first described in the 'Edinburgh Journal of Science,' vol. i. p. 178. The

[s] 'Philosophical Transactions,' vol. li. p. 100.
[t] 'Philosophical Magazine,' 3rd Series, vol. xxvi. p. 372.

difference between the judgment of observers at the
Greenwich Observatory usually varies from $\frac{1}{100}$ to $\frac{1}{3}$ of
a second, or even a little more, and remains pretty con-
stant for the same observers[u]. In some observers it has
amounted to seven or eight-tenths of a second[x]. De
Morgan appears to have entertained the opinion that
this source of error was essentially incapable of elimination or correction[y]. But it seems clear that this personal
error might be determined absolutely with any desirable
degree of accuracy by test experiments, consisting in
making an artificial star move at a considerable distance
and recording by electricity the exact moment of its
passage over the wire. This method has in fact been
successfully employed in Leyden, Paris, and Neuchatel[z].

Newton employed the pendulum for making experiments on the impact of balls. Two balls were hung in
contact, and one of them, being drawn aside through a
measured arc, was then allowed to strike the other, the
arcs of vibration giving sufficient data for calculating the
distribution of energy at the moment of impact. The
resistance of the air was an interfering cause which he
estimated very simply by causing one of the balls to
make several complete vibrations and then marking the
reduction in the length of the arcs, a proper fraction
of which reduction was added to each of the other observed arcs of vibration[a].

In the modern use of the pendulum, to measure
terrestrial gravity, it is not found convenient to annul

[u] 'Greenwich Observations for 1866,' p. xlix.

[x] 'Penny Cyclopædia,' art. *Transit*, vol. xxv. pp. 129, 130.

[y] Ibid. art. *Observation*, p. 390.

[z] 'Nature,' vol. i. pp. 85, 337. See references to the Memoirs describing the method.

[a] 'Principia,' Book I. Law III. Corollary VI. Scholium. Motte's
translation, vol. i. p. 33.

the resistance of the air by operating in a vacuum. Consequently this resistance has to be ascertained by appropriate and tedious series of experiments, which should be made if possible upon each pendulum employed.

The exact definition of the standard of length is one of the most important, as it is one of the most difficult questions in physical science, and the different practice of different nations introduces wholly needless confusion. Were all standards constructed so as to give the true length at a fixed uniform temperature, for instance the freezing-point, then any two standards could be compared without the interference of temperature by bringing them both to exactly the same fixed temperature. Unfortunately the French metre is defined by a bar of platinum at 0°C, while our yard is defined by a bronze bar at 62°F. It is quite impossible, then, to make a comparison of the yard and metre without the introduction of a correction, either for the expansion of platinum or bronze, or both. Bars of metal differ too so much in their rates of expansion according to their molecular condition that it is dangerous to infer from one bar to another.

When we come to use instruments with great accuracy there are many minute sources of error which must be guarded against. If a thermometer has been graduated when perpendicular, it will read somewhat differently when laid down, as the pressure of a column of mercury is removed from the bulb. The reading may also be somewhat altered if it has recently been raised to a higher temperature than usual, if it be placed under a vacuous receiver, or if the tube be unequally heated as compared with the bulb. For these minute causes of error we may have to introduce troublesome corrections, unless we adopt the simple mode of using the thermometer in circumstances of position, &c. exactly similar to those in which it was graduated [c]. There is no end to the number of minute corrections which may ultimately be required. A very large number of experiments on gases, standard weights and measures, &c. depend upon the height of the barometer ; but when experiments in different parts of the world are compared together we ought to take into account the varying force of gravity, which even between London and Paris makes a difference of ·008 inch of mercury.

The measurement of quantities of heat is a matter of great difficulty, because there is no known substance impervious to heat, and the problem is therefore as difficult as to measure liquids in porous vessels. To determine the latent heat of steam we must condense a certain amount of the steam in a known weight of water, and then observe the rise of temperature of the water. But while we are carrying out the experiment, part of the heat will have escaped by radiation or conduction from the condensing vessel or calorimeter. We may indeed reduce the loss of heat by using vessels with double sides and bright surfaces, surrounded with swan's-down wool or other non-conducting materials ; and we may also avoid raising the temperature of the water much above that of the surrounding air. Yet we cannot by any such means render the loss of heat inconsiderable. Rumford ingeniously proposed to reduce the loss to zero by commencing the experiment when the temperature of the calorimeter is as much below that of the air as it is at the end of the experiment above it. Thus the vessel will first gain and then lose by radiation and conduction, and these opposite errors will approximately balance each other. But Regnault has shown that the loss and gain do not proceed by exactly the same laws, so that in very accurate investigations Rumford's method is not sufficient. There

[c] Balfour Stewart, ' Elementary Treatise on Heat,' p. 16.

remains the method of correction which was beautifully carried out by Regnault in his determination of the latent heat of steam. He employed two calorimeters, made in exactly the same way and alternately used to condense a certain amount of steam, so that while one was measuring the latent heat, the other calorimeter was engaged in determining the corrections to be applied, whether on account of radiation and conduction from the vessel or on account of heat reaching the vessel by means of the connecting pipes[d].

4. *Method of Compensation.*

There are many cases in which a cause of error cannot conveniently be rendered null, and is yet beyond the reach of the third method, that of calculating the requisite correction from independent observations. The magnitude of an error may be subject to continual variations, on account of change of weather, or other fickle circumstances beyond our control. It may either be impracticable to observe the variation of those circumstances in sufficient detail, or, if observed, the calculation of the amount of error may be subject to doubt. In these cases, and only in these cases, it will be desirable to invent some artificial mode of counterpoising the variable error against an equal error subject to exactly the same variation.

We cannot weigh any object with great accuracy unless we make a correction for the weight of the air displaced by the object, and add this to the apparent weight. In very accurate investigations relating to standard weights, it is usual to note the barometer and thermometer at the time of making a weighing, and, from the measured bulks of the objects compared, to calculate the weight of air

d Graham's 'Chemical Reports and Memoirs,' Cavendish Society, pp. 247, 268, &c.

displaced ; the third method in fact is adopted. To make
all the calculations in the frequent weighings requisite in
chemical analysis would be exceedingly laborious, hence
the correction is usually neglected. But when the chemist
wishes to weigh a quantity of gas contained in a glass
globe for the purpose of determining its specific gravity,
the correction becomes of much importance. Hence
chemists avoid at once the error, and the labour of correcting it, by attaching to the opposite scale of the balance
a sealed glass globe of exactly equal capacity to that
containing the gas to be weighed, noting only the difference of weight when the globe is full and empty. The
correction, being exactly the same for both globes, may be
entirely neglected [e].

A device of nearly the same kind is employed in the
construction of galvanometers which measure the force of
an electric current by the deflection of a suspended
magnetic needle. The resistance of the needle is partly
due to the directive influence of the earth's magnetism,
and partly to the torsion of the thread. But the former
force may often be inconveniently great as well as
troublesome to determine for different inclinations. Hence
it is customary to connect together two exactly equal
needles, with their poles pointing in opposite directions,
one needle being within and another without the coil of
wire. As regards the earth's magnetism, the needles are
now *astatic* or indifferent, the tendency of one needle
being exactly balanced by that of the other.

An elegant instance of the elimination of a disturbing
force by compensation is found in Faraday's researches
upon the magnetism of gases. To observe the magnetic
attraction or repulsion of a gas seems impossible unless we
enclose the gas in an envelope, probably best made of

[e] Regnault's 'Cours Élémentaire de Chimie,' 1851, vol. I. p. 141.

glass. But any such envelope is sure to be more or less
affected by the magnet, so that it becomes difficult to
distinguish between three forces which enter the problem ;
namely, the magnetism of the gas in question, that of the
envelope, and that of the surrounding atmospheric air.
Faraday avoided all difficulties by employing two exactly
equal and similar glass tubes connected together, and so
suspended from the arm of a torsion balance that the
tubes were in similar parts of the magnetic field. One
tube being filled with nitrogen and the other with oxygen,
it was found that the oxygen seemed to be attracted and
the nitrogen repelled. The suspending thread of the
balance was then turned until the force of torsion restored
the tubes to their original places, where the magnetism of
the tubes as well as that of the surrounding air, being
exactly the same and in the opposite direction upon the
two tubes, could not produce any interference. The force
thus required to restore the tubes was measured by the
amount of torsion of the thread, and it indicated correctly
the comparative attractive powers of oxygen and nitrogen.
The oxygen was then withdrawn from one of the tubes,
and a second experiment made, so as to compare a vacuum
with nitrogen. No force was now required to maintain
the tubes in their places, so that nitrogen was found to
be, approximately speaking, indifferent to the magnet,
that is, neither magnetic nor diamagnetic, while oxygen
was proved to be positively magnetic [f]. It required the
highest experimental skill on the part of Faraday and
Tyndall, to distinguish between what is apparent and real
in magnetic attraction and repulsion.

Experience alone can absolutely decide when a com-
pensating arrangement is conducive to accuracy. As a
general rule mechanical compensation is the last resource,

[f] Tyndall's 'Faraday,' pp. 114–15.

and in the more accurate observations it is likely to introduce more uncertainty than it removes. A multitude of instruments involving mechanical compensation have been devised, but they are usually of an unscientific character[g], because the errors compensated can be more accurately determined and allowed for. But there are exceptions to this rule, and it seems to be proved that in the delicate and tiresome operation of measuring a base line, invariable bars, compensated for expansion by heat, give a very accurate result, the observation of their varying temperature and the calculation of the corrections being an uncertain and tedious work[h].

We thus see that the choice of one or other mode of eliminating a simple error depends entirely upon circumstances and the object in view; but we may safely lay down the following conclusions. First of all, seek to avoid the source of error altogether if it can be conveniently done; if not, make the experiment so that the error may be as small, but more especially as constant, as possible. If the means are at hand for determining its amount by calculation from other experiments and principles of science, allow the error to exist and make a correction in the result. If this cannot be accurately done or involves too much labour for the purposes in view, then throw in a counteracting error which shall as nearly as possible be of equal amount in all circumstances with that to be eliminated. There yet remains, however, one important method, that of Reversal, which will form an appropriate transition to the succeeding chapters on the Method of Mean Results and the Law of Error.

[g] See, for instance, the Compensated Sympiesometer, ' Philosophical Magazine,' 4th Series, vol. xxxix. p. 371.

[h] Grant, ' History of Physical Astronomy,' pp. 146, 147.

5. *Method of Reversal.*

The fifth method of eliminating error is most potent and satisfactory whenever it can be applied, but it requires that we shall be able to reverse the apparatus and mode of procedure, so as to make the interfering cause act alternately in opposite directions. If we can get two experimental results, one of which is as much too great as the other is too small, the error is equal to half the difference, and the true result is the mean of the two apparent results. It is an unavoidable defect of the chemical balance, for instance, that the points of suspension of the pans cannot be fixed at exactly equal distances from the centre of suspension of the beam. Hence two weights which seem to balance each other will never be quite equal in reality. The difference is detected by reversing the weights, and it may be estimated by adding sufficient small weights to the deficient side to restore equilibrium, and then taking as the true weight the geometric mean of the two apparent weights of the same object. If the difference is small the arithmetic mean, that is half the sum, may be substituted for the geometric mean, from which it will not appreciably differ.

This method of reversal is most extensively employed in practical astronomy. The apparent elevation of a heavenly body is observed by a telescope moving upon a divided circle, upon which the inclination of the telescope is read off. Now this reading will be erroneous if the circle and the telescope have not accurately the same centre. But if we read off at the same time both ends of the telescope, the one reading will be about as much too small as the other is too great, and the mean will be nearly free from error. In practice the observation is differently conducted, but the principle is the same; the telescope is fixed to the circle, which moves with it, and the angle through which it moves is read off at three, six, or more points, disposed of at equal intervals round the circle. The older astronomers, down even to the time of Flamsteed, were accustomed to use portions only of a divided circle, generally quadrants, and Römer made a vast improvement when he introduced the complete circle.

The transit circle, employed to determine the meridian passage of heavenly bodies, is so constructed that the telescope and the axis bearing it, in fact the whole moving part of the instrument, can be taken out of the bearing sockets and turned over, so that what was formerly the western pivot becomes the eastern one, and *vice versâ.* It is impossible that the instrument could have been so perfectly constructed, mounted, and adjusted that the telescope should point exactly to the meridian, but the effect of the reversal is that it will point as much to the west in one position as it does to the east in the other, and the mean result of observations in the two positions must be free from such cause of error.

The accuracy with which the inclination of the compass needle can be determined depends almost entirely on the method of reversal. The dip needle consists of a bar of magnetized steel, suspended like the beam of a delicate balance on a slender axis passing through the centre of gravity of the bar, so that it is at liberty to rest in that exact degree of inclination in the magnetic meridian which the magnetism of the earth induces. The inclination is read off upon a vertical divided circle, but to avoid any error in the centring of the needle and circle, both ends are read, and the mean of the results is taken. The whole instrument is now turned carefully round through 180°, which gives two new readings, in

which any error due to the wrong position of the zero
of the division will be reversed. As the axis of the
needle may not be exactly horizontal, it is now reversed
in the same manner as the transit instrument, the end of
the axis which formerly pointed east being made to point
west, and a new set of readings is taken.

Finally, error may arise from the axis not passing
accurately through the centre of gravity of the bar, and
this error can only be detected and eliminated on reversing the magnetic poles of the bar by the application
of a strong magnet. The error is thus made to act in
opposite directions. To ensure all possible accuracy each
reversal ought to be combined with each other reversal,
so that the needle will be observed in eight different
positions by sixteen different readings, the mean of the
whole of which will give the required inclination free
from all eliminable errors [k].

There are certain cases of experiment in which a
disturbing cause can with much ease be made to act in
opposite directions, in alternate observations, so that the
mean of the results will be free from disturbance. Thus
in direct experiments upon the velocity of sound in
passing through the air between stations two or three
miles apart, the wind is a cause of error. It will be well,
in the first place, to choose a time for the experiment
when the air is very nearly at rest, and the disturbance
slight, but if at the same moment signal sounds be made
at each station and observed at the other, two sounds will
be passing in opposite directions through the same body
of air and the wind will accelerate one sound almost
exactly as much as it retards the other [l]. Again, in
trigonometrical surveys the apparent height of a point

[k] Quetelet, 'Sur la Physique du Globe,' p. 174. Jamin, 'Cours de Physique,' vol. i. p. 504.

[l] Herschel, On Sound, ' Encyclopædia Metropolitana,' p. 748.

will be affected by atmospheric refraction and the curvature of the earth. But if in the case of two points the apparent elevation of each as seen from the other be observed, the corrections will be the same in amount, but reversed in direction, and the mean between the two apparent differences of altitude will give the true difference of level [m].

In the next two chapters we really pursue the Method of Reversal into more complicated applications.

[m] Hutton, 'Philosophical Transactions,' abridgment, vol. xiv. p. 422.

CHAPTER XVI.

THE METHOD OF MEANS.

ALL results of the measurement of continuous quantity can only be approximately true. Were this assertion doubted, it could readily be proved by direct experience. For if any person, using an instrument of the greatest precision, makes and registers successive observations in an unbiassed manner, it will almost invariably be found that the results differ from each other. When we operate with sufficient care we cannot perform so simple an experiment as weighing an object in a good balance without getting discrepant numbers. Only the rough and careless experimenter will think that his observations agree, but in reality he will be found to overlook the differences. The most elaborate researches, such as those undertaken in connexion with standard weights and measures, always render it apparent that complete coincidence is out of the question, and that the more accurate our modes of observation are rendered, the more numerous are the sources of minute error which become apparent. We may look upon the existence of error in all measurements as the normal state of things. It is absolutely impossible to eliminate separately the multitude of small disturbing influences, except by balancing them off against each other. And even in drawing a mean it is to be expected that we shall come near the truth rather than exactly to it. In the measurement of continuous quantity, absolute coincidence, if it even occurs or seems to occur, must be purely casual, and is no indication of precision. It is one of the most embarrassing things we can meet when experimental results agree too closely. Such coincidences should raise our suspicion that the apparatus in use is in some way restricted in its operation, so as not really to give the true result at all[a], or that the actual results have not been faithfully recorded by the assistant in charge of the apparatus.

If then we cannot get twice over exactly the same result, the question arises, How can we ever attain the truth or select the result which may be supposed to approach most nearly to it? The quantity of a certain phenomenon is expressed in several numbers which differ from each other; no more than one of them at the most can be true, and it is more probable that they are all false. It may be suggested, perhaps, that the observer should select the one observation which he judged to be the best made, and there will often doubtless be a feeling that one or more results were satisfactory, and the others less trustworthy. This seems to have been the course adopted by some of the early astronomers. Flamsteed when he had made several observations of a star probably chose in an arbitrary manner that which seemed to him nearest to the truth[b].

When Horrocks selects for his estimate of the sun's semidiameter a mean between the results of Kepler and Tycho he professes not to do it from any regard to the idle adage, 'Medio tutissimus ibis,' but because he thought it from his own observations to be correct[c]. But this method will not apply at all when the observer has

---

[a] Thomson and Tait, 'Treatise on Natural Philosophy,' vol. i. p. 309.

[b] Baily's 'Account of Flamsteed,' p. 376.

[c] 'The Transit of Venus across the Sun,' by Horrocks, London, 1859, p. 146.

made a number of measurements which are equally good in his opinion, and it is quite apparent that in using an instrument or apparatus of considerable complication the observer will not necessarily be able to judge whether slight causes have affected its operation or not.

In this question, as indeed throughout inductive logic, we deal only with probabilities. There is no infallible mode of arriving at the absolute truth, which lies beyond the reach of human intellect, and can only be the distant object of our long continued and painful approximations. Nevertheless there is a mode pointed out alike by common sense and the highest mathematical reasoning, which is more likely than any other, as a general rule, to bring us near the truth. The ἄριστον μέτρον, or the *aurea mediocritas*, was highly esteemed in the ancient philosophy of Greece and Rome ; but it is not probable that any of the ancients should have been able clearly to analyse and express the reasons why they advocated the *mean* as the safest course. But in the last two centuries this apparently simple question of the mean has been found to afford a field for the exercise of the utmost mathematical skill. Roger Cotes, the editor of the 'Principia,' appears to have had some insight into the value of the mean ; but profound mathematicians such as De Moivre, Daniel Bernouilli, Laplace, Lagrange, Gauss, Quetelet, De Morgan, Airy, Leslie, Ellis and others have hardly exhausted the subject.

Several uses of the Mean Result.

The elimination of errors of unknown sources, is almost always accomplished by the simple arithmetical process of taking the *mean*, or, as it is often called, the *average* of several discrepant numbers. To take an average is to add the several quantities together, and divide by the number of quantities thus added, which gives a quotient lying among, or in the *middle* of, the several quantities. Before however inquiring fully into the grounds of this procedure, it is essential to observe that this one arithmetical process is really applied in at least three different cases, for different purposes, and upon different principles, and we must take great care not to confuse one application of the process with another. A *mean result*, then, may have any one of the following significations.

(1) It may give a merely representative number, expressing the general magnitude of a series of quantities, and serving as a convenient mode of comparing them with other series of quantities. Such a number is properly called *The fictitious mean* or *The average result*.

(2) It may give a result approximately free from disturbing quantities, which are known to affect some results in one direction, and other results equally in the opposite direction. We may say that in this case we get a *Precise mean result*.

(3) It may give a result more or less free from unknown and uncertain errors; this we may call the *Probable mean result*.

Of these three uses of the mean the first is entirely different in nature from the two last, since it does not yield an approximation to any natural quantity, but furnishes us with an arithmetic result comparing the aggregate of certain quantities with their number. The third use of the mean rests entirely upon the theory of probability, and will be more fully considered in a later part of this chapter. The second use is closely connected, or even identical with, the Method of Reversal already described (p. 410), but it will be convenient to enter somewhat fully on all the three employments of the same arithmetical process.

The significations of the terms Mean and Average.

Much confusion exists in the popular, or even the scientific employment of the terms *mean* and *average*, and they are commonly taken as synonymous. It is desirable to ascertain carefully what significations we ought to attach to them. The English word *mean* is exactly equivalent to *medium*, being derived perhaps, through the French *moyen*, from the latin *medius*, which again is undoubtedly kindred with the Greek μεσος. Etymologists believe, too, that this Greek word is connected with the preposition μετα, the German *mitte*, and the true English *mid* or *middle*; so that after all the *mean* is a technical term identical in its root with the more popular equivalent *middle*.

If we inquire what is a mean in a mathematical point of view, the true answer is that there are several or many kinds of means. The old arithmeticians recognised at least ten kinds, which are stated by Boethius, and even an eleventh was added by Jordanus[d].

The *arithmetic mean* is the one by far the most commonly denoted by the term, and that which we may understand it to signify in the absence of any qualification. It is the sum of any series of quantities divided by their number, and may be represented by the formula $\frac{1}{2}(a+b)$. But there is also the *geometric mean*, which is the square root of the product, $\sqrt{a \times b}$, or that quantity the logarithm of which is the arithmetic mean of the logarithms of the quantities. There is also the *harmonic mean*, which is the reciprocal of the arithmetic mean of the reciprocals of the quantities. Thus if a and b be the

---

[d] De Morgan, Supplement to the 'Penny Cyclopædia,' art. *Old Appellations of Numbers.*

quantities, as before, their reciprocals are $\frac{1}{a}$ and $\frac{1}{b}$, the mean of which is $\frac{1}{2}\left(\frac{1}{a}+\frac{1}{b}\right)$, and the reciprocal again is $\frac{2ab}{a+b}$. Other kinds of means might no doubt be invented for particular purposes, and we might apply the term, as De Morgan pointed out [e], to any quantity a function of which is equal to a function of two or more other quantities, and is such, that the interchange of these latter quantities among themselves will make no alteration in the value of the function. Symbolically, if $\phi\,(y, y, y \ldots.)$ $= \phi\,(x_1, x_2, x_3 \ldots.)$, then y is a kind of mean of the quantities x_1, x_2, &c.

The geometric mean is necessarily adopted in certain cases. Thus when we estimate the work done against a force which varies inversely as the square of the distance from a fixed point, the mean force is the geometric mean between the forces at the beginning and end of the path [f]. When in an imperfect balance, we reverse the weights to eliminate error, the true weight will be the geometric mean of the two apparent weights of the one body (see p. 410).

In almost all the calculations of statistics and commerce the geometric mean ought, strictly speaking, to be used. Thus if a commodity rises in price 100 per cent. and another remains unaltered, the mean rise of price is not 50 per cent. because the ratio 150 : 200 is not the same as 100 : 150. The mean ratio is as unity to $\sqrt{1\cdot00 \times 2\cdot00}$ or 1 to 1·41. The difference between the three kinds of mean in such a case, as I have elsewhere shown [g], is very considerable, being as follows—

[e] 'Penny Cyclopædia,' art. *Mean.*

[f] Thomson and Tait, 'Treatise on Natural Philosophy,' vol. i. p. 366.

[g] 'Journal of the Statistical Society,' June 865, vol. xxviii. p. 296.

Arithmetic mean 50 per cent.
Geometric   „   41   „
Harmonic   „   33   „

In all calculations concerning the average rate of progress of a community, or any of its operations, the geometric mean should be employed. For if a quantity increases 100 per cent. in 100 years, it would not on the average increase 10 per cent. in each ten years, as the 10 per cent. would at the end of each decade be calculated upon larger and larger quantities, and give at the end of 100 years much more than 100 per cent., in fact as much as 159 per cent. The true mean rate in each decade would be $\sqrt[10]{2}$ or about 1·07, that is, the increase would be about 7 per cent. in each ten years. But when the quantities differ but little, the arithmetic and geometric means are approximately the same. Thus the arithmetic mean of 1·000 and 1·001 is 1·0005, and the geometric mean is about 1·0004998, the difference being of an order inappreciable in almost all scientific or practical processes. Even in the comparison of standard weights by Gauss' method of transposition the arithmetic mean may usually be substituted for the geometric mean which is the true result.

Regarding the mean in the absence of express qualification to the contrary as the common arithmetic mean, we must still distinguish between its two uses where it defines with more or less accuracy and probability a really existing quantity, and where it acts as a mere representative of other quantities. If I make many experiments to determine the specific gravity of a homogeneous piece of gold there is a certain definite ratio which I wish to approximate to, and the mean of my separate results will, in the absence of any reasons to the contrary, be the most probable approximate result. When we determine on the other hand the mean density of the earth, it is exceedingly unlikely that there is any part of the earth exactly of that density, and, as the crust is only about half the mean density, there must be other parts of greater density. I may also determine the mean specific gravity of a body composed of iron and gold, so that there will certainly be no portion possessing the mean density.

The very different signification of the word 'mean' in these two uses has been fully explained by M. Quetelet[h], and the importance of the distinction has moreover been pointed out by Sir John Herschel in reviewing his work[i]. It is much to be desired that scientific men would mark the difference by using the word *mean* only in the former sense when it denotes approximation to a definite existing quantity; and *average*, when the mean is only a fictitious quantity, used for the convenience of thought and expression. The etymology of this word 'average' is somewhat obscure; but according to De Morgan[k] it comes from *averia*, 'havings or possessions,' especially applied to farm stock. By the accidents of language *averagium* came to mean the labour of farm horses to which the lord was entitled, and it probably acquired in this manner the notion of distributing a whole into parts, a sense in which it was very early applied to maritime averages or contributions of the other owners of cargo to those whose goods have been thrown overboard or used for the safety of the vessel.

[h] 'Letters on the Theory of Probabilities,' transl. by Downes, Part ii.

[i] Herschel's 'Essays,' &c. pp. 404, 405.

[k] 'On the Theory of Errors of Observations,' 'Cambridge Philosophical Transactions,' vol. x. Part ii. 416.

On the Fictitious Mean or Average Result.

Although the average when employed in its proper sense of a fictitious mean, represents no really existing quantity, it is yet of the highest scientific importance, as enabling us to conceive in a single result a multitude of complex details. It enables us to make a hypothetical simplification of a problem, and avoid complexity without committing error. Thus the aggregate weight of a body is the sum of the weights of the indefinitely small particles, each acting at a different place, so that the simplest mechanical problem concerning a body really resolves itself, strictly speaking, into an infinite number of distinct problems. We owe to Archimedes the first introduction of the beautiful idea that one point might be discovered in a gravitating body such that the weight of all the particles might be regarded as concentrated in that point, and yet the behaviour of the whole body would be exactly represented by the behaviour of this heavy point. This Centre of Gravity may be within the body, as in the case of a sphere, or it may be in empty space, as in the case of a ring. Any two bodies, whether connected or separate, may be conceived as having a centre of gravity ; that of the sun and earth, for instance, lying within the sun and only 267 miles from its centre.

Although we most commonly use the notion of a centre or average point with regard to gravity, the same notion is applicable to many other cases. Terrestrial gravity is only one case of approximately parallel forces, so that the centre of gravity is but a special case of the more general Centre of Parallel Forces. Wherever a number of forces of whatever amount act in parallel lines, it is possible to discover a point at which the algebraic sum of the forces may be imagined to act with exactly the same effect. Water in a cistern presses against the side with a pressure varying according to the depth, but always in a direction perpendicular to the side. We may then conceive the whole pressure as exerted on one point, which will be one-third from the bottom of the cistern, and may be called the Centre of Pressure. The Centre of Oscillation of a pendulum, discovered by Huyghens, is that point at which the whole weight of the pendulum may be considered as concentrated, without altering the time of oscillation (see p. 370). Similarly when one body strikes another the Centre of Percussion is that point in the striking body at which all its mass might be concentrated without altering the effect of the stroke. Mathematicians have also described the Centre of Gyration, the Centre of Conversion, the Centre of Friction, &c.

We ought however carefully to distinguish between those circumstances in which an invariable centre can be assigned, and those in which it cannot. In perfect strictness, there is no such thing as a true invariable centre of gravity. As a general rule a body is capable of possessing an invariable centre only for perfectly parallel forces, and gravity never does act in absolutely parallel lines. Thus, as usual, we find that our conceptions are only hypothetically correct, and only approximately applicable to real circumstances. There are indeed certain geometrical forms, called *Centrobaric*[1], such that bodies of that shape would attract each other exactly as if the mass were concentrated at the centre of gravity, whether the forces act in a parallel manner or not. Newton shewed that uniform spheres of matter have this property, and this truth proved of the greatest importance in simplifying his calculations. But it is after all a purely hypothetical truth, because we can nowhere meet with, nor can we construct, a perfectly spherical

[1] Thomson and Tait, 'Treatise on Natural Philosophy,' vol. i. p. 394.

and homogeneous body. The slightest irregularity or protrusion from the surface will destroy the rigorous correctness of the assumption. The spheroid, on the other hand, has no invariable centre at which its mass may always be regarded as concentrated. The point at which its resultant attraction acts will move about according to the distance and position of the other attracting body, and it will only coincide with the centre as regards an infinitely distant body whose attractive forces may be considered as acting in parallel lines.

Physicists speak familiarly of the pole of a magnet, and the term may be used with convenience. But, if we attach any real and definite meaning to it, the pole is not the end of the magnet, nor is it any one fixed point within, but the variable point from which the resultant of all the forces exerted by the particles in the whole bar upon exterior magnetic particles may be considered as acting. The pole is, in short, a Centre of Magnetic Forces; but as those forces are really never parallel, this centre will vary in position according to the relative place of the object attracted. Only when we regard the magnet as attracting a very distant, or, strictly speaking, infinitely distant particle, does the centre become a fixed point, situated in short magnets approximately at one sixth of the whole length from each end of the bar. We have in the above instances of centres or poles of force sufficient examples of the mode in which the Fictitious Mean or Average is employed in physical science.

The Precise Mean Result.

We now turn to that mode of employing the mean result which is analogous to the method of reversal, but which is brought into practice in a most extensive manner throughout many branches of physical science. We find

the simplest possible case in the determination of the
latitude of a place by observations of the Pole-star.
Tycho Brahe suggested that if the elevation of any circumpolar star were observed at its higher and lower
passages across the meridian, half the sum of the elevations
would be the latitude of the place, which is equal to the
height of the pole. Such a star is as much above the
pole at its highest point, as it is below at its lowest, so
that the mean must necessarily give the height of the
pole itself free from doubt, except as regards incidental
errors of observation. The Pole-star is usually selected
for the purpose of such observations because it describes
the smallest circle, and is thus on the whole least affected
by atmospheric refraction.

Whenever several causes are in action, each of which
at one time increases and at another time decreases the
joint effect by equal quantities, we may apply this method
and disentangle the effects. Thus the solar and lunar
tides roll on in almost complete independence of each
other. When the moon is new or full the solar tide coincides, or nearly so, with that caused by the moon, and the
joint effect is the sum of the separate effects. When the
moon is in quadrature, or half full, the two tides are
acting in opposition, one raising and the other depressing
the water, so that we observe only the difference of the
effects. We have in fact—

Spring tide = lunar tide + solar tide
Neap tide = lunar tide − solar tide.

We have only then to add together the heights of the
maximum spring tide and the minimum neap tide, and
half the sum is the true height of the lunar tide. Half
the difference of the spring and neap tides on the other
hand gives the solar tide.

Effects of very small amount may with great approach
to certainty be detected among much greater fluctuations,

provided that we have a series of observations sufficiently numerous and long continued to enable us to balance all the larger effects against each other. For this purpose the observations should be continued over at least one complete cycle, in which the effects run through all their variations, and return exactly to the same relative position as at the commencement. If casual or irregular disturbing causes exist, we should probably require many such cycles of results to render their effect inappreciable. We obtain the desired result by taking the mean of all the observations in which a cause acts positively, and the mean of all in which it acts negatively. Half the difference of these means will be the desired quantity, provided indeed that no other effect happens to vary in the same period.

Since the moon causes so considerable a movement of the ocean, it is evident that its attraction must have some effect upon the atmosphere. The laws of these tides were investigated by Laplace, but as it would be impracticable by theory to calculate their amount, we can only determine them by observation, as Laplace predicted that they would one day be determined [m]. But the oscillations of the barometer thus caused are far smaller than the oscillations due to several other causes. Storms, hurricanes, or changes of weather produce movements of the barometer sometimes as much as a thousand times as great as the tide in question. There are also regular daily, yearly, or other fluctuations, all greater than the desired quantity. To detect and measure the atmospheric tide it was desirable that observations should be made in a place as free as possible from irregular disturbances. On this account several long series of observations were made at St. Helena, where the barometer is far more regular in its movements than in a continental climate. The effect of the moon's attraction was then detected by taking the mean of all the readings when the moon was on the meridian and the similar mean when she was on the horizon. The difference of these means was found to be only ·00365, yet it was possible to discover even the variation of this tide according as the moon was nearer to or further from the earth, though this difference was only ·00056 inch [n]. It is quite evident that such minute effects could never be discovered in a purely empirical manner. Having no information but the series of observations before us, we could have no clue as to the mode of grouping them which would give so small a difference. In applying this method of means in an extensive manner we must generally then have à priori knowledge as to the periods at which a cause will act in one direction or the other.

We are sometimes able to eliminate fluctuations and take a mean result by purely mechanical arrangements. The daily variations of temperature, for instance, become imperceptible one or two feet below the surface of the earth, so that a thermometer placed with its bulb at that depth would give very nearly the true daily mean temperature. At a depth of twenty feet even the yearly fluctuations would become nearly effaced, and the thermometer would stand a little above the true mean temperature of the locality. In registering the rise and fall of the tide by a tide-guage, it is desirable to avoid the oscillations arising from surface waves, which is very readily accomplished by placing the float which marks the level of the water in a cistern communicating by a small hole with the sea. Only a general rise or fall of the level is then perceptible, just as in the marine barometer the narrow tube prevents any casual fluctuations and allows only a continued change of pressure to manifest itself.

[n] Grant, 'History of Physical Astronomy,' p. 163.

Determination of the Zero point by the Method
of Means.

There are a number of important observations in which
one of the chief difficulties consists in defining exactly the
zero point from which we are to measure.   We can point
a telescope with great precision to a star and can measure
the angle through which the telescope is raised or lowered
to a second of arc ; but all this precision will be useless
unless we can know exactly where the centre point of
the heavens is from which we measure, or, what comes to
the same thing, the horizontal line 90° distant from it.
Since the true horizon has reference to the figure of the
earth at the place of observation, we can only determine
it by the direction of gravity, as marked either by the
plumb-line or the surface of a liquid.   The question resolves itself then into the most accurate mode of observing
the direction of gravity, and as the plumb-line has long
been found hopelessly inaccurate, astronomers generally
employ the surface of mercury in repose as the criterion
of horizontality.   They ingeniously observe the direction
of the surface by making a star the index.   From the
Laws of Reflection it follows that the angle between the
direct ray from a star and that reflected from a surface
of mercury will be exactly double the angle between the
surface and the direct ray from the star.   Hence the
horizontal or zero point is the mean between the apparent
place of any star or other very distant object and its
reflection in mercury.

A plumb-line is perpendicular, or a liquid surface is horizontal only in an approximate sense ; for any irregularity
of the surface of the earth, a mountain, or even a house
must cause some deviation by its attracting power.   To
detect such deviation might seem very difficult, because
every other plumb-line or liquid surface would be equally

affected by the very principles of gravity. Nevertheless it can be detected ; for if we place one plumb-line to the north of a mountain, and another to the south, they will be about equally deflected in opposite directions, and if by observations on the same star we can measure the angle between the plumb-lines, half the inclination will be the deviation of either, after allowance has been made for the inclination due to the difference of latitude of the two places of observation. By this mode of observation applied to the mountain Schehallien the deviation of the plumb-line was accurately measured by Maskelyne, and thus a comparison instituted between the attractive forces of the mountain and the whole globe, which led to a very probable estimate of the earth's average density.

In some cases it is actually better to determine the zero point by the average of equally diverging quantities than by direct observations. Thus in delicate weighings by a chemical balance it is requisite to ascertain exactly the point at which the beam comes to rest, and when standard weights are being compared the position of the beam is ascertained by a carefully divided scale viewed through a microscope. But when the beam is just coming to rest, friction, small impediments or other accidental causes may readily obstruct it, because it is near the point at which the force of stability becomes infinitely small. Hence it is found better to let the beam vibrate and observe the terminal points of the vibrations. The mean between two extreme points will nearly indicate the position of rest. Friction and the resistance of air tend to reduce the vibrations, so that this mean will be erroneous by half the amount of this effect during a half vibration. But by taking several observations we may determine this retardation and allow for it. Thus if a, b, c be the terminal points of three excursions of the beam from the zero of the scale, then $\frac{1}{2}(a+b)$ will be about as much erroneous in one direction as $\frac{1}{2}(b+c)$ in the other, so that the mean of these two means, or what is the same, $\frac{1}{4}(a+2b+c)$, will be exceedingly near to the point of rest [o]. A still closer approximation may be made by taking four readings and reducing them by the formula $\frac{1}{6}(a+2b+2c+d)$.

The accuracy of Baily's experiments, directed to determine the density of the earth, entirely depended upon this mode of observing oscillations. The balls whose gravitation was measured were so delicately suspended by a torsion balance that they never came to rest. The extreme points of the oscillations were observed both when the heavy leaden attracting ball was on one side and on the other. The difference of the mean points when the leaden ball was on the right hand and that when it was on the left hand gave double the amount of the deflection.

A most beautiful instance of the mode of avoiding the use of a zero point is to be found in Mr. E. J. Stone's observations on the radiated heat of the fixed stars. The great difficulty in these observations arose from the comparatively great amounts of heat which were sent into the telescope from the atmosphere, and which were sufficient almost entirely to disguise the feeble heat rays of a star. But Mr. Stone fixed at the focus of his telescope a double thermo-electric pile of which the two parts were reversed in order. Now any disturbance of temperature which acted upon both piles uniformly produced no effect upon the galvanometer needle, and when the rays of the star were made to fall alternately upon one pile and the other, the total amount of the deflection represented double the heating power of the star. Thus Mr. Stone was able to detect with much certainty a heating effect of the star Arcturus, which even when concentrated by the telescope amounted only to $\frac{2}{100}$th of a degree

[o] Gauss, Taylor's 'Scientific Memoirs,' vol. ii. p. 43, &c.

Fahrenheit, and which represents a heating effect of the direct ray of only about 0°·00000137 Fahrenheit, equivalent to the heat which would be received from a three-inch cubic vessel full of boiling water at the distance of 400 yards [P]. It is probable that Mr. Stone's arrangement of the pile might be usefully employed in other delicate thermometric experiments subject to considerable disturbing influences.

Determination of Maximum Points.

We employ the method of means in a certain number of observations directed to determine the moment at which a phenomenon reaches its highest point in quantity. In noting the place of a fixed star at a given time there is no difficulty in ascertaining the point to be observed, for a star in a good telescope presents an exceedingly small disc. In observing a nebulous body which from a bright centre fades gradually away on all sides, it will not be possible to select with certainty the middle point. In many such cases the best method is not to select arbitrarily the supposed middle point, but points of equal brightness on either side, and then take the mean of the observations of these two points for the centre. As a general rule, a variable quantity in reaching its maximum increases at a less and less rate, and after passing the highest point begins to decrease by insensible degrees. The maximum may indeed be defined as that point at which the increase or decrease is insensibly small. Hence it will usually be the most indefinite point in the whole course, and if we can accurately measure the phenomenon we shall best determine the place of the maximum by determining points on either side at which the ordinates are equal. There is

[P] 'Proceedings of the Royal Society,' vol. xviii. p. 159 (Jan. 13, 1870): 'Philosophical Magazine' (4th Series), vol. xxxix. p. 376.

moreover this advantage in the method that several points may be determined with the corresponding ones on the other side, and the mean of the whole taken as the true place of the maximum. But this method entirely depends upon the existence of symmetry in the curve, so that of two equal ordinates one shall be as far on one side of the maximum as the other is on the other side. The method fails when other laws of variation prevail.

In tidal observations great difficulty is encountered in fixing the moment of high water, because the rate at which the water is then rising or falling is almost imperceptible. Dr. Whewell proposed, therefore, to note the time at which the water passes a fixed point somewhat below the maximum both in rising and falling, and take the mean time as that of high water. But this mode of proceeding unfortunately does not give a correct result, because the tide follows different laws in rising and in falling. There is a difficulty again in selecting the highest spring tide, another object of much importance in tidology. Laplace discovered that the tide of the second day preceding the conjunction of the sun and moon is nearly equal to that of the fifth day following; and, believing that the increase and decrease of the tides proceeded in a nearly symmetrical manner, he decided that the highest tide would occur about thirty-six hours after the conjunction, that is half-way between the second day before and the fifth day after[q].

This method is also employed in determining the time of passage of the middle or densest point of a stream of meteors. The earth takes two or three days in passing completely through the November stream; but astronomers need for their calculations to have some definite point fixed within a few minutes if possible. When near to the middle they observe the numbers of meteors which

q Airy 'On Tides and Waves,' Encycl. Metrop. pp. 364*–366*.

come within the sphere of vision in each half hour or quarter hour, and then, assuming that the law of variation is symmetrical, they select a moment for the passage of the whole body equidistant between times of equal frequency.

The eclipses of Jupiter's satellites are not only of great interest as regards the motions of the satellites themselves, but used to be, and perhaps still are, of importance in determining longitudes, because they are events occurring at fixed moments of absolute time, and visible in all parts of the planetary system at the same time, allowance being made for the interval occupied by the light in travelling. But as is excellently explained by Sir John Herschel[r], the moment of the event is wanting in definiteness, partly because the long cone of Jupiter's shadow is surrounded by a penumbra, and partly because the satellite has itself a sensible disc, and takes a certain time in entering the shadow. Different observers using different telescopes would usually select different moments for that of the eclipse. But it is evident that the increase of light in the emersion will proceed according to a law exactly the reverse of that observed in the immersion, so that if an observer notes the time of both events with the same telescope, he will be as much too soon in one observation as he is too late in the other, and the mean moment of the two observations will represent with considerable accuracy the time when the satellite is in the middle of the shadow. The personal error of judgment of the observer is thus eliminated, provided that he takes care to act at the emersion as he did at the immersion.

[r] 'Outlines of Astronomy,' 4th edition, § 538.

CHAPTER XVII.

THE LAW OF ERROR.

To bring error itself under law might seem beyond human power. He who errs surely diverges from law, and it might well be deemed hopeless to suppose that out of error we can draw truth. One of the most remarkable achievements of the human intellect is the establishment of a general theory which not only enables us among discrepant results to approximate to the truth, but to assign the degree of probability which fairly attaches to this conclusion. It would be a gross misapprehension indeed to suppose that this law is necessarily the best guide under all circumstances. Every measuring instrument and every form of experiment may have its own special law of error; there may in one instrument be a tendency in one direction and in another in the opposite direction. Every process has its peculiar liabilities to mistake and disturbance, and we are never relieved from the necessity of vigilantly providing against such special difficulties. The general Law of Error is the best guide only when we have exhausted all other means of approximation, and still find discrepancies, which are due to entirely unknown causes. We must treat such residual differences in some way or other, since they will occur in all accurate experiments, and as their peculiar nature and origin is assumed to be unknown, there is no reason why we should treat them differently in different cases. Accordingly the ultimate Law of Error must be a uniform and general one.

It is perfectly recognised by mathematicians that in each special case a special Law of Error may apply, and should be discovered and adopted if possible. 'Nothing can be more unlikely than that the errors committed in all classes of observations should follow the same law[a],' and the special Laws of Error which will apply to certain instruments, as for instance the repeating circle, have been investigated by M. Bravais[b]. He concludes that every partial and distinct cause of error gives rise to a curve of possibility of errors, which may have any form whatever,— a curve which we may either be able or unable to discover, and which in the first case may be determined by considerations *à priori*, on the peculiar nature of this cause, or which may be determined *à posteriori* by observation. Whenever it is practicable and worth the labour, we ought to investigate these special conditions of error; nevertheless, when there are a great number of different sources of minute error, the general resultant will always tend to obey that general law which we are about to consider.

Establishment of the Law of Error.

Mathematicians agree far better as to the nature of the ultimate Law of Error than they do as to the manner in which it can be deduced and proved. They agree that among a number of discrepant results of observation, that mean quantity is probably the most nearly approximate to the truth which makes the sum of the squares of the errors as small as possible. But there are at least three different ways in which this principle has been arrived at respectively by Gauss, by Laplace, by Quetelet and by Sir John Herschel. Gauss proceeds much upon assump-

[a] 'Philosophical Magazine,' 3rd Series, vol. xxxvii. p. 324.
[b] 'Letters on the Theory of Probabilities,' by Quetelet, transl. by O. G. Downes, Notes to Letter XXVI. pp. 286–295.

THE PRINCIPLES OF SCIENCE.

tion; Herschel rests upon geometrical considerations; while Laplace and Quetelet regard the Law of Error as a development of the doctrine of combinations ; that of Gauss may be first noticed.

The Law of Error expresses the comparative probability of errors of various magnitude, and partly from experience, partly from à *priori* considerations, we may readily lay down certain conditions to which the law will certainly conform. It may fairly be assumed as a first principle to guide us in the selection of the law, that large errors will be far less frequent and probable than small ones. We know that very large errors are almost impossible, so that the probability must rapidly decrease as the amount of the error increases. A second principle is that positive and negative errors shall be equally probable, which may certainly be assumed, because we are supposed to be devoid of any knowledge as to the causes of the residual errors. It follows that the probability of the error must be a function of an even power of the magnitude, that is of the square, or the fourth power, or the sixth power, otherwise the probability of the same amount of error would vary accordingly as the error was positive or negative. The even powers x^2, x^4, x^6, &c., are always intrinsically positive, whether x be positive or negative. There is no à *priori* reason why one rather than another of these even powers should be selected. Gauss himself allows that the fourth or sixth powers would fulfil the conditions as well as the second[c], but in the absence of any theoretical reasons we should prefer the second power, because it leads to formulæ of great comparative simplicity. Did the Law of Error necessitate the use of the higher powers of the error, the complexity of

[c] 'Méthode des Moindres Carrés.' 'Mémoires sur la Combinaison des Observations, par Ch. Fr. Gauss. Traduit en Français par J. Bertrand,' Paris, 1855, pp. 6, 133, &c.

the necessary calculations would much reduce the utility of the theory.

By a process of reasoning, which it would be undesirable to attempt to follow in detail in this place, it is shown that, under these conditions, the most probable result of any series of recorded observations is that which makes the sum of the squares of the errors the least possible. Let a, b, c, &c., be the results of observation, and x the quantity selected as the most probable, that is the most free from unknown errors : then we must determine x so that $(a - x)^2 + (b - x)^2 + (c - x)^2 + \dots$ shall be the least possible quantity. Thus we arrive at the celebrated *Method of Least Squares,* as it is usually called, which appears to have been first distinctly put in practice by Gauss in 1795, while Legendre first published in 1806 an account of the process in his work, entitled, 'Nouvelles Méthodes pour la détermination des Orbites des Comètes.' It is worthy of notice, however, that Roger Cotes had long previously recommended a method of equivalent nature in his tract, 'Estimatio Erroris in Mixta Mathesi[d].'

Herschel's Geometrical Proof.

A second method of demonstrating the Principle of Least Squares was proposed by Sir John Herschel, and although only applicable to geometrical notions, it is remarkable as showing that from whatever point of view we regard the subject, the same principle will be detected. After assuming that some general law must exist, and that it is subject to the general principles of probability, he supposes that a ball is dropped from a high point with the intention that it shall strike a given mark on a horizontal plane. In the absence of any known causes of deviation it will either strike that mark, or, as

[d] De Morgan, 'Penny Cyclopædia,' art. *Least Squares.*

is infinitely more probable, diverge from it by an amount which we must regard as error of unknown origin. Now, to quote the words of Sir J. Herschel[c], 'the probability of that error is the unknown function of its square, i.e. of the sum of the squares of its deviations in any two rectangular directions. Now, the probability of any deviation depending solely on its magnitude, and not on its direction, it follows that the probability of each of these rectangular deviations must be the same function of *its* square. And since the observed oblique deviation is equivalent to the two rectangular ones, supposed concurrent, and which are essentially independent of one another, and is, therefore, a compound event of which they are the simple independent constituents, therefore its probability will be the product of their separate probabilities. Thus the form of our unknown function comes to be determined from this condition, viz., that the product of such functions of two independent elements is equal to the same function of their sum. But it is shown in every work on algebra that this property is the peculiar characteristic of, and belongs only to, the exponential or antilogarithmic function. This, then, is the function of the square of the error, which expresses the probability of committing that error. That probability decreases, therefore, in geometrical progression, as the square of the error increases in arithmetical.'

Laplace's and Quetelet's Proof of the Law of Error.

However much presumption the modes of determining the Law of Error, already described, may give in favour of the law usually adopted, it is difficult to feel that the

c 'Edinburgh Review,' July 1850, vol. xcii. p. 17   Reprinted 'Essays,' p. 399. This method of demonstration is discussed by Boole, 'Transactions of Royal Society of Edinburgh,' vol. xxi. pp. 627–630.

arguments are satisfactory and conclusive. The law adopted is chosen rather on the grounds of convenience and plausibility, than because it can be seen to be the true and necessary law. We can however approach the subject from an entirely different point of view, and yet get to the same result.

Let us assume that a particular observation is subject to four chances of error, each of which will increase the result one inch if it occurs. Each of these errors is to be regarded as an event independent of the rest and we can therefore assign, by the theory of probability, the comparative probability and frequency of each conjunction of errors. From the Arithmetical Triangle (pp. 208, 213) we learn that the ways of happening are as follows :—

No error at all	1 way.
Error of 1 inch	4 ways.
Error of 2 inches . . .	6 ways.
Error of 3 inches . . .	4 ways.
Error of 4 inches . . .	1 way.

We may infer that the error of two inches is the most likely to occur, and will occur in the long run in six cases out of sixteen. Errors of one and three inches will be equally likely, but will occur less frequently ; while no error at all, or one of four inches will be a comparatively rare occurrence. If we now suppose the errors to act as often in one direction as the other, the effect will be to alter the average error by the amount of two inches, and we shall have the following results :—

Negative error of 2 inches . . .	1 way.
Negative error of 1 inch . . .	4 ways.
No error at all	6 ways.
Positive error of 1 inch . . .	4 ways.
Positive error of 2 inches . . .	1 way.

We may now imagine the number of causes of error

increased and the amount of each error decreased, and the arithmetical triangle will always give us the proportional frequency of the resulting errors. Thus if there be five positive causes of error and five negative causes, the following table shows the comparative numbers of aggregate errors of various amount which will be the result :—

Direction of Error.	Positive Error.		Negative Error.
Amount of Error.	5, 4, 3, 2, 1	0	1, 2, 3, 4, 5
Number of such Errors.	1, 10, 45, 120, 210	252	210, 120, 45, 10, 1

It is plain that from such numbers I can ascertain the probability of any particular amount of error under the conditions supposed. Thus the probability of a positive error of exactly one inch is $\frac{210}{1024}$, in which fraction the numerator is the exact number of combinations giving one inch positive error, and the denominator the whole number of possible errors of all magnitudes. I can also, by adding together the appropriate numbers, get the probability of an error not exceeding a certain amount. Thus the probability of an error of three inches or less, positive or negative, is a fraction whose numerator is the sum of $45 + 120 + 210 + 252 + 210 + 120 + 45$, and the denominator, as before, giving the result $\frac{1002}{1024}$.

We may see at once that, according to these principles, the probability of small errors is far greater than of large ones: thus the odds are 1002 to 22, or more than 45 to 1, that the error will not exceed three inches; and the odds are 1022 to 2 against the occurrence of the greatest possible error of five inches. The existence of no error at all is the most likely event; but a small error, such as that of one inch positive, is little less likely.

If any case should arise in which the observer knows the number and magnitude of the independent errors which may occur, he ought certainly to calculate from the Arithmetical Triangle the special Law of Error which would apply. But the general law, of which we are in search, is to be used in the dark, when we have no knowledge whatever of the sources of error. To assume any special number of causes of error is then an arbitrary proceeding, and mathematicians have chosen the least arbitrary course of imagining the existence of an infinite number of infinitely small errors, just as, in the inverse method of probabilities, an infinite number of infinitely improbable hypotheses were submitted to calculation (p. 296).

The reasons in favour of this choice are of several different kinds.

1. It cannot be denied that there may exist infinitely numerous causes of error in any act of observation.

2. The resulting law on the hypothesis of a large finite, or even a moderate finite number of causes of error, does not appreciably differ from that given by the hypothesis of infinity.

3. We gain by the hypothesis of infinity a general law capable of ready calculation, and applicable by uniform rules to all problems.

4. This law, when tested by comparison with extensive series of observations, is strikingly verified, as will be shown in a later section.

When we imagine the existence of any large number of causes of error, for instance one hundred, the numbers of combinations become impracticably large, as may be seen to be the case from a glance at the Arithmetical Triangle (p. 208), which proceeds only up to the seventeenth line. M. Quetelet, by suitable abbreviating processes, succeeded in calculating out a table of probability of errors on the hypothesis of one thousand distinct causes[f]; but mathematicians have generally proceeded on the hypothesis of infinity, and then, by some of the beautiful devices of analysis, have substituted a general law of easy treatment. In mathematical works upon the subject, it is shown that the standard Law of Error is expressed in the formula

$$y = Y \epsilon^{-cx^2},$$

in which x is the amount of the error, Y the maximum ordinate of the curve of error, and c a number constant for each series of observations, and expressing the general amount of the tendency to error, but varying between one series of observations and another, while ϵ is the peculiar constant, $2\cdot71828 \ldots\ldots$ the base of the Naperian logarithms. To show the close correspondence of this general law with the special law which might be derived from the supposition of any moderate number of causes of error, I have in the accompanying figure drawn a

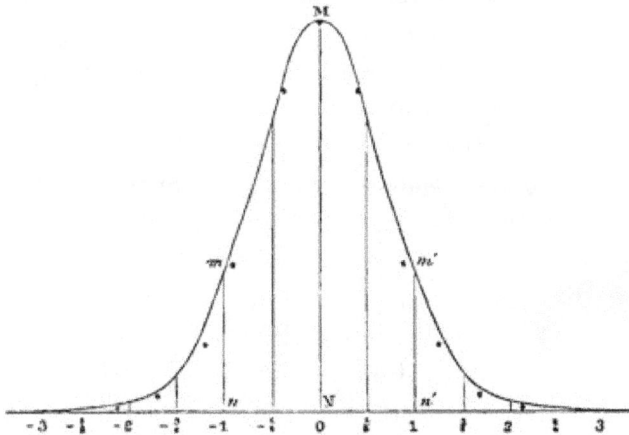

curved line representing accurately the variation of y when x in the above formula is taken equal to 0, $\frac{1}{2}$, 1, $\frac{3}{2}$, 2, &c., positive or negative, the arbitrary quantities Y and c

[f] 'Letters on the Theory of Probabilities,' Letter XV. and Appendix, note pp. 256–266.

being both assumed equal to unity, in order to simplify the calculations. In the same figure are inserted eleven dots, whose heights above the base line are proportional to the numbers in the eleventh line of the Arithmetical Triangle, thus representing the comparative probabilities of errors of various amounts arising from ten equal causes of error. It is apparent that the correspondence of the general and the special Law of Error is almost as close as can be exhibited in the figure, and the assumption of a greater number of equal causes of error would render the correspondence far more close.

It may be explained that the ordinates, for instance NM, nm, $n'm'$, represent values of y in the equation expressing the Law of Error. The occurrence of any one definite amount of error is infinitely improbable, because an infinite number of such ordinates might be drawn. But the probability of an error occurring between certain definite limits is finite, and is represented by a portion of the *area* of the curve. Thus the probability that an error, positive or negative, not exceeding unity will occur, is represented by the area M$mnn'm'$, in short, by the area standing upon the line nn'. Since every observation must either have some definite error or none at all, it follows that the whole area of the curve should be considered as the unit expressing certainty, and the probability of an error falling between particular limits will then be expressed by the ratio which the area of the curve between those limits bears to the whole area of the curve.

Derivation of the Law of Error from Simple Logical Principles.

It is worthy of notice that this Law of Error, abstruse though the subject may seem, is really founded upon the simplest principles. It arises entirely out of the difference between permutations and combinations, a subject upon which I may seem to have dwelt with unnecessary prolixity in previous pages (pp. 200–216). The order in which we add quantities together does not affect the amount of the sum, so that if there be three positive and five negative causes of error in operation, it does not matter in which order they are considered as acting. They may be indifferently intermixed in any arrangement, and yet the result will be the same. The reader should not fail to notice how laws or principles which appeared to be absurdly simple and evident when first noticed, reappear in the most complicated and mysterious processes of scientific method. The fundamental Laws of Identity and Difference gave rise to the Logical Abecedarium, which, after abstracting the character of the differences, led to the Arithmetical Triangle (p. 214). The Law of Error is defined by an infinitely high line of that triangle, and the law proves that the mean is the most probable result, and that divergencies from the mean become much less probable as they increase in amount. Now the comparative greatness of the numbers towards the middle of each line of the Arithmetical Triangle is entirely due to the indifference of order in space or time, which was first prominently pointed out as a condition of logical relations, and the symbols indicating them (pp. 40–42), and which was afterwards shown to attach equally to numerical symbols, the derivatives of logical terms (pp. 180, 181).

Verification of the Law of Error.

The theory of error which we have been considering rests entirely upon an assumption, namely that when known sources of disturbances are allowed for, there yet remain an indefinite, possibly an infinite number of other minute sources of error, which will as often produce excess as deficiency. Granting this assumption, the Law of Error must be as it is usually taken to be, and there is no more need to verify empirically than to test the truth of one of Euclid's propositions mechanically, after we have proved it theoretically. Nevertheless, it is an interesting occupation to verify even the propositions of geometry in an approximate manner, and it is still more instructive to inquire whether a large number of observations will be found to justify our assumption of the Law of Error.

Encke has given an excellent instance of the correspondence of theory with experience, in the case of certain observations of the difference of Right Ascension of the sun and two stars, namely a Aquilæ and a Canis minoris. The observations were 470 in number, and were made by Bradley and reduced by Bessel, who found the probable error of the final result to be only about one-fourth part of a second ($0''\cdot2637$). He then compared the number of errors of each magnitude from $\frac{1}{10}$th part of a second upwards, as actually given by the observations, with what should occur according to the Law of Error.

The results were as follow[g]:—

Magnitude of the errors in parts of a second.	Number of errors of each magnitude according to	
	Observation.	Theory.
0·0 to 0·1	94	95
·1 „ ·2	88	89
·2 „ ·3	78	78
·3 „ ·4	58	64
·4 „ ·5	51	50
·5 „ ·6	36	36
·6 „ ·7	26	24
·7 „ ·8	14	15
·8 „ ·9	10	9
·9 „ 1·0	7	5
above „ 1·0	8	5

[g] Encke, 'On the Method of Least Squares,' Taylor's 'Scientific Memoirs,' vol. ii. pp. 338, 339.

The reader will remark that the correspondence is remarkably close, except as regards larger errors, which are excessive in practice. It is one objection, indeed, to the theory of error, that, being expressed in a continuous mathematical function, it contemplates the possible existence of errors of every magnitude, such indeed as could not practically occur ; yet in this case the theory seems to under-estimate the number of large errors.

Another excellent comparison of the law with observation has been made by Quetelet, who has investigated the errors of 487 determinations in time of the Right Ascension of the Pole-star, made at Greenwich during the four years 1836–39. These observations, although carefully corrected for all known causes of error, as well as for nutation, precession, &c., are yet of course found to differ, and being classified as regards intervals of one-half second of time, and then proportionately increased in number, so that their sum may be one thousand, give the following results as compared with what theory would lead us to expect [h] :—

Magnitude of error in tenths of a second.	Number of errors		Magnitude of error in tenths of a second.	Number of errors	
	by Observation.	by Theory.		by Observation.	by Theory.
0·0	168	163	—	—	—
+ 0·5	148	147	− 0·5	150	152
+ 1·0	129	112	− 1·0	126	121
+ 1·5	78	72	− 1·5	74	82
+ 2·0	33	40	− 2·0	43	46
+ 2·5	10	19	− 2·5	25	22
+ 3·0	2	10	− 3·0	12	10
—	—	—	− 3·5	2	4

In this instance the correspondence is also satisfactory, but the divergence between theory and fact is in the opposite direction to that discovered in the former com-

[h] Quetelet, ‘Letters on the Theory of Probabilities,’ translated by Downes, Letter XIX. p. 88. See also Galton’s ‘ Hereditary Genius,’ p. 379.

parison, the larger errors being less frequent than theory would indicate.

We may also regard the experiments enumerated in the chapter on Probabilities (p. 238), as forming an empirical verification of the theory of error.

Remarks on the General Law of Error.

The mere fact that the Law of Error allows of the possible existence of errors of every assignable amount shows that it is only approximately true. We may fairly say that in measuring a mile it would be impossible to commit an error of a hundred miles, and the length of life would never allow of our committing an error of one million miles. Nevertheless the general Law of Error would assign a probability for an error of that amount or more, but so small a probability as to be utterly inconsiderable, and almost inconceivable. All that can, or in fact need, be said in defence of the law is, that it may be made to represent the errors in any special case to a very close approximation, and that the probability of large and practically impossible errors, as given by the law, will be so small as to be entirely inconsiderable. And as we are dealing with error itself, and our results pretend to nothing more than approximation and probability, an indefinitely small error in our process of approximation is of no importance whatever.

The Probable Mean Result as defined by the Law of Error.

One immediate result of the Law of Error, as thus stated, is that the mean result is the most probable one ; and when there is only a single variable this mean is found by the familiar arithmetical process. An unfortunate error has crept into several works which allude to this subject. Mr. Mill, in treating of the ' Elimination of Chance,' remarks in a note [i] that ' the mean is spoken of as if it were exactly the same thing with the average. But the mean, for purposes of inductive inquiry, is not the average, or arithmetical mean, though in a familiar illustration of the theory the difference may be disregarded.' He goes on to say that, according to mathematical principles, the most probable result is that for which the sums of the squares of the deviations is the least possible. In Bowen's ' Treatise on Logic' (p. 439), we find the Method of Least Squares mentioned as ' a mode of finding the most probable result in those cases in which the arithmetical mean is not an applicable expedient for determining the probability.' It seems probable that these and other writers were misled by Dr. Whewell's remarks on the subject ; for he says [k] that ' The Method of Least Squares is in fact a Method of Means, but with some peculiar characters. . . . The method proceeds upon this supposition ; that all errors are not equally probable, but that small errors are more probable than large ones.' He adds that this method ' removes much that is arbitrary in the Method of Means.' It is strange to find a mathematician like Dr. Whewell making such remarks, when there is no doubt whatever that the Method of Means is only an application of the Method of Least Squares. They are, in fact, the same method, except that the latter method may be applied to cases where two or more quantities have to be determined at the same time. Many authorities might be quoted to this effect, but it will be sufficient to mention Lubbock and Drinkwater, who say [l], ' If only one quantity has to be

[i] 'System of Logic,' bk. iii. chap. 17, § 3.   5th ed. vol. ii. p. 56.

[k] 'Philosophy of the Inductive Sciences,' 2nd ed. vol. ii. pp. 408, 409.

[l] ' Essay on Probability,' by J. W. Lubbock and J. E. Drinkwater, Useful Knowledge Society, 1833. p. 41.

determined, this method evidently resolves itself into taking the mean of all the values given by observation.' Encke, again, distinctly says[m], that the expression for the probability of an error 'not only contains in itself the principle of the arithmetical mean, but depends so immediately upon it, that for all those magnitudes for which the arithmetical mean holds good in the simple cases in which it is principally applied, no other law of probability can be assumed than that which is expressed by this formula.'

It can be shown, too, in a moment that the mean is the result which gives the least sum of squares of errors. For if a, b, c, &c., be the results of observation and x the selected mean result, the sum of squares of the errors is $(a-x)^2 + (b-x)^2 + (c-x)^2 + $ &c., which is at a minimum when its differential coefficient $2(a-x+b-x+c-x+$ &c.$)=0$. From this equation we immediately obtain, denoting by n the number of separate results, a, b, c, &c., $x = (a+b+c+ \dots) \frac{1}{n}$, or the ordinary arithmetic mean.

Weighted Observations.

It is to be distinctly understood that when we take the mean of certain numerical results as the most probable number aimed at, we regard all the different results as equally good and probable in themselves. The theory of error expresses no preference for any one number over any other. If, then, an observer has reason to suppose that some results are not so trustworthy as others, he must take account of this difference in drawing the mean. By the method of *weighting* observations this difference of value is easily allowed for. Astronomers are in the habit

---

[m] Taylor's 'Scientific Memoirs,' vol. ii. p. 333.

of recording with an observation the value or degree of confidence with which they regard it, freely estimated according to the impression of success or failure in accuracy *immediately after the observation.* This value is usually expressed in a decimal scale, so that 10 denotes the highest degree of satisfaction with the result, and 1 the least degree. Before taking the mean of the observations each number is multiplied by its weight or value, and the sum of the products is divided by the sum of the weights. Thus if a, b, c, &c., be the observed numbers, and w, w', w'', &c., the weights, then the most probable mean is $\frac{aw + bw' + cw'' + \ldots}{w + w' + w'' + \ldots}$. This formula, it will be observed, is identical in form with that for finding the centre of gravity of particles of different weights arranged in a straight line. When we regard w, w', w'', &c., as all equal, it becomes identical with the formula for the ordinary mean. This method of weighting observations, now of much importance in astronomical and other very exactly quantitative investigations, appears to have been first proposed by Roger Cotes, the editor of the 'Principia,' as pointed out by De Morgan[n].

The practice of giving weights would open the way to much error and abuse, if the weights were assigned when the mean was being drawn, and when the divergence of some results from the others would be likely to become the guide. As a general rule the weights must be assigned at the moment of observation, and afterwards rigidly maintained, and they must be assigned not from regard to the apparent intrinsic accuracy of the result, but the extrinsic circumstances which seem to render it valuable. An observed result, in short, must be discredited, not because it is divergent, but because there were other reasons to suppose that it would be divergent.

[n] 'Penny Cyclopædia,' art. *Least Squares.*

The Probable Error of Mean Results.

When we draw any conclusion from the numerical results of observations we ought not to consider it sufficient, in cases of importance, to content ourselves with finding the simple mean and treating it as true. We ought also to ascertain what is the degree of confidence we may place in this mean, and our confidence should be measured by the degree of concurrence of the observations from which it is derived. In some cases the mean may be so close to the correct result that we may consider it as approximately certain and accurate. In other cases it may really be worth little or nothing. The Law of Error enables us to give exact expression to the degree of confidence proper in any case ; for it shows how to calculate the probability of a divergence of any amount from the mean, and we can thence ascertain the probability that the mean in question is within a certain distance from the true number. The *probable error* is taken by mathematicians to mean the limits within which it is as likely as not that the truth will fall. Thus if 5·45 be the mean of all the determinations of the density of the earth, and ·20 be approximately the probable error, the meaning is that the probability of the real density of the earth falling between 5·25 and 5·65 is ½. Any other limits might have been selected at will. We might readily calculate the limits within which it was one hundred or one thousand to one that the truth would fall ; but there is a general convention to take the even odds, one to one, as the quantity of probability of which the limits are to be estimated.

Many books on the subject of probability give rules for making the calculations, but as, in the gradual progress of science, all persons ought to be more familiar with these processes, I propose to repeat the rules here and illustrate their use. The calculations, when made in strict accordance with the directions, involve none but arithmetic operations.

Rules for finding the probable error of a mean result :—

1. Draw the mean of all the observed results.

2. Find the excess or defect, that is, the error of each result from the mean.

3. Square each of these reputed errors.

4. Add together all these squares of the errors.

5. Take the square root of this sum.

6. Divide the square root by the number of results.

7. Multiply the quotient by 0·67449 (or approximately by 0·674, or even 0·67), a natural constant number derived from the Law of Error in a manner which is described in mathematical works upon the subject.

Suppose, for instance, that five measurements of the height of a hill, by the barometer or otherwise, have given the numbers of feet as 293, 301, 306, 307, 313; we want to know the probable error of the mean, namely 304. Now the differences between this mean and the above numbers, *paying no regard to direction,* are 11, 3, 2, 3, 9; their squares are 121, 9, 4, 9, 81, and the sum of the squares consequently 224. Taking the square root of this sum by the common arithmetic process, or by logarithms, we obtain 14·966, and dividing by five, the number of observations, we have 2·99, which has only to be multiplied by ·67 to yield us 2·019. This number is so close to 2, that we may call the probable error equal to two. Thus the probability is one-half, or the odds are even, that the true height of the mountain lies between 302 and 306 feet. We have thus an exact measure of the degree of credibility of our mean result, which mean indicates the most likely point for the truth to fall upon.

The reader should observe that as the object in these calculations is only to gain a notion of the degree of confidence with which we view the mean, there is no real use in carrying the calculations to any great degree of precision ; and whenever the neglecting of decimal fractions, or even the slight alteration of a number will much abbreviate the computations, it may be fearlessly done, except in cases of high importance and precision. It has been stated that the voyages of the Great Britain steamship to Melbourne from Liverpool, up to May, 1871, have been thirteen in number, with the following durations in days : 62, 63, 59, 60, 58, 61, 57, 57, 57, 57, 56, 63, 55. The mean duration of the voyages is 58·85 days, which is the most probable length of any similar future voyage ; but to calculate the probable error, we may take the mean to be 59 days. The sum of the squares of the errors is only 88, and the probable error thence calculated 0·49 day, or, say half a day. It is as likely as not, then, that any particular voyage will be not less than 58½ days, nor more than 59½ days.

The experiments of Benzenberg to detect the revolution of the earth, by the deviation of a ball from the exact perpendicular line in falling down a deep pit, have been cited by Encke[o] as an interesting illustration of the Law of Error. The mean deviation was 5·086 lines, and its probable error was calculated by Encke to be not more than ·950 line, that is, the odds were even that the true result lay between 4·136 and 6·036. As the deviation should, according to astronomical theory be 4·6 lines, which lies well within the limits, we may consider that the experiments are consistent with the Copernican system of the universe.

It will of course be understood that the probable error has regard only to the differences of the results from which the mean is drawn, and takes no account of constant errors. The true result accordingly will often fall far beyond the limits of probable error.

[o] Taylor's ‘Scientific Memoirs,’ vol. ii. pp. 330, 347, &c.

The Rejection of the Mean Result.

We ought always to bear in mind that the mean of any series of observations is the best, that is, the most probable approximation to the truth, only in the entire absence of any knowledge to the contrary. The selection of the mean rests entirely upon the probability that wholly unknown causes of error will in the long run fall as often in one direction as the opposite, so that in drawing the mean they will balance each other. If we have any presumption to the contrary, any reason to suppose that there exists a tendency to error in one direction rather than the other, then to choose the mean would be to ignore that tendency. Thus we may certainly approximate to the length of the circumference of a circle, by measuring the perimeters of inscribed and circumscribed polygons of an equal and large number of sides. The correct length of the circular line undoubtedly lies between the lengths of the two perimeters, but it does not follow that the mean is the best approximation. It may in fact be shown upon mathematical principles that the circumference of the circle is *very nearly* equal to the perimeter of the inscribed polygon, together with one-third part of the difference between the inscribed and circumscribed polygons of the same number of sides. Having this knowledge we ought of course to act upon it, instead of upon vague grounds of probability.

We may often perceive that a series of measurements tends towards an extreme limit rather than towards a mean. Thus in endeavouring to obtain a correct estimate of the apparent diameter of the brightest fixed stars, we should find a continuous diminution in estimates as the powers of observation increased. Kepler assigned to Sirius an apparent diameter of 240 seconds; Tycho Brahe made it 126; Gassendi 10 seconds; Galileo, Hevelius, and J. Cassini, 5 or 6 seconds. Halley, Michell, and subsequently Sir W. Herschel came to the conclusion that the brightest stars in the heavens could not have real discs of a second, and were probably much less in diameter. It would of course be absurd to take the mean of quantities which differ more than 240 times; and as the tendency has always been to smaller estimates, there is a considerable indication in favour of the smallest[p].

In the case of many experiments and measurements we shall know on which side there is a tendency to error. Thus the readings of a thermometer always tend to rise as the age of the instrument increases, and no drawing of means will correct this result. Barometers, on the other hand, are always likely to read too low instead of too high, owing to the imperfection of the vacuum, or the action of capillary attraction. If the mercury be perfectly pure and no considerable error be due to the measuring apparatus, the best barometer will be that which gives the highest result.

When we have reasonable grounds for supposing that certain experimental results are liable to grave errors, we should exclude them in drawing a mean. If we want to find the most probable approximation to the velocity of sound in air, it would be absurd to go back to the old experiments which made the velocity from 1200 to 1474 feet per second; for we know that the old observers did not guard against errors arising from wind and other causes. Old chemical experiments are absolutely valueless as regards quantitative results. The old chemists found the atmosphere to differ in composition nearly ten per cent. in different places, whereas modern accurate experimenters find very slight variations. Any method of measurement which we know to avoid a source of error is far to be preferred to others which trust to probabilities

p Quetelet, 'Letters,' &c. p. 116.

for the elimination of the error. As Flamsteed says [q], 'One good instrument is of as much worth as a hundred indifferent ones.' But an instrument is good or bad only in a comparative sense, and no instrument gives invariable and truthful results. Hence we must always ultimately fall back upon general probabilities for the selection of the final mean, when our other precautions are exhausted.

Very difficult questions sometimes arise when one or more results of a method of experiment diverge widely from the mean of the rest. Are we or are we not to exclude them in adopting the supposed true mean result of the method. The drawing of a mean result rests, as I have frequently explained, upon the assumption that every error acting in one direction will probably be balanced by other errors acting in an opposite direction. If then we know or can possibly discover any causes of error not agreeing with this assumption, we shall be justified in excluding results which seem to be affected by this cause.

In reducing large series of astronomical observations, it is not uncommon to meet with numbers differing from others by a whole degree or half a degree, or some considerable integral quantity. These are errors which could hardly arise in the act of observation or in instrumental irregularity; but they might readily be accounted for by misreading of figures or mistaking of division marks. It would be absurd to trust to chance that such mistakes would balance each other in the long run, and it is therefore better to correct arbitrarily the supposed mistake, or better still, if new observations can be made, to strike out the divergent numbers altogether. When results come sometimes too great or too small in a regular manner, we should suspect that some part of the instrument slips through a definite space, or that a definite cause of error enters at times, and not at others. We

q Baily, 'Account of Flamsteed,' p. 56.

should then make it a point of prime importance to discover the exact nature and amount of such an error, and either prevent its occurrence for the future or else introduce a corresponding correction. In many researches the whole difficulty will consist in this detection and avoidance of sources of error. Thus Professor Roscoe found that the presence of phosphorus caused serious and almost unavoidable errors in the determination of the atomic weight of vanadium[r]. Sir John Herschel, in reducing his observations of double stars at the Cape of Good Hope, was perplexed by an unaccountable difference of the angles of position as measured by the Seven-feet Equatorial and the Twenty-feet Reflector Telescopes, and after a careful investigation was obliged to be contented with introducing a correction experimentally determined[s].

Even the most patient and exhaustive investigations will sometimes fail to disclose any reason why some results diverge in an unusual and unexpected manner from others. The question again recurs—Are we arbitrarily to exclude them? The answer should be in the negative as a general rule. The mere fact of divergence ought not to be taken as conclusive against a result, and the exertion of arbitrary choice would open the way to the most fatal influence of bias, and what is commonly known as the 'cooking' of figures. It would amount in most cases to judging fact by theory instead of theory by fact. The apparently divergent number may even prove in time to be the true one. It may be an exception of that peculiarly valuable kind which upsets our false theories, a real exception, exploding apparent coincidences, and opening the way to a wholly new view of the subject. To establish this position for the divergent fact will of course require additional research; but in the meantime we should give it a fair

---

[r] Bakerian Lecture, 'Philosophical Transactions' (1868), vol. clviii. p. 6.

[s] 'Results of Observations at the Cape of Good Hope,' p. 283.

weight in our mean conclusions, and should bear in mind the discrepancy as one demanding attention. To neglect a divergent result is to neglect the possible clue to a great discovery.

Method of Least Squares.

When two or more unknown quantities are so involved that they cannot be separately determined by the single Method of Means, we can yet obtain their most probable amounts by the Method of Least Squares, without more difficulty than arises from the length of the arithmetical computations. If the result of each observation gives an equation between two unknown quantities of the form

$$ax + by = c$$

then, if the observations were free from error, we should only need two observations giving two equations; but, for the attainment of greater accuracy, we may take a series of observations, and then reduce the equations so as to give only a pair with average coefficients. This reduction is effected by, firstly, multiplying the coefficients of each equation by the first coefficient, and adding together all the similar coefficients thus resulting for the coefficients of a new equation; and secondly, by repeating this process, and multiplying the coefficients of each equation by the coefficient of the second term. Thus meaning by (sum of a^2) the sum of all quantities of the same kind, and having the same place in the equations as a^2, we may briefly describe the two resulting mean equations as follows :—

(sum of a^2) . x + (sum of ab) . y = (sum of ac),
(sum of ab) . x + (sum of b^2) . y = (sum of bc).

When there are three or more unknown quantities the process is exactly the same in nature, and we only need additional mean equations to be obtained by multiplying by the third, fourth, &c., coefficients. As the numbers are in any case only approximate, it is usually quite un-necessary to make the computations with any great degree of accuracy, and places of decimals may therefore be freely cut off to save arithmetical work. The mean equations having been computed, their solution by the ordinary methods of algebra gives the most probable values of the unknown quantities.

Works upon the Theory of Probability and the Law of Error.

Regarding the Theory of Probability and the Law of Error as constituting, perhaps, the most important subjects of study for any one who desires to obtain a complete comprehension of logical and scientific method as actually applied in physical investigations, I will briefly indicate the works in one or other of which the reader will best pursue the study.

The best popular, and at the same time profound English work on the subject is De Morgan's 'Essay on Probabilities and on their Application to Life Contingencies and Insurance Offices,' published in the 'Cabinet Cyclopædia,' and to be obtained from Messrs. Longman. No mathematical knowledge beyond that of common arithmetic is required in reading this work. Quetelet's 'Letters,' already often referred to, also form a most interesting and excellent popular introduction to the subject, and the mathematical notes are also of value. Sir George Airy's brief treatise 'On the Algebraical and Numerical Theory of Errors of Observation and the Combination of Observations,' contains a complete explanation of the Law of Error and its practical applications. De Morgan's treatise 'On the Theory of Probabilities' in the 'Encyclopædia Metropolitana,' presents an abstract of the more abstruse investigations of Laplace, together with a multitude of pro-

found and original remarks concerning the theory generally.
In Lubbock and Drinkwater's work on 'Probability,' in the
Library of Useful Knowledge, we have a very concise but
good statement of a number of important problems. The
Rev. W. A. Whitworth has given, in an interesting little
work entitled 'Choice and Chance,' a number of good illustrations of the calculations both in the theories of Combinations and Probabilities. In Mr. Todhunter's admirable
History we have an exhaustive critical account of almost all
writings upon the subject of probability down to the cul-
mination of the theory in Laplace's works. In spite of the
existence of these and some other good English works, there
seems to be a want of an easy and yet pretty complete
introduction to the study of the theory of probabilities.

Among French works the 'Traité Élémentaire du Calcul des Probabilités,' by S. F. Lacroix, of which several
editions have been published, and which is not difficult
to obtain, forms probably the best elementary treatise.
Poisson's 'Recherches sur la Probabilité des Jugements,'
(Paris, 1837), commences with an admirable investigation
of the grounds and methods of the theory. While Laplace's great 'Théorie Analytique des Probabilités' is of
course the 'Principia' of the subject, his 'Essai Philosophique sur les Probabilités' is a popular discourse, and
is one of the most profound and interesting essays ever
published. It should be familiar to every student of
logical method, and has lost little or none of its importance by lapse of time.

Detection of Constant Errors.

The Method of Means is absolutely incapable of eliminating any error which is always the same, and which
always lies in one direction. We sometimes require to be
aroused from a false feeling of security, and to be urged

to take suitable precautions against such occult errors. 'It is to the observer,' says Gauss[t], 'that belongs the task of carefully removing the causes of constant errors,' and this is quite true when the error is absolutely constant. When we have made a number of determinations with a certain apparatus or method of measurement, there is a great advantage in altering the arrangement, or even devising some entirely different method of getting estimates of the same quantity. The reason obviously consists in the improbability that exactly the same constant error will affect two or more different methods of experiment. If a discrepancy is found to exist, we shall at least be aware of the existence of error, and can take measures for finding in which way it lies. If we can try a considerable number of methods, the probability becomes considerable that errors constant in one method will be balanced or nearly so by errors of an opposite effect in the others. Suppose that there be three different methods each affected by an error of equal amount. The probability that this error will in all fall in the same direction is only $\frac{1}{4}$; and with four methods similarly $\frac{1}{8}$. If each method be affected, as is always the case by several independent sources of error, the probability becomes very great that in the mean result of all the methods some of the errors will partially compensate the others. In this case, as in all others, when human foresight and vigilance has exhausted itself, we must trust the theory of probability.

In the determination of a zero point, of the magnitude of the fundamental standards of time and space, in the personal equation of an astronomical observer, we have instances of such fixed errors; but as a general rule a change of procedure is likely to reverse the character of the error, and many instances may be given of the value of this precaution.

[t] Gauss, translated by Bertrand, p. 25.

If we measure over and over again the same angular magnitude by the same divided circle, maintained in exactly the same position, it is evident that the same mark in the circle will be the criterion in each case, and any error in the position of that mark will equally affect all our results. But if in each measurement we use a different part of the circle, a new mark will come into use, and as the error of each mark can hardly be in the same direction, the average result will be nearly free from errors of division. It will be still better to use more than one divided circle.

Even when we have no clear perception of the points of our apparatus at which fixed error is likely to enter, we may with advantage vary the construction of our apparatus with the hope that we shall accidentally detect some latent imperfection. Baily's purpose in repeating the experiments of Michell and Cavendish on the density of the earth, was not merely to follow the same course and verify the previous numbers, but to try whether variations in the size and substance of the attracting balls, the mode of suspension, the temperature of the surrounding air, &c., would yield different results. He performed no less than 62 distinct series, comprising 2153 experiments, and he carefully classified and discussed the results so as to disclose the utmost differences. Again, in experimenting upon the resistance of the air to the motion of a pendulum, Baily employed no less than 80 pendulums of various forms and materials, in order to ascertain exactly upon what conditions the resistance depends. Regnault, in his exact researches upon the dilatation of gases made arbitrary changes in the magnitude of parts of his apparatus. He thinks that if, in spite of such modification the results are unchanged, the errors are probably of inconsiderable amount[u]; but in reality it is

[u] Jamin, 'Cours de Physique,' vol. ii. p. 60.

always possible, and usually likely, that we overlook sources of error which a future generation will detect. Thus the pendulum experiments of Baily and Sabine were directed to ascertain the nature and amount of a correction for air resistance, which had been entirely misunderstood in the experiments upon which was founded the definition of the standard yard, by means of the seconds pendulum in the Act of 5th George IV. c. 74. It has already been mentioned that a considerable error was discovered in the determination of the standard metre as the ten-millionth part of the distance from the pole to the equator (p. 368).

We shall return in the second volume to the further consideration of the methods by which we may as far as possible secure ourselves against permanent and undetected sources of error. In the meantime, having completed the consideration of the special methods requisite for treating quantitative phenomena, we must return to our principal subject, and endeavour to trace out the course by which the physicist, from observation and experiment, collects the materials of natural knowledge, and then proceeds by hypothesis and inverse calculation to educe from them the laws of nature.

END OF THE FIRST VOLUME.

www.ingramcontent.com/pod-product-compliance
Lightning Source LLC
Chambersburg PA
CBHW020901210326
41598CB00018B/1742